Exercise, Autophagy and Chronic Diseases

Ning Chen
Editor

Exercise, Autophagy and Chronic Diseases

Springer

Editor
Ning Chen
Tianjiu Research and Development Center
for Exercise Nutrition and Foods, Hubei
Key Laboratory of Exercise Training and
Monitoring, College of Health Science
Wuhan Sports University
Wuhan, China

ISBN 978-981-16-4524-2 ISBN 978-981-16-4525-9 (eBook)
https://doi.org/10.1007/978-981-16-4525-9

This Springer imprint is published by the registered company Springer Nature Singapore Pte Ltd.
The registered company address is: 152 Beach Road, #21-01/04 Gateway East, Singapore 189721, Singapore

Abstract *As the gradual increase in life expectancy, non-communicable chronic diseases such as metabolic diseases, neurodegenerative diseases, cardiovascular diseases, and aging-related diseases present a quick increasing trend. Physical inactivity is a critical inducer for these non-communicable chronic diseases. Consequently, regular exercise training has long been recognized as a green and environment-friendly strategy for health promotion, and a reasonable and effective remedy for a wide variety of chronic diseases. Meanwhile, regular exercise at appropriate intensity and duration, as an inducer of autophagy that is an evolutionarily conserved degradation pathway for the elimination and clearance of denatured or aggregated proteins and aging or damaged cellular organelles to maintain cellular homeostasis in the body, can afford a series of health benefits and execute the prevention, treatments, and rehabilitation of these chronic diseases through inducing autophagy or optimizing the functional status of autophagy.*

This book covers the signaling pathways for regulating the functional status of

autophagy upon acute and chronic exercise interventions for the prevention, treatments, and rehabilitation of common and frequently occurring chronic diseases such as metabolic diseases including diabetes, obesity and non-alcohol fatty liver disease, neurodegenerative diseases including brain aging, Alzheimer's disease, Parkinson's disease, and cardiovascular diseases including hypertension, atherosclerosis, cardiomyopathy, and myocardial ischemia-reperfusion injury, as well as sarcopenia, which partially uncovers the underlying mechanisms of exercise-mediated autophagy for intervening these chronic diseases and puts forward the potential application prospects. Meanwhile, the precision exercise mimetic pills using autophagy as the target have also proposed for health promotion, and the prevention and treatments of these chronic diseases, which will be beneficial to people with disability for exercise performance, or people without willing to exercise. This book will facilitate to the development of new preventative and therapeutic strategies for health promotion and chronic diseases. Meanwhile, it will provide the theoretical guidance and practical references for persons who need physical fitness or have chronic diseases, and clinical physicians, as well as scientists in the fields of exercise science and medicine.
Keywords *Autophagy; Exercise intervention; Metabolic diseases; Neurodegenerative diseases; Cardiovascular diseases; Sarcopenia; Exercise mimetics*

Preface

As the development of modern scientific technology and gradual increase in life expectancy, non-communicable chronic diseases such as metabolic diseases, neurodegenerative diseases, cardiovascular diseases, and aging-related diseases present a quick increasing trend. Lifestyles play critical roles in the initiation and progression of these non-communicable chronic diseases. Physical inactivity, as one of the unhealthy lifestyles, is also a critical inducer for these non-communicable chronic diseases. Consequently, regular exercise training has been gained public recognition as a green and environment-friendly strategy for health promotion, and a reasonable and effective remedy for a wide variety of chronic diseases. Although increasing volume of studies have confirmed regular exercise-induced benefits to chronic diseases, its underlying accurate mechanisms are not fully understood or elucidated, thereby leading to the difficulties in exploring optimal exercise interventions for these chronic diseases and corresponding extension of novel and effective interventional strategies. Recently, regular exercise at appropriate intensity and duration, as an inducer of autophagy that is an evolutionarily conserved degradation pathway for the elimination and clearance of denatured or aggregated proteins and aging or damaged cellular organelles to maintain cellular homeostasis in the body, can afford a series of health benefits and execute the prevention, treatment, and rehabilitation of these chronic diseases through inducing autophagy or optimizing the functional status of autophagy. Meanwhile, the clarified regulators for modulating the signal pathways of exercise-mediated functional status of autophagy can be the useful targets for the interventions of these chronic diseases, or can be used to develop the novel and promising exercise mimetic pills for these chronic diseases.

This book aims to establish a bridge between exercise-mediated functional status of autophagy and above-mentioned non-communicable chronic diseases for elucidating and clarifying the corresponding signal pathways and underlying mechanisms, which will be beneficial for optimizing exercise interventions, extending interventional strategies, or developing novel and effective exercise mimetic pills for the prevention, treatments, and rehabilitation of common and frequently occurring chronic diseases. This book covers 13 chapters including: (1) molecular processes and regulation of autophagy; (2) acute and chronic exercise on autophagy; (3) the beneficial roles of exercise-mediated autophagy in T2DM; (4) exercise-induced autophagy and obesity; (5) exercise-mediated autophagy and non-alcoholic fatty liver disease; (6) exercise-mediated autophagy and brain aging; (7) exercise-mediated autophagy and Alzheimer's disease; (8) exercise-induced autophagy and Parkinson's disease; (9) exercise-mediated autophagy in cardiovascular diseases; (10) exercise-induced autophagy in the prevention and treatment of sarcopenia; (11) prospective advances in exercise-induced autophagy on health; (12) exercise mimetic pills for chronic diseases based on autophagy; and (13) exercise-mediated functional status of autophagy is beneficial to health. Meanwhile, the precision exercise mimetic pills using autophagy as the target have also proposed for health promotion, and the prevention and treatments of these chronic diseases, which will be beneficial to people with disability for exercise performance, or people without willing to exercise.

The authors of this book have multiple research backgrounds with the involvement in the studies in the fields of Biochemistry, Molecular Physiology, Molecular Exercise Physiology, Exercise Science and Clinical Medicine from Wuhan Sports University, Huazhong University of Science and Technology, Georgia Gwinnett College, and the University of Massachusetts Boston. Most of these authors are involved in the studies on autophagy for the regulation of chronic diseases for several years with large volume of first-hand experimental data. As a practical guide book, the chief emphasis is not only to offer the in-depth discussion on the signal pathways for regulating the functional status of autophagy for the prevention, treatments, and rehabilitation of chronic diseases, but also to provide the optimization of exercise interventional strategies for chronic diseases, as well as to contribute to develop exercise mimetic pills for the persons with disability for exercise performance, or the persons without willing to exercise. Therefore, this book will facilitate to the development of new preventative and therapeutic strategies for health promotion and chronic diseases. Meanwhile, this book will provide the theoretical guidance and practical references for the persons who need physical fitness or have chronic diseases, and clinical physicians, as well as scientists in various fields of exercise science, fundamental and clinical medicine, exercise fitness, and functional foods.

Wuhan, China

Ning Chen

Acknowledgments

This work was supported by the National Natural Science Foundation of China (Nos. 31571228, 31771318, and 32071176); the Key Special Project of Disciplinary Development, Hubei Superior Discipline Group of Physical Education and Health Promotion; Hubei Superior Discipline Group of Exercise and Brain Science; and the Outstanding Youth Scientific and Technological Innovation Team (T201624) from the Education Department of Hubei Province; as well as the Chutian Scholar Program and Innovative Start-up Foundation from Wuhan Sports University to Ning Chen.

Contents

Chapter 1
Molecular Processes and Regulation of Autophagy

Mohammad Nasb, Michael Kirberger, and Ning Chen

1 Introduction

The molecular processes involved in autophagy have attracted much attention in the past few years, and our understanding of its underlying mechanism has increased dramatically. Autophagy is a highly conserved homeostatic process involving the degradation of cytoplasmic elements, including intracellular pathogens, toxic protein aggregates, and damaged organelles [1]. It also plays a pivotal role in tumor suppression, cellular response to stress, and lifespan extension [1, 2]. Autophagy significantly contributes to cellular homeostasis via the catabolism of carbohydrates (glycophagy), iron (ferritinophagy), and lipids (lipophagy); the clearance of abnormal protein aggregates (aggrephagy); and the elimination of damaged organelles including mitochondria (mitophagy), endoplasmic reticulum (ER-phagy), and ribosomes (ribophagy). Moreover, autophagy plays an important role in intracellular signaling pathways for controlling replicative senescence, differentiation, and proliferation [2]. Alternatively, autophagy may have relevance to some neurodegenerative conditions or myopathies [3]. These activities are critically organized by the intervention of a series of major components, including autophagy-related genes (ATGs). The classical autophagy model describes the destruction of intracellular

M. Nasb
Department of Rehabilitation Medicine, Tongji Medical College, Huazhong University of Science and Technology, Wuhan, China

Department of Physical Therapy, College of Health Science, Albaath University, Homs, Syria

M. Kirberger
School of Science and Technology, Georgia Gwinnett College, Lawrenceville, GA, USA

N. Chen (✉)
Tianjiu Research and Development Center for Exercise Nutrition and Foods, Hubei Key Laboratory of Exercise Training and Monitoring, College of Health Science, Wuhan Sports University, Wuhan, China

N. Chen (ed.), *Exercise, Autophagy and Chronic Diseases*,
https://doi.org/10.1007/978-981-16-4525-9_1

proteins under starvation conditions, while basal autophagy usually occurs under nutrient-rich conditions [4]. Basal autophagy may play a critical role in quality control of intracellular proteins and organelles in each cell [5]. The degradation of eukaryotic cells usually occurs in two ways: autophagy and proteasomal degradation. Autophagy is primarily responsible for degradation of long-lived cytosolic proteins and organelles, whereas specific short-lived proteins are degraded via the ubiquitin–proteasome system (UPS) [6]. Chaperone-mediated autophagy (CMA) is a process used for the degradation of cytosolic proteins with a basic pentapeptide motif, although CMA has limited degradation capacity. The other major modes of autophagy are microautophagy and macroautophagy. These two pathways include dynamic membrane rearrangements that terminate at lysosomes/vacuoles, which are obviously distinct from proteasomal degradation [7].

2 Molecular Mechanism of Autophagy: A Historical Interest

While the autophagy process was first recognized in mammalian cells roughly 50 years ago, the understanding of its molecular mechanism only began in the past decade [8]. This understanding is principally based on the discovery of ATGs in yeast, followed by the recognition of homologs in other higher eukaryotes. One subset of ATG proteins is vital for the formation of autophagosomes (APs). This subset is termed “core” molecular machinery [9].

The Nobel Prize in Physiology and Medicine 2016 was awarded to cell biologist Yoshinori Ohsumi for his early distinct discovery and description of the autophagy machinery and components, especially yeast autophagy-related genes [10]. This topic was more formally articulated by Christian De Duve, a Belgian cytologist and biochemist, who first used the term “autophagy” in 1963 to describe the appearance of single- or double-membrane vesicles enveloping cytoplasmic cargoes or cellular organelles for degradation during lysosome fusion at various stages [11].

Our comprehension of autophagy regarding the mechanistic and pathophysiological grounds, which have been strongly conserved throughout evolution, has increased dramatically over the last several decades [2, 12]. This knowledge includes a better understanding of the importance of dietary or pharmacological interventions for stimulating or suppressing autophagy as distinct treatments for various disorders, including infectious [13], cardiovascular [14], pulmonary [15], rheumatic [16], metabolic [17], neurodegenerative [18], autoimmune [19], and malignant diseases [20], in addition to aging [21]. However, at present, neither the Food and Drug Administration (FDA) nor any comparable agency from any countries has approved any autophagy-modulating drugs [22, 23], suggesting a present barrier in translating the preclinical data from a model organism to clinical therapeutic interventions. Therefore, improved understanding of the underlying mechanisms of autophagy could be helpful for researchers and clinicians to overcome this barrier.

3 The Molecular Mechanisms of Autophagy in Mammals

3.1 *Autophagy Induction*

Autophagy can be activated as a response to several stimuli, including decreased cellular energy levels (reduction in ATP), extra- or intracellular stress (i.e., hypoxia or oxidative stress), withdrawal of growth factors (e.g., insulin-like growth factors), nutrient deficiency (including glucose and amino acids), and pathogenic infections [24].

The deprivation of growth factors and amino acids has the greatest capacity to induce autophagy, which is connected with the mammalian target of the rapamycin complex 1 (mTORC1) signal pathway [25]. The mammalian target of rapamycin (mTOR) is an important and highly conserved threonine–serine protein kinase that plays a regulatory role in cellular metabolism through the recognition of nutrient signals. Under nutrient-enriched conditions, mTORC1 phosphorylates ATG13 and Unc-51-like kinase 1 (ULK1) [26, 27]. The ATG1/ULK1 complex plays an integral role in starvation-induced autophagy and is involved in the initiation of autophagy. Merging signals received by upstream sensors such as AMPK and mTOR, and then transducing them, can activate the downstream signal pathway of autophagy [28] (Fig. 1.1).

The suppression of mTORC1 due to starvation conditions causes the dephosphorylation of ATG13, which contributes to the formation of a complex between ULK1, ATG13, ATG101, and focal adhesion kinase (FAK) family of interacting protein, with a molecular weight of 200 kDa (FIP200) (Fig. 1.2). The ULK1 protein is the only protein kinase of the core ATG and can significantly modify and recruit downstream ATG proteins, including, but not limited to, ATG13 and FIP200, with the aim of triggering autophagy [29].

It is noteworthy that earlier studies have revealed the collaboration of mTOR and AMPK to regulate the cellular energy and nutrient signals for preserving cellular homeostasis through ULK1 phosphorylation at defined serine residues [30]. During the conditions of energy insufficiency, ULK1 is phosphorylated by AMPK to inhibit the reaction between ULK1 and mTOR, thereby stimulating the initiation of autophagy [30, 31].

3.2 *Autophagosome Formation*

When the ULK1 complex is created, the class III phosphatidylinositol 3-kinase (PI3K) complex is allocated to the isolation membrane as the autophagosomal precursor, where it plays a crucial role in the initiation of sequestering of cargoes [32]. The PI3K complex is essential for nucleation of the isolation membrane and phagophore assembly. The chief component of the PI3K complex is the vacuolar protein sorting 34 (VPS34). Meanwhile, ULK1 recruits the catalytic subunit of the

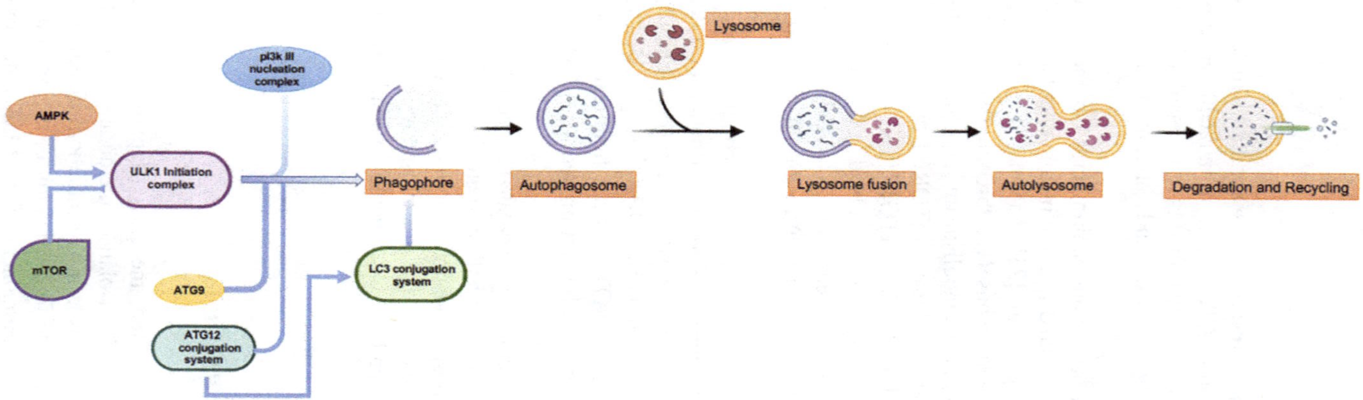

Fig. 1.1 Molecular process of autophagy, with schematic description of major regulatory machinery and process. The maintained metabolic sensors and longevity determinants including mTOR and AMPK are the key controllers of the autophagy machinery, as mTOR and AMPK are involved in inhabitation and activation of the autophagy process, respectively. Once autophagy is activated, the double membranes engulfed the cytoplasmic components, initiation by forming a cup-shaped structure named phagophore, and followed by growing of the phagophores into the double-membrane vesicles, named autophagosomes, that fuse with acidic lysosomes in order to form the autolysosomes for cargo degradation

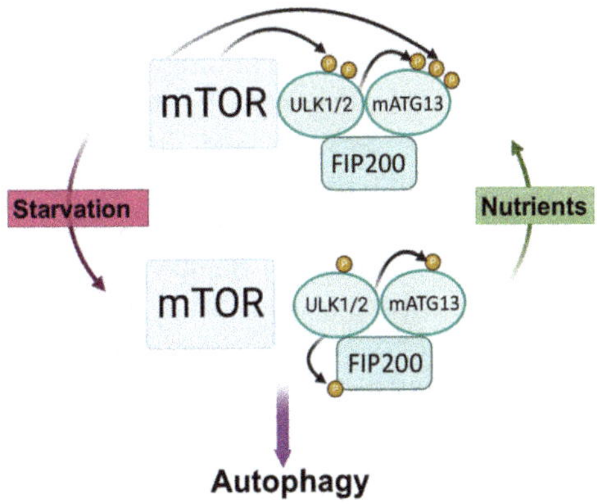

Fig. 1.2 Schematic description of autophagy core molecular machinery in mammals [9]

PI3K complex to generate phosphatidylinositol 3-phosphate (PI3P) at the initiation site. PI3P is pivotal for the formation of the APs, and is considered as a biomarker of the AP membrane [33], while there remains some debates regarding the origin of the isolation membrane. Several studies suggest that, in mammals, the endoplasmic reticulum provides the materials to initiate the formation of the isolation membrane [1, 34–36].

Evidence to support this is based on the availability of a PI3P-rich area in the endoplasmic reticulum membrane with a Ω-like shape, called the omegasome [37]. The omegasome is the site of the PI3P downstream effector, double-FYVE-containing protein1 (DFCP1), which is a biomarker for the initiation of the isolation membrane [38–40]. Morphological observations have demonstrated that the majority of isolation membrane initiation locations are found in an adjacent locality of the endoplasmic reticulum microdomains [40–42]. After the localization of ULK1 and PI3K complexes to the endoplasmic reticulum-derived omegasomes, different ATG proteins are enrolled to assist the expansion of the isolation membrane. Besides enrolling the PI3K complex, the ULK1 complex plays a critical role in recruiting ATG9 to the isolation membrane [28]. Among the core ATG proteins, ATG9 is the transmembrane protein shuttle that moves between the isolation membrane and other membrane components during the formation of ATG9 vesicles. This transference may be reliant on proteins WD repeat domain phosphatidylinositide-interacting 2 (WIPI2) and WIPI2-interacting protein ATG2 [43]. Thus, ATG9 vesicles transport to assist the expansion of the isolation membrane. The mitochondrial membrane is also an important contributor to the expansion of the isolation membrane [44]. There are two significant ubiquitin-like (Ubl) conjugation systems, the ATG12 system and the microtubule-associated protein 1 light chain 3 (LC3) system, which are important during expansion. Both systems involve Ubl proteins. ATG12 is consecutively triggered through ubiquitin-conjugating enzyme (E2)-like enzyme ATG10, and ubiquitin-activating enzyme (E1)-like enzyme ATG7, and subsequently combines with ATG5 to form an ATG12-ATG5-ATG16L1 complex on the isolation membrane [45].

Earlier studies have reported that the enrollment of ATG12-ATG5-ATG16L1 requires WIPI2 to facilitate the conjugation of LC3 with PI3P at the location of the generating AP [46], suggesting a major role of the WIPI family in building the conjugation system. Prior to conjugation, the LC3 precursor is cleaved by the cysteine protease ATG4, thus exposing the C-terminal glycine. Next, the E1-like enzyme ATG7 and the E2-like enzyme ATG3 can activate cytosolic LC3-I for the conjugation with phosphatidylethanolamine (PE), thus producing the membrane-associated LC3-PE, which is LC3-II [47]. Unlike ATG12 conjugation, the final phase of ATG3 conjugation requires a ubiquitin (Ub) ligase ATG12-ATG5-ATG16L1 complex [48]. Studies have shown that ATG12 localizes only on the isolation membrane, while LC3-II is located on the autophagosomal membrane during the entire AP formation process and during tethering or fusing with lysosomes [49].

3.3 Fusion Between Autophagosome and Lysosome

The underlying mechanism of the fusion between autophagosome and lysosome is still not completely understood. Soluble *N*-ethylmaleimide-sensitive factor attachment protein receptor (SNARE) proteins, and RAS-related GTP-binding protein (RAB) GTPases, are intracellular membrane-trafficking-associated protein families that have regulatory roles in the fusion process of autophagosomes and lysosomes [31]. Moreover, several studies have revealed that the microtubule system is vital for trafficking mature APs from various starting locations to perinuclear sites [50], where the endosomes and lysosomes fuse together. Furthermore, the class III PI3K complex II, where ATG14L can be converted into UV irradiation resistance-associated gene (UVRAG), is involved in controlling the delivery of APs to lysosomes [31]. In addition to fusing with lysosomes directly, APs can also first fuse with endosomes and then traffic to lysosomes to form autolysosomes [51]. After the formation of the autolysosome, lysosome-derived hydrolases degrade the cargoes and then recycle materials into the cytoplasm for energy generation or biosynthesis [52].

4 Molecular Regulation of Autophagy Machinery

The autophagy machinery is primarily regulated by two pivotal signaling pathways as a response to environmental and intracellular stresses, which are mTOR-dependent and mTOR-independent pathways.

4.1 The mTOR-Dependent Signaling Pathway

The mTOR signal pathway is considered a crucial regulator of autophagy, cell metabolism, and mitochondrial metabolism (Fig. 1.3) [53]. Multiple studies have revealed that mTOR activation is also critical for cell survival during ischemia [54–

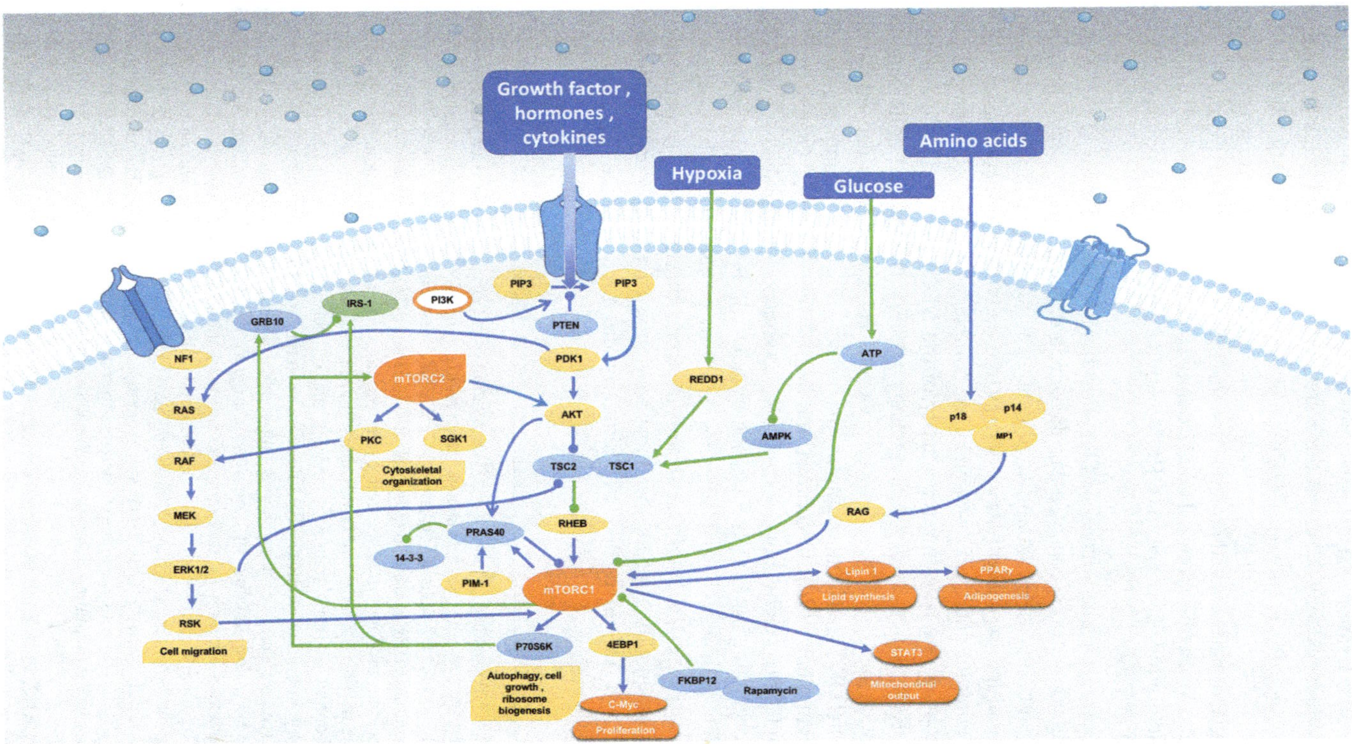

Fig. 1.3 mTOR signaling pathway. Cytokines, growth factors, and hormones can bind to their receptors on the cell surface, which activates phosphatidylinositol 3-kinase (PI3K). PI3K then phosphorylates phosphatidylinositol 4,5-bisphosphate (PIP2) and forms phosphatidylinositol 3,4,5-trisphosphate (PIP3), which in turn activates phosphoinositide-dependent protein kinase 1 (PDK1), followed by the downstream effector AKT. Next, AKT phosphorylates tuberous sclerosis complex 2 (TSC2), and TSC2 attaches to TSC1 to form the TSC1-TSC2 complex, thereby suppressing the Ras homolog

56], but mTOR also plays a vital role in controlling other biological processes, such as immune cell activation. mTOR is usually observed in the form of mTORC1 and mTORC2 complexes (Fig. 1.3), and there are some similarities in cellular roles of two complexes. The mTORC1 complex mainly acts as a leading regulator of proliferation and cell growth by promoting anabolic routes that involve the biosynthesis of lipids, organelles, and proteins, and through reducing catabolic routes, such as autophagy. Meanwhile, the mTORC2 complex controls cell survival and cytoskeletal structure [57, 58]. The mTORC1 complex consists of five components, as summarized in Table 1.1 [57].

The receptor tyrosine kinase (RTK)/phosphoinositide 3-kinase (PI3K)/AKT/mTORC1 signaling axis is considered as the standard, and the most environmentally responsive, signal-mediated controller of canonical autophagy. In order to initiate autophagy, multiple signaling pathways converge on two protein complexes: PI3K lipid kinase complex and ULK1 protein kinase complex. The mTORC1 complex can regulate autophagy by adjusting the activity of the ULK1 protein kinase complex. When mTORC1 is activated, the autophagy process is inhibited following the phosphorylation of ULK1 by mTORC1. Various cellular stresses can inhibit the activity of mTORC1, which promotes the activity of ULK1 complex, thereby stimulating autophagy [26]. In contrast, the availability of growth factors such as insulin, and the activation of PI3K complex established by RTK autophosphorylation, initiates the phosphorylation of insulin receptor substrates (IRSs), which then enrolls and activates PI3K to produce phosphatidylinositol (3,4,5)-trisphosphate (PIP3) [60]. A serine–threonine kinase, PDK1, is recruited by PIP3 and subsequently triggers the phosphorylation of the pleckstrin homology domain (PHD) of AKT at Thr308 [61]. Similarly, the stimulus of the PI3K signaling

---

Fig. 1.3 (continued) enriched in brain (RHEB) activation of mTORC1. Proline-rich AKT substrate of 40 kDa (PRAS40) also plays a role as a negative regulator of mTORC1, and it is sequestered by 14-3-3. PIM-1 promotes the phosphorylation of PRAS40, separating it from mTORC1, while the phosphorylation of 4EBP1 is controlled by mTORC1. Then, mTORC1 activates both 4EBP1 and p70S6K to regulate protein synthesis, cell cycle, progression, and proliferation. The p70S6K1 protein inhibits mTOR signaling. mTORC1 activation plays a stimulatory role in lipid synthesis, mitochondrial output via signal transducer and activator of transcription 3 (STAT3) signaling, and adipogenesis by lipin1 stimulation [59]. Glucose, hypoxia, and stress inhibit mTOR signaling pathways via TSC activation, or through direct interactions with mTORC1 in the glucose signaling pathway. However, the signaling of amino acids triggers the mTORC1 signaling via forming regulator complex in addition to activating the recombination-activating gene (RAG) complex. The mTORC2 participates in modulating survival and cell growth. Akt activation relies on the phosphorylation at two sites, which are Thr308 and Ser473 via another activator termed PDK2. The upstream of AKT through mTORC2 can subsequently activate mTORC1. The effector protein kinase C (PKC), a partner of the mitogen-activated protein kinase (MAPK)/extracellular signal-regulated kinase (ERK) signaling pathway, can be activated and phosphorylated by mTORC2. The signaling pathways of RAS, RAF, and MAPK can also stimulate mTOR signaling. The activation of ERK1 and ERK2, in addition to p90 ribosomal protein S6 kinase (RSKs), can suppress TSC2, thereby triggering mTORC1 or phosphorylation of RAPTOR

Table 1.1 Five components of mTORC1

Component abbreviation	Component name
mTOR	Mammalian target of rapamycin
Raptor	Regulatory-associated protein of mTOR
mLST8	Mammalian lethal with Sec13 protein 8
PRAS40	Proline-rich AKT substrate 40 kDa
DEPTOR	DEP domain-containing mTOR-interacting protein

pathway results in an increase in PIP3, which leads to the suppression of autophagy through the phosphorylation and activation of AKT via PDK1, while the excessive activation of the phosphatidylinositol (PI) phosphatase PTEN stimulates autophagy by inhibiting the corresponding pathway [57].

The mTORC1 complex is controlled via various signals arising from several cytokines, growth factors, cellular energy levels, hypoxia, Wnt proteins, and DNA damage throughout the classical PI3K-AKT-TSC (tuberous sclerosis complex)1/2-RHEB signaling pathway, which subsequently regulates autophagy [57, 62]. Through interaction with the mTOR catalytic domain, active GTP-loaded RHEB triggers mTORC1 on the lysosome, and TSC1/2 controls mTORC1 negatively by switching RHEB to the inactive GDP-bound state. Furthermore, AKT similarly offers a signal to mTORC1 in a TSC1/2-independent fashion by dissociating and phosphorylating PRAS40, an mTORC1 inhibitor, from RAPTOR [63]. RTK-dependent RAS-MAPK signaling pathway molecules such as ERK and ribosomal protein S6K can also phosphorylate TSC2, thereby inhibiting the TSC1/2 complex, in addition to activating RHEB-mediated mTORC1 [57].

Likewise, hypoxic conditions or alterations in energy states can also regulate the activity of mTORC1 using the AMP-activated protein kinase (AMPK)/TSC signaling pathway, which affects the autophagy process [64]. AMPK plays an important role as a cellular energy sensor, capable of detecting the differences in ATP/AMP ratios [64]. In the course of metabolic stress, several molecules, such as Ca^{2+}/calmodulin (CaM)-dependent protein kinase-β (CaMKKβ) [65], transforming growth factor-β-activated kinase 1 (TAK1), and liver kinase B125, can phosphorylate and activate AMPK [66]. When energy conditions are low, the activated AMPK will phosphorylate TSC2, which in turn is phosphorylated via glycogen synthase kinase 3 (GSK-3) [67, 68]. Consequently, AMPK signaling promotes the tuberous sclerosis complex 1 (TSC1) complex-mediated inhibition of mTORC1, thereby inducing autophagy. Moreover, there is a strong connection between the core signaling pathway of autophagy and AMPK, which can be observed when ULK1 is directly phosphorylated by AMPK [30, 69], stimulating autophagy signaling. Furthermore, AMPK can also control mTOR signaling by phosphorylating Raptor, which then connects to 14-3-3 protein, thereby leading to the allosteric inhibition of mTORC1 [70]. AMPK also has a role as the mediator during hypoxia-induced autophagy. The impairment of mitochondrial respiration due to hypoxia, or resulting from a low ATP/AMP ratio, can subsequently trigger AMPK and at the same time inhibit mTORC1 [71].

In addition to AMPK-mediated activation, mTORC1 can also be inhibited by the induction of regulation in development and DNA damage response 1 (REDD1) protein, which activates TSC1/2 [72]. Moreover, hypoxia conditions stimulate mitophagy through hypoxia-inducible factor 1α (HIF-1α)-dependent transcriptional activation of Bcl-2/adenovirus E1B 19 kDa protein-interacting protein 3 (BNIP3), which interrupts the interaction between Beclin1 and Bcl-2 [73]. Under the same hypoxic conditions, the tumor suppressor promyelocytic leukemia (PML) protein hinders mTOR signaling by physically interacting with mTOR [74], which subsequently terminates the interaction between mTOR and the positive regulator RHEB [73]. On the other hand, during genotoxic stress conditions, the tumor suppressor p53 protein stimulates the autophagy process by inhibiting mTORC1 via AMPK- and TSC1/2-dependent manner [75]. The signaling of mTORC1 is quite sensitive to an amino acid shortage, which activates mTORC1 and stimulates the lysosomal nutrient-sensing (LYNUS) mechanism before finally involving multiple protein complexes [76–78].

Many of the relevant protein complexes have been identified and characterized, including the vacuolar H^+-adenosine triphosphatase (ATPase) complex, heterodimeric Rag GTPase, RHEB, and pentameric regulator [78]. The connection within the mTORC1 signaling pathway and amino acid sensing is complicated as mTORC1 senses intralysosomal and cytosolic amino acids through separate mechanisms. The translocation of mTORC1 to the lysosomal surface as a response to intralysosomal amino acids is stimulated by the nucleotide state of Rag GTPase, which contains RagA and RagB bound to RagC or RagD [79]. Binding of Raptor to the active nucleotide-bound states of the Rags facilitates connection of mTORC1 to the lysosomal surface, where the RHEB activator is observed [80]. Furthermore, the vATPase-dependent fashion also plays a role in controlling amino acids within the lysosomal lumen because the inhibition of the vATPase can increase the concentrations of most metabolites at the lysosomal lumen [81, 82]. Among the mTORC1-regulating amino acids, glutamine and leucine have received the most attention [83]. Glutamine can enter cells via solute carrier family protein 1 member 5/alanine/serine/cysteine-preferring transporter 2 (SLC1A5/ASCT2) complex, while leucine can move into cells in the presence of glutamine through the heterodimeric SLC7A5/L-amino acid transporter 1 (LAT1)/SLC3A2 antiporter [57]. The results of a previous study have documented that leucine flux toward the lysosomal lumen is assisted via SLC7A5-SLC3A2 complex in the SLC family proteins [84], which acts as a plasma membrane antiporter that can carry leucine to the cells in exchange for glutamine [85]. Moreover, a similar study has demonstrated that overexpression of the lysosomal-associated transmembrane protein 4b facilitates lysosomal localization of the SLC7A5-SLC3A243 complex and consequently a leucine flux transported toward the lysosomes [86]. The mTORC1 complex also reportedly can sense the existence of arginine [87]. It has been proposed that arginine activates mTORC1 through a similar mechanism as leucine, including Rag GTPases, where SLC38A9, the lysosomal amino acid transporter, plays a role as an amino acid sensor [87–89]. Intralysosomal arginine sensor stimulation can lead to the interaction of SLC38A9 with Rag–regulator–vATPase complex, in addition to activation of mTORC1 signaling on the lysosomal membrane [89]. In contrast, arginine cooperates with growth factor signaling to block the interaction between RHEB and TSC1/2 on the lysosomes, which additionally facilitates the separation of TSC2 from lysosomes, thus allowing RHEB to activate mTORC1 [90]. Additionally, as a feedback response to *S*-adenosylmethionine (either SAM, SAMe, or AdoMet), SAmTOR is recognized as an mTORC1 signaling controller through interactions with GTPase-activating protein (GAP) for RagA/B and GATOR1 (GAP activity toward Rag 1). The detection of arginine and cytosolic leucine via mTORC1 is controlled through a novel GATOR signal pathway, which consists of two subcomplexes (GATOR1 and GATOR2). GATOR1 is a negative regulator of mTORC1 signaling by acting as a GAP for RagA/B, and a chief negative regulator of the amino-acid-sensing pathway, whereas GATOR2 is a positive regulator of mTORC1 signaling through acting upstream of or in parallel to GATOR1 [91].

The newly recognized KICSTOR complex connects GATOR1 with the lysosomes, in addition to regulating the mTORC1 signaling pathway via nutrient sensing [58]. Arginine sensor CASTOR153 (cellular arginine sensor for mTORC1) and leucine sensor Sestrins [92] are nonlysosomal amino acid sensors that can regulate mTORC1 via hydrolysis of RagA/RagB GTPases by GATOR1.

Sestrin2, a physiological controller of GATOR, can physically interact with GATOR2, thus leading to the release of the GATOR1. GATOR1 then inhibits RagB GTPase and subsequently suppresses the activation of mTORC1 under conditions of amino acid deficiency [57]. The cytosolic arginine sensor CASTOR1 can also stimulate mTORC1, via direct interaction with the GATOR1/2-Rag signaling pathway. Similar to Sestrin2, CASTOR1 can bind to and inhibit GATOR2 in the absence of arginine, thereby leading to the prohibited activation of the mTORC1 [93].

In the presence of amino acids, the folliculin-interacting protein 1 (FNIP1) complex plays a role as a powerful GAP for RagC/D, which can activate mTORC1 [57, 93]. Folliculin (FLCN) can also control the functions of lysosomes by stimulating mTORC1-dependent phosphorylation and cytoplasmic sequestration of transcription factor EB (TFEB) [94]. Under nutrient-rich conditions, TFEB is correlated with Rag GTPases on the lysosomes and phosphorylated on Ser142/211 via mTORC1 [78]. The mTORC1 complex can also activate interactions between phosphorylated TFEB and the 14-3-3 protein family, in addition to stabilizing TFEB in the cytosol, which subsequently inhibits its nuclear translocation [95]. Conversely, under starvation conditions, both Rag GTPase and mTORC1 are inhibited. The mTORC1 complex dissociates from the surface of the lysosome, and TFEB is dephosphorylated [96]. As a result, the phosphorylated TFEB/14-3-3 complexes are totally missed, and TFEB can move from the surface of lysosomes toward the nucleus.

In order to activate mTORC1, glutamine can be inserted into the cells using SLC1A5 (also called ASCT2). Leucine is imported into the cells by exchanging with glutamine through the heterodimeric SLC7A5-SLC3A2 antiporter [57]. The mTORC1 complex can also control mTORC2 signaling between the PI3K-AKT and mTORC1 signal pathways through a negative feedback loop [97]. Unlike mTORC1, mTORC2 signaling is not sensitive to the availability of nutrients. However, mTORC2 contributes to the regulation of autophagy via a transcription factor that is in charge of the signaling pathway of forkhead box protein O3 (FOXO3) [98]. The mTORC2 complex phosphorylates AKT at Ser473 [99]. Likewise, AKT activation requires the phosphorylation at Ser473 via mTORC2 and at Ser308 via PDK1. The mTORC2 is helpful for phosphorylating AKT, which in turn phosphorylates FOXO3. The phosphorylated FOXO3 combines with 14-3-3 protein to remain in the cytoplasm, in addition to preventing the activation of autophagy-related gene transcription [57]. FOXO also can inhibit mTORC1 by producing its negative control, Sestrin3. Conversely, mTORC2 can inhibit FOXO activity by stimulating glucocorticoid-regulated kinase 1 (SGK162) and AKT [98]. The inhibition of AKT subsequently causes the activation and dephosphorylation of FOXO1/3a, which regulates the expression of autophagy genes.

4.2 mTOR-Independent Signaling Pathways

4.2.1 Calcium Signaling in Autophagy

There are several alternative mTOR-independent signaling pathways that also regulate autophagy. For example, Ca^{2+}, MAPK/JNK, AMPK, HIF-15-007, microRNAs, and reactive oxygen species (ROS) also perform regulatory roles in autophagy. The most puzzling intracellular autophagy-regulating pathway is Ca^{2+} channels, which can lead to both activation and inhibition of Ca^{2+}-mediated autophagy [100, 101]. The debate over the multiple roles of Ca^{2+} channels in the regulation of autophagy emerges from the ambiguity of the subsequent intracellular calcium ion signaling and the cellular situation. Normally, Ca^{2+} signaling is maintained through differences in localization close to the plasma membrane (PM) or organelles, such as the mitochondria, Golgi, lysosomes, nucleus, and endoplasmic reticulum, with different types of machinery for regulation and activation [102, 103]. In a stable cellular environment, an unprompted, basal inositol triphosphate (IP3)-mediated Ca^{2+} signal arises from endoplasmic reticulum to mitochondria, which facilitates ATP production and mitochondrial bioenergetics [104]. Because of the existence of membrane contact sites, the release of Ca^{2+} from the endoplasmic reticulum leads to an increase in cytosolic Ca^{2+} and, at the same time, the increase in Ca^{2+} concentration in other organelles such as lysosomes and mitochondria [105, 106]. Although lysosomal Ca^{2+} can be released, Ca^{2+} also can be stored by several signaling pathways. The most significant channels are transient receptor potential (TRP) and two-pore segment channel superfamily cation channels [107]. Under conditions of cellular stress, the increased cytosolic Ca^{2+} signals stimulate autophagy due to low energy generation [108]. In this way, cells can change the Ca^{2+} signaling mode from inhibition to induction of autophagy [100].

According to several previous studies, basal autophagy can be preserved at a minimum level via L-type calcium ion channel-dependent and IP3R channel-dependent mechanisms [109], while the inhibition of IP3R can cause the induction of autophagy via reduced production of ATP and stimulated AMPK [110]. There is also a positive feedback loop associated with the alteration of calcium ion concentrations near the plasma membrane, the production of IP3R, calcium ion-dependent cysteine protease, and the activation of phospholipase Cε and the cAMP.

When growth factors are released, cellular stimulation can lead to the activation of calpain through intracellular Ca^{2+}, which induces the production of cAMP and activates adenylyl cyclase. It has been proposed that active calpain can mediate activity of the complex consisting of Beclin1 and ATG5 to suppress the initiation of autophagy [111, 112]. Similarly, the increased level of intracellular cAMP can also lead to the activation of a guanine nucleotide exchange factor and exchange protein directly activated by *cAMP* (EPAC). EPAC can activate both PLCε and Rap2B. The activated PLCε facilitates the hydrolysis of PI(4,5)P2 to produce IP3, which in turn inhibits autophagy [109]. The tetrameric endoplasmic reticulum-resident calcium ion-IP3R channel discharges calcium ions from ER to cytosol as a response to the second messenger IP3 [113]. Under conditions promoting cell growth, mitochondria can harvest Ca^{2+} through IP3R/voltage-dependent anion channel (VDAC) complexes, thus promoting the production of ATP in mitochondria, and leading to the suppression of basal autophagy through AMPK [100, 114]. On the other hand, the absence of Ca^{2+} leads to reduced production of ATP and AMPK activation, which subsequently initiates autophagy through an mTOR-independent pathway [57].

In addition, previous studies have shown that VDAC1 sited in the outer mitochondrial membrane is a target for Parkin-mediated K63 and K27-linked polyubiquitination and can cause the activation of mitophagy [115]. Conversely, VDAC2, also located in outer mitochondrial membrane, can hinder autophagy by facilitating the interaction between Beclin1 and the Bcl-2 family member Bcl-XL that forms a triple complex [116]. Moreover, IP3R plays a role as a scaffold protein in the interaction between Beclin1 and Bcl-2, and also in forming the tripartite complex consisting of Beclin1, Bcl-2, and IP3R, under Ca^{2+}-releasing actions [117]. As IP3R independently interacts with Bcl-2 and Beclin1, the reduced level of free Beclin1 inhibits autophagy. Furthermore, the ryanodine receptors (RyRs) and IP3Rs are similarly involved in controlling autophagy [118]. The inhibition of RyR causes an increase in autophagic flux, independently from mTOR signaling pathway [119]. Additionally, the high level of cytosolic Ca^{2+} can lead to the inhibition of mTOR activity and cause an increase in autophagic flux by stimulating the activity of upstream AMPK activator, CaMKKβ [120]. On the other hand, AMPK can avoid mTORC1 inhibition and stimulate the autophagy process through phosphorylation of the ULK1 complex [30]. Cells are extremely sensitive to the elevated intraluminal Ca^{2+} concentration that can cause endoplasmic reticulum stress and the unfolded protein response (UPR) [121, 122]. ER stress can also occur via the sarcoendoplasmic reticulum Ca^{2+} ATPase (SERCA) blocker thapsigargin, thus causing the elevation of sustained cytosolic Ca^{2+} and chronic depletion of ER Ca^{2+} stores, and, consequently, triggering impaired protein folding in the ER [122].

Under low or transient ER stress conditions, ER-resident chaperones such as glucose-regulating protein 78 (GRP78) or immunoglobulin heavy-chain-binding protein (BiP) can stimulate autophagy [121], whereas severe or persistent ER stress can lead to the suppression of autophagy and the initiation of cell death. With depletion of calcium ions in the ER, the unfolded proteins are collected atypically in ER and cells, thereby leading to a filling mechanism with adequate endoplasmic reticulum calcium ions through the UPR, a concerted program activated by three endoplasmic reticulum stress sensors, including enabling transcription factor 6 (ATF6), protein kinase RNA-like endoplasmic reticulum kinase (PERK), and inositol-requiring enzyme 1α (IRE1α) [123].

With onset of ER stress, BiP preferentially attaches to misfolded proteins in the ER, thus leading to the activation of stress sensors. The accumulation of unfolded proteins in the ER can trigger IRE1α oligomerization in the ER membrane, in addition to the cytosolic domain of IRE1α autophosphorylation [124]. The RNase domain of IRE1α permits the translation of X-box-binding protein 1 (XBP-1) mRNA into the XBP-1 protein. The XBP-1 protein can then react with the agents of upregulating genes elaborated in the unfolded protein response, and ER-assisted degradation, to stimulate cytoprotection and re-establish protein homeostasis [124]. To control ER stress-induced autophagy, IRE1α stimulates XBP-1-forkhead box O1 (XBP1-FOXO1) interactivity [125] and tumor necrosis factor (TNF) receptor-associated factor 2 apoptosis signal-regulating kinase 1-JNK cascade [124]. Likewise, XBP-1 mRNA linking activates autophagy via transcriptional triggering of Beclin1 [126]. During ER stress, PERK inhibits the influx of recently created proteins toward the stressed endoplasmic reticulum and phosphorylates eukaryotic translation initiation factor 2 subunit α (eIF2α). The phosphorylated eIF2α prevents its recycling into the active GTP-binding protein, which is necessary to initiate polypeptide chain synthesis, and permits the preferential translation of the UPR-dependent genes. The significant target genes driven by activating transcription factor 4 (ATF4) are those for the growth arrest and DNA damage-inducible 34 (GADD34) protein and the transcription factor C/EBP homologous protein (CHOP) [127]. The UPR-dependent signaling pathway of PERK and eIF2α can stimulate ATG12 expression and trigger transcriptional control of ATG genes via ATF4 [124, 128].

4.2.2 Calcium-Releasing Channel Control in Autophagy Mediated by TRPML

Previous studies have suggested that the elevation of nicotinic acid adenine dinucleotide phosphate (NAADP) and second messenger cyclic ADP ribose (cADPR) can stimulate transient receptor potential (TRP) channels to mediate the release of endolysosomal calcium ions [107, 129]. The hetero-multimeric TRP channels include members from different subfamilies: TRPC (canonical), TRPM (melastatin), TRPML (mucolipin), TRPN (no mechanoreceptor potential C), TRPA (ankyrin), TRPV (vanilloid), and TRPP (polycystin) [107]. The functional role of TRP channels of receptor-operated calcium entry (ROCE) and store-operated calcium entry (SOCE) remains unclear. SOCE is a very common calcium ion entry signal pathway, and it can be activated in response to the depletion of ER calcium ion stores [130]. The chief mediators for SOCE are the ER calcium sensor stromal interaction molecule 1 (STIM1), and the plasma membrane calcium release-activated calcium modulator 1 (ORAI1), which can construct the calcium ion-selective opening of the calcium store-operated channel [131, 132]. SOCE is a vital cellular mechanism involved in refilling endoplasmic reticulum calcium ion stores, thereby triggering calcium ion-mediated pro- or antisurvival mechanisms and corresponding gene transcription. The diminution of calcium ions from the endoplasmic reticulum can cause STIM1 to transfer into the plasma membrane, where it attaches to ORAI1 and activates it, thus leading in turn to calcium entry [133]. Transient receptor potential canonical 1 (TRPC1) protein has been identified as a molecular regulator of the SOCE. The reduced availability of STIM1 severely diminishes the current of calcium ions and endogenous TRPC1-mediated SOCE, whereas the co-expression of exogenous STIM1 with TRPC1 can stimulate the SOCE [134, 135]. TRPC1 interacts with STIM1 to reduce and then fill the calcium ion stores in the ER, which triggers the separation of STIM1 from TRPC1 and the inhabitation of the TRPC1 role [135–137]. The transient receptor potential ion channel mucolipin (TRPML) includes TRPML1 (MCOLN1/mucolipin1), TRPML2 (MCOLN2), and TRPML3 (MCOLN3) [107]. TRPML1 is present in several tissues, while TRPML2 is generally found only in immune cells, specifically in the spleen, lymphatic nodes, and thymus. TRPML3 is present in several types of tissues, such as the colon, intestine, skin, and lung [138]. In mammals, the TRPML channels are activated on the internal side of the plasma membrane by an increase in phosphoinositide in the intracellular membrane, PI(3,5)P2 (phosphatidylinositol 3,5-bisphosphate) [139]. In contrast, TRPML1 is antagonized via increased PI(4,5)P2 (phosphatidylinositol 4,5-bisphosphate) [140] in the plasma membrane, thus increasing the activity of TRPML channels, once located at the lysosomal membrane.

The TRPML1 channel is mainly located in the late endolysosomal compartment. It is involved in various intracellular procedures, such as the regulation of autophagy and lysosomal biogenesis [107, 141]. TRPML1 is able to target the plasma membrane via a biosynthetic route from either the Golgi apparatus or lysosomal exocytosis [141]. Previous studies have defined three classes of proteins that can significantly react with TRPML1. First, the lysosomal-associated protein transmembrane (LAPTM) is known to interact with TRPML1 on the lysosomes and endosomes [142]. Second, the TRPML1 amino terminus can also interact with calcium ions dependently with apoptosis-linked gene 2 (ALG-2) [143]. Third, TRPML1 is believed to interact with chaperone heat shock cognate 70 (HSC70), in addition to other members of the chaperone-mediated autophagy (CMA) complex. Although the physiological importance of the first two classes of interactions is yet to be clarified, TRPML1 is believed to control CMA through its interaction with HSC70 [144]. Moreover, TRPML1 is co-localized with the endoplasmic reticulum calcium ion sensor STIM1 in differentiated motoneuron-like cell line (NSC-34), although this particular interaction remains questionable [145].

During malnutrition, the overexpression of TRPML1 results in a substantial increase in autophagic flux, while the silencing of TRPML1 decreases PI3P-positive vesicles, suggesting that TRPML1 may additionally be involved in the early stage of autophagy [106]. Furthermore, TRPML1 can also regulate the location of autophagous lysosomes. Under starvation conditions, TRPML1 is also involved in the centripetal transport of lysosomes [146].

Calcium-binding protein ALG-2 is discharged by the PI(3,5)P2/TRPML1-mediated calcium ions that fuse to TRPML1, which then involves the dynein–dynactin complex, in order to affect the transport of lysosomes [143, 146]. TRPML1 can also sense pH changes, which may be a possible explanation for starvation-induced reduction in the concentration of PI(3,5)P2, together with alterations in cytosolic pH [106, 146]. TRPML1 has also been identified as a vital factor in activating the transcription factor TFEB [77, 96]. Under conditions of nutrient deficiency, the lysosomal calcium ion release channel TRPML1 can adjust the action of CaM-dependent calcineurin (CaN) phosphatase in close range to lysosomes [77, 106, 147]. In order to facilitate the nuclear translocation, CaN binds and dephosphorylates TFEB at Ser211 and Ser142 [106]. The dephosphorylated TFEB binds to coordinated lysosomal expression and regulation (CLEAR) sites, and can control autophagy by regulating lysosomal transcription and autophagy genes [76, 77, 96]. Because TRPML1 is a transcriptional target of TFEB (to make an RNA copy of its sequence), a positive feedback loop is established to enhance the TRPML1/TFEB response [77, 106, 147]. The TRPML1/TFEB signaling pathway is also involved in ROS production. Here, TRPML1 plays a role as a ROS level sensor in the cell, in addition to having the capability to be activated under the excessive ROS condition, thus inducing autophagy and mitochondrial damage [148, 149].

TRPML3 is found in various intracellular compartments, including lysosomes, endosomes, and APs. TRPML3 furthermore has a regulatory role in autophagy [107, 141], since its overexpression in HeLa cells has been shown to activate autophagy [150]. Under starvation conditions, TRPML3 overexpression can activate the formation of APs. The TRPML3 channel is recruited to APs, which provide the calcium ions that are vital for fusion and fission during autophagy [150]. Additionally, the overexpression of the calcium ion channel TRPML3 in the plasma membrane stimulates autophagy [150]. Moreover, TRPML3 particularly binds to GATE-16, an ATG8 homolog essential for AP maturation in mammalian cells [151], yet the relationship between GATE-16, TRPML3, and AP maturation remains unclear. Similarly, the role of TRPML2 in autophagy is also still ambiguous [138].

4.2.3 The mTOR-Independent TRPML1 Channel

The mTOR-independent signal pathway is not fully independent from mTOR activity, suggesting that the calcium ion TRPML channel release is necessary for enabling mTORC1, which subsequently suppresses autophagy [107, 152]. The knockdown of lysosomal calcium ion channel TRPML1 reportedly inhibits the activity of mTORC1 [153]. The outcome of TRPML1 knockdown is inverted via thapsigargin, which is consistent with the fact that mTORC1 can activate cytosolic calcium ion signals. Previous studies have revealed that mTOR promptly targets and deactivates the TRPML1 channel, thereby suppressing autophagy via the phosphorylation of serine residues, including Ser576 and Ser572, in TRPML1 [154]. Interestingly, TRPML1, AMPK, and mTOR interactions are vital for regulating autophagy. During starvation conditions, AMPK suppresses mTOR activity, which subsequently modifies TRPML1 actions through a feedback loop. Conversely, under nutrient-rich conditions, the activity of AMPK is not observed in a feedback loop, and likewise, the negative regulation of mTOR is notably diminished. Under such conditions, the TRPML1 channel is phosphorylated and regulated directly via mTOR kinase [154].

The energy inputs and growth factors produce signals to modulate the mTORC1-activating model [155, 156] throughout the TSC1/2-Rheb axis. In this context, amino acids play a pivotal role by adjusting the nucleotide state of the Rag GTPases, in addition to stimulating the translocation of mTORC1 to lysosomes, where the interaction and activation are related to lysosomal-localized GTP bound to Rheb [60, 157]. Additionally, mTOR is also recognized as an atypical CaM-dependent kinase, since both CaM and calcium ions are needed for mTORC1 activity [158]. During starvation conditions, with a deficiency of amino acids, a major upregulation of TRPML1 activity occurs [147]. At the time that mTORC1 separates from the surface of lysosomes, calcium release occurs across the TRPML1 channel, which could play a role in activating mTORC1.

The results of another study suggest that the released lysosomal calcium via increasing TRPML1-mediated local levels of calcium ions subsequent to translocation of mTORC1 to lysosomes, and stimulating calcium ions binding with CaM, attaches and stimulates the kinase of the mTORC1 complex [153]. This suggests an additional function of lysosomal localization of mTORC1, involving the detection of stimulation from localized lysosomal calcium [153].

4.2.4 The Regulation of Autophagy Via miRNA, ROS, and JNK-Beclin1

The Melastatin1 (TRPM1) gene, a TRP channel member, encodes two transcriptions. The first one is microRNA such as miR-211 encoded via the sixth intron, and related to regulating the expression of TRPM1. The second one is the TRPM1 channel protein encoded by each exon [107, 159, 160]. Under stress conditions, such as starvation, miR-211 promptly targets the RICTOR, a part of mTORC2, which subsequently suppresses the mTORC1 signaling pathway and activates the translocation of microphthalmia-associated transcription factor (MITF) to the nucleus. The axis of MITF-miR-211 finalizes the autophagy amplification loop system, which can regulate autophagic activity [161]. In addition, miR-155 can also hinder autophagy by inhibiting numerous ATG genes, including ATG14, ATG5, ATG3, FOXO3, LC3, and ULK1, at both protein and mRNA levels [162]. Although miR-155 is able to suppress autophagy by mTOR signals [163], its interactions with downstream targets create multiple autophagy-inhibiting events.

Previous studies have also shown that miR-30b binds directly to Beclin1 and significantly affects the nucleation of the isolation membrane [162]. Similarly, miR-376b [164] and miR-216b [165, 166] inhibit autophagy through downregulation of Beclin1. Under hypoxic conditions, the expression of miR-210 is enabled by hypoxia-inducible factor 1-alpha (HIF-1α), which specifically regulates the Bcl-2-dependent autophagy signal pathway [167].

Activated autophagy due to starvation conditions can be reduced by miR-183 through targeting of UV radiation resistance-associated gene (UVRAG) [168], whereas autophagy triggered by cisplatin can be inhibited by miR-152 directly targeting ATG14, which is considered a chief part of vesicle nucleation [169]. Alternatively, during glucose starvation, the miR-90-295 cluster can suppress autophagic cell death in melanoma cells by inhibiting ATG7 and ULK1 [170], while the inhibition of miR-25 can trigger autophagic cell death via upregulated expression of ULK1 [170].

During intracellular oxidative stress, ROS production increases significantly, which activates autophagy [171]. It is conspicuous that the initiation and execution phases of autophagy are linked to the effects of ROS [172]. These signal pathways, mainly involving transcriptional processes, occur in the nucleus, while posttranscriptional processes occur in the cytoplasm. Additionally, these transcriptional controlling processes include the activation of nuclear factor E2-related factor 2 (NRF2), HIF-1, FOXO3, and p53, and the production of proteins, as well as the adjustment of the autophagy process in the cytoplasm at the site of exposure to ROS [173]. The amount of ROS could contribute to the adjustment during the initiation process of autophagy through several signaling pathways, including ROS-TP53-induced glycolysis and apoptosis regulator (TIGAR) autophagy, ROS-NRF2-p62 autophagy, ROS-FOXO3-LC3/BNIP3 autophagy, and ROS-HIF-1α-BNIP3/NIX autophagy [173]. Moreover, ROS stimulates the ER PERK stress sensor and sequentially activates the expression of autophagy genes [174]. ROS also suppresses the deconjugation activity of ATG4 protease in the LC3 recycling route, and thus supports sustained AP production [175]. An earlier study reported that, under hypoxia conditions, HIF-1α stimulates EC109 cell autophagy by positively regulating the p27-E2F1 signal pathway [176]. Through the activation of mitochondria-selective autophagy, ROS production can be reduced by HIF-1α [177]. Additionally, the redox-sensitive protein, AMPK, is expressed differently in lysosomes/endosomes, mitochondria, and ER [30]. A particular mitochondria-associated endoplasmic reticulum membrane (MAM) microdomain-oriented AMPK has been developed, with the capability to sense the level of intracellular ATP, in addition to phosphorylating Beclin1 and initiating autophagy in an mTORC1-independent manner [178].

Even though ROS can activate AMPK indirectly through an increase in AMP [179], some studies have documented that ROS can also regulate AMPK activity directly via a posttranslational modification [180]. At least one previous study has proposed that the $\alpha1$ isoform of AMPK is extremely important for AP maturation and fusion with lysosomes [181]. Under stress conditions such as nutritional deficiency or oxidative stress, the AMPK/JNK signal pathway is responsible for the regulation of autophagy [182]. The phosphorylation of c-Jun by JNK can trigger the translocation of c-Jun toward the nucleus. The translocated c-Jun controls the expression of Bcl-2 by phosphorylation [183]. Thus, JNK releases Beclin1 from the Beclin2 complex and stimulates the formation of AP. Furthermore, JNK adjusts autophagy through FOXO transcription of ATG genes [183].

5 Conclusion

Autophagy is a vital process, which is responsible for homeostasis and controlling the quality of cellular components. During the past decade, broad advances have been achieved regarding an understanding of molecular machinery and the network-regulating autophagy, beginning with yeast and concluding with mammals. Hitherto, the majority of autophagy studies have focused on the role of mTOR as a chief regulator of the autophagy process in yeast and mammalian cells. Furthermore, significant progress has been made toward understanding the function of nutrient sensing and controlling the mTORC1 via RHEB and TSC1/2 at the lysosome level. Despite these accomplishments, our comprehension of the signaling network is still not complete and many fundamental questions have yet to be solved, including, but not limited to, the cellular localization of mTORC1, the way of rapamycin to inhibit mTORC1 function, how RHEB triggers mTORC1, and the localization of mTORC1 and TSC1/TSC2 complexes in the cytoplasm during periods of amino acid deficiency or enrichment.

Some studies have described the components involved in controlling mTORC1 through amino acids, but the underlying mechanism for sensing amino acids is still not clearly understood. The latest studies have reported that signal transduction via mTORC1/2 is essential for the regulation of autophagy, but have not adequately addressed how the mTORC1/2 signal pathways can be merged.

The roles of ion channels in autophagy are also quite vague. The majority of studies report that the roles of ion channels in the regulation of autophagy have focused on calcium ion channels such as TRPML, but the roles of the other channels, such as potassium and sodium ion channels, remain obscure. Despite the important progress accomplished in calcium ion channel biology, the roles of intracellular calcium signaling in regulating autophagy are unclear. Answers to these questions will improve our understanding of autophagy and help us to determine how this information can be applied to promote new research aimed at preventing and/or curing diseases.

References

1. Mortimore GE, Miotto G, Venerando R et al (1996) Autophagy. In: Biology of the lysosome. Plenum Press, New York, pp 93–135

2. Choi AM, Ryter SW, Levine B (2013) Autophagy in human health and disease. N Engl J Med 368:651–662
3. Levine B, Kroemer G (2008) Autophagy in the pathogenesis of disease. Cell 132:27–42
4. Poüs C, Codogno P (2011) Lysosome positioning coordinates mTORC1 activity and autophagy. Nat Cell Biol 13:342–344
5. Mizushima N, Hara T (2006) Intracellular quality control by autophagy: how does autophagy prevent neurodegeneration? Autophagy 2:302–304
6. Korolchuk VI, Menzies FM, Rubinsztein DC (2009) A novel link between autophagy and the ubiquitin-proteasome system. Autophagy 5:862–863
7. Hayat M (2017) Overview of autophagy. In: Autophagy: cancer, other pathologies, inflammation, immunity, infection, and aging. Elsevier, Amsterdam, pp 1–122
8. Mizushima N (2018) A brief history of autophagy from cell biology to physiology and disease. Nat Cell Biol 20:521–527
9. Yang Z, Klionsky DJ (2010) Mammalian autophagy: core molecular machinery and signaling regulation. Curr Opin Cell Biol 22:124–131
10. Tsukada M, Ohsumi Y (1993) Isolation and characterization of autophagy-defective mutants of Saccharomyces cerevisiae. FEBS Lett 333:169–174
11. Yang Z, Klionsky DJ (2010) Eaten alive: a history of macroautophagy. Nat Cell Biol 12:814–822
12. Noda NN, Inagaki F (2015) Mechanisms of autophagy. Annu Rev Biophys 44:101–122
13. Deretic V, Saitoh T, Akira S (2013) Autophagy in infection, inflammation and immunity. Nat Rev Immunol 13:722–737
14. Shirakabe A, Kobayashi N, Hata N et al (2016) The serum heart-type fatty acid-binding protein (HFABP) levels can be used to detect the presence of acute kidney injury on admission in patients admitted to the non-surgical intensive care unit. BMC Cardiovasc Disord 16:1–12
15. Ryter SW, Nakahira K, Haspel JA et al (2012) Autophagy in pulmonary diseases. Annu Rev Physiol 74:377–401
16. Rockel JS, Kapoor M (2016) Autophagy: controlling cell fate in rheumatic diseases. Nat Rev Rheumatol 12:517
17. Lim Y-M, Lim H, Hur KY et al (2014) Systemic autophagy insufficiency compromises adaptation to metabolic stress and facilitates progression from obesity to diabetes. Nat Commun 5:1–14
18. Menzies FM, Fleming A, Rubinsztein DC (2015) Compromised autophagy and neurodegenerative diseases. Nat Rev Neurosci 16:345–357
19. Zhong Z, Sanchez-Lopez E, Karin M (2016) Autophagy, NLRP3 inflammasome and autoinflammatory/immune diseases. Clin Exp Rheumatol 34:12–16
20. Galluzzi L, Pietrocola F, Bravo-San Pedro JM et al (2015) Autophagy in malignant transformation and cancer progression. EMBO J 34:856–880
21. Lapierre LR, Kumsta C, Sandri M et al (2015) Transcriptional and epigenetic regulation of autophagy in aging. Autophagy 11:867–880
22. Poklepovic A, Gewirtz DA (2014) Outcome of early clinical trials of the combination of hydroxychloroquine with chemotherapy in cancer. Autophagy 10:1478–1480
23. Vakifahmetoglu-Norberg H, Xia H-G, Yuan J (2015) Pharmacologic agents targeting autophagy. J Clin Invest 125:5–13
24. He C, Klionsky DJ (2009) Regulation mechanisms and signaling pathways of autophagy. Annu Rev Genet 43:67–93
25. Sengupta S, Peterson TR, Laplante M et al (2010) mTORC1 controls fasting-induced ketogenesis and its modulation by ageing. Nature 468:1100–1104
26. Hosokawa N, Hara T, Kaizuka T et al (2009) Nutrient-dependent mTORC1 association with the ULK1–Atg13–FIP200 complex required for autophagy. Mol Biol Cell 20:1981–1991
27. Lee E-J, Tournier C (2011) The requirement of uncoordinated 51-like kinase 1 (ULK1) and ULK2 in the regulation of autophagy. Autophagy 7:689–695

28. Wong P-M, Puente C, Ganley IG et al (2013) The ULK1 complex: sensing nutrient signals for autophagy activation. Autophagy 9:124–137
29. Jung CH, Jun CB, Ro S-H et al (2009) ULK-Atg13-FIP200 complexes mediate mTOR signaling to the autophagy machinery. Mol Biol Cell 20:1992–2003
30. Kim J, Kundu M, Viollet B et al (2011) AMPK and mTOR regulate autophagy through direct phosphorylation of Ulk1. Nat Cell Biol 13:132–141
31. Bento CF, Renna M, Ghislat G et al (2016) Mammalian autophagy: how does it work? Annu Rev Biochem 85:685–713
32. Yang YP, Liang ZQ, Gu ZL et al (2005) Molecular mechanism and regulation of autophagy 1. Acta Pharmacol Sin 26:1421–1434
33. Hamad MNM (2017) Autophagy: the powerful of immune response. MedCrave Group LLC, Edmond, OK
34. Mizushima N (2004) Methods for monitoring autophagy. Int J Biochem Cell Biol 36:2491–2502
35. Ueno T, Muno D, Kominami E (1991) Membrane markers of endoplasmic reticulum preserved in autophagic vacuolar membranes isolated from leupeptin-administered rat liver. J Biol Chem 266:18995–18999
36. Noda T, Suzuki K, Ohsumi Y (2002) Yeast autophagosomes: de novo formation of a membrane structure. Trends Cell Biol 12:231–235
37. Mizushima N, Yoshimori T, Ohsumi Y (2011) The role of Atg proteins in autophagosome formation. Annu Rev Cell Dev Biol 27:107–132
38. Axe EL, Walker SA, Manifava M et al (2008) Autophagosome formation from membrane compartments enriched in phosphatidylinositol 3-phosphate and dynamically connected to the endoplasmic reticulum. J Cell Biol 182:685–701
39. Park Y-E, Hayashi YK, Bonne G et al (2009) Autophagic degradation of nuclear components in mammalian cells. Autophagy 5:795–804
40. Hayashi-Nishino M, Fujita N, Noda T et al (2009) A subdomain of the endoplasmic reticulum forms a cradle for autophagosome formation. Nat Cell Biol 11:1433–1437
41. Ylä-Anttila P, Vihinen H, Jokitalo E et al (2009) 3D tomography reveals connections between the phagophore and endoplasmic reticulum. Autophagy 5:1180–1185
42. Uemura T, Yamamoto M, Kametaka A et al (2014) A cluster of thin tubular structures mediates transformation of the endoplasmic reticulum to autophagic isolation membrane. Mol Cell Biol 34:1695–1706
43. Orsi A, Razi M, Dooley H et al (2012) Dynamic and transient interactions of Atg9 with autophagosomes, but not membrane integration, are required for autophagy. Mol Biol Cell 23:1860–1873
44. Mcewan DG, Dikic I (2010) Not all autophagy membranes are created equal. Cell 141:564–566
45. Lystad AH, Carlsson SR, Simonsen A (2019) Toward the function of mammalian ATG12-ATG5-ATG16L1 complex in autophagy and related processes. Autophagy 15:1485–1486
46. Dooley HC, Razi M, Polson HE et al (2014) WIPI2 links LC3 conjugation with PI3P, autophagosome formation, and pathogen clearance by recruiting Atg12-5-16L1. Mol Cell 55:238–252
47. Kabeya Y, Mizushima N, Ueno T et al (2000) LC3, a mammalian homologue of yeast Apg8p, is localized in autophagosome membranes after processing. EMBO J 19:5720–5728
48. Wible DJ, Chao H-P, Tang DG et al (2019) ATG5 cancer mutations and alternative mRNA splicing reveal a conjugation switch that regulates ATG12-ATG5-ATG16L1 complex assembly and autophagy. Cell Discov 5:1–19
49. Ichimura Y, Kirisako T, Takao T et al (2000) A ubiquitin-like system mediates protein lipidation. Nature 408:488–492
50. Jahreiss L, Menzies FM, Rubinsztein DC (2008) The itinerary of autophagosomes: from peripheral formation to kiss-and-run fusion with lysosomes. Traffic 9:574–587

51. Berg TO, Fengsrud M, Strømhaug PE et al (1998) Isolation and characterization of rat liver amphisomes: evidence for fusion of autophagosomes with both early and late endosomes. J Biol Chem 273:21883–21892
52. Perera RM, Zoncu R (2016) The lysosome as a regulatory hub. Annu Rev Cell Dev Biol 32:223–253
53. Laplante M, Sabatini DM (2012) mTOR signaling in growth control and disease. Cell 149:274–293
54. Wang P, Guan Y-F, Du H et al (2012) Induction of autophagy contributes to the neuroprotection of nicotinamide phosphoribosyltransferase in cerebral ischemia. Autophagy 8:77–87
55. Xie R, Wang P, Ji X et al (2013) Ischemic post-conditioning facilitates brain recovery after stroke by promoting Akt/mTOR activity in nude rats. J Neurochem 127:723–732
56. Chen H, Qu Y, Tang B et al (2012) Role of mammalian target of rapamycin in hypoxic or ischemic brain injury: potential neuroprotection and limitations. Rev Neurosci 23:279–287
57. Saxton RA, Sabatini DM (2017) mTOR signaling in growth, metabolism, and disease. Cell 168:960–976
58. Sabatini DM (2017) Twenty-five years of mTOR: uncovering the link from nutrients to growth. Proc Natl Acad Sci 114:11818–11825
59. Gitto SB, Altomare DA (2015) Recent insights into the pathophysiology of mTOR pathway dysregulation. Res Rep Biol 6:1–16
60. Lane JD, Korolchuk VI, Murray JT et al (2017) mTORC1 as the main gateway to autophagy. Essays Biochem 61:565–584
61. Ziemba BP, Pilling C, Calleja VR et al (2013) The PH domain of phosphoinositide-dependent kinase-1 exhibits a novel, phospho-regulated monomer–dimer equilibrium with important implications for kinase domain activation: single-molecule and ensemble studies. Biochemistry 52:4820–4829
62. Martin TD, Chen X-W, Kaplan RE et al (2014) Ral and Rheb GTPase activating proteins integrate mTOR and GTPase signaling in aging, autophagy, and tumor cell invasion. Mol Cell 53:209–220
63. Vander Haar E, Lee S-I, Bandhakavi S et al (2007) Insulin signalling to mTOR mediated by the Akt/PKB substrate PRAS40. Nat Cell Biol 9:316–323
64. Meijer AJ, Codogno P (2007) AMP-activated protein kinase and autophagy. Autophagy 3:238–240
65. Nakanishi A, Hatano N, Fujiwara Y et al (2017) AMP-activated protein kinase–mediated feedback phosphorylation controls the Ca2+/calmodulin (CaM) dependence of Ca2+/CaM-dependent protein kinase kinase β. J Biol Chem 292:19804–19813
66. Herrero-Martín G, Høyer-Hansen M, García-García C et al (2009) TAK1 activates AMPK-dependent cytoprotective autophagy in TRAIL-treated epithelial cells. EMBO J 28:677–685
67. Inoki K, Ouyang H, Zhu T et al (2006) TSC2 integrates Wnt and energy signals via a coordinated phosphorylation by AMPK and GSK3 to regulate cell growth. Cell 126:955–968
68. Inoki K, Zhu T, Guan K-L (2003) TSC2 mediates cellular energy response to control cell growth and survival. Cell 115:577–590
69. Egan DF, Shackelford DB, Mihaylova MM et al (2011) Phosphorylation of ULK1 (hATG1) by AMP-activated protein kinase connects energy sensing to mitophagy. Science 331:456–461
70. Gwinn DM, Shackelford DB, Egan DF et al (2008) AMPK phosphorylation of raptor mediates a metabolic checkpoint. Mol Cell 30:214–226
71. Papandreou I, Lim A, Laderoute K et al (2008) Hypoxia signals autophagy in tumor cells via AMPK activity, independent of HIF-1, BNIP3, and BNIP3L. Cell Death Differ 15:1572–1581
72. Cam H, Easton JB, High A et al (2010) mTORC1 signaling under hypoxic conditions is controlled by ATM-dependent phosphorylation of HIF-1α. Mol Cell 40:509–520
73. Song Y, Du Y, Zou W et al (2018) Involvement of impaired autophagy and mitophagy in Neuro-2a cell damage under hypoxic and/or high-glucose conditions. Sci Rep 8:1–14

74. Salsman J, Stathakis A, Parker E et al (2017) PML nuclear bodies contribute to the basal expression of the mTOR inhibitor DDIT4. Sci Rep 7:1–16
75. Agarwal S, Bell CM, Rothbart SB et al (2015) AMP-activated protein kinase (AMPK) control of mTORC1 is p53-and TSC2-independent in pemetrexed-treated carcinoma cells. J Biol Chem 290:27473–27486
76. Settembre C, De Cegli R, Mansueto G et al (2013) TFEB controls cellular lipid metabolism through a starvation-induced autoregulatory loop. Nat Cell Biol 15:647–658
77. Settembre C, Di Malta C, Polito VA et al (2011) TFEB links autophagy to lysosomal biogenesis. Science 332:1429–1433
78. Settembre C, Fraldi A, Medina DL et al (2013) Signals from the lysosome: a control centre for cellular clearance and energy metabolism. Nat Rev Mol Cell Biol 14:283–296
79. Sancak Y, Bar-Peled L, Zoncu R et al (2010) Ragulator-Rag complex targets mTORC1 to the lysosomal surface and is necessary for its activation by amino acids. Cell 141:290–303
80. Powis K, De Virgilio C (2016) Conserved regulators of Rag GTPases orchestrate amino acid-dependent TORC1 signaling. Cell Discov 2:1–16
81. Zhang T, Wang R, Wang Z et al (2017) Structural basis for Ragulator functioning as a scaffold in membrane-anchoring of Rag GTPases and mTORC1. Nat Commun 8:1–10
82. Abu-Remaileh M, Wyant GA, Kim C et al (2017) Lysosomal metabolomics reveals V-ATPase-and mTOR-dependent regulation of amino acid efflux from lysosomes. Science 358:807–813
83. Jewell JL, Kim YC, Russell RC et al (2015) Differential regulation of mTORC1 by leucine and glutamine. Science 347:194–198
84. Milkereit R, Persaud A, Vanoaica L et al (2015) LAPTM4b recruits the LAT1-4F2hc Leu transporter to lysosomes and promotes mTORC1 activation. Nat Commun 6:1–9
85. Nicklin P, Bergman P, Zhang B et al (2009) Bidirectional transport of amino acids regulates mTOR and autophagy. Cell 136:521–534
86. Cormerais Y, Massard PA, Vucetic M et al (2018) The glutamine transporter ASCT2 (SLC1A5) promotes tumor growth independently of the amino acid transporter LAT1 (SLC7A5). J Biol Chem 293:2877–2887
87. Wang S, Tsun Z-Y, Wolfson RL et al (2015) Lysosomal amino acid transporter SLC38A9 signals arginine sufficiency to mTORC1. Science 347:188–194
88. Rebsamen M, Pochini L, Stasyk T et al (2015) SLC38A9 is a component of the lysosomal amino acid sensing machinery that controls mTORC1. Nature 519:477–481
89. Wyant GA, Abu-Remaileh M, Wolfson RL et al (2017) mTORC1 activator SLC38A9 is required to efflux essential amino acids from lysosomes and use protein as a nutrient. Cell 171:642–654, e612
90. Carroll B, Maetzel D, Maddocks OD et al (2016) Control of TSC2-Rheb signaling axis by arginine regulates mTORC1 activity. elife 5:e11058
91. Bar-Peled L, Chantranupong L, Cherniack AD et al (2013) A tumor suppressor complex with GAP activity for the Rag GTPases that signal amino acid sufficiency to mTORC1. Science 340:1100–1106
92. Wolfson RL, Chantranupong L, Saxton RA et al (2016) Sestrin2 is a leucine sensor for the mTORC1 pathway. Science 351:43–48
93. Saxton RA, Chantranupong L, Knockenhauer KE et al (2016) Mechanism of arginine sensing by CASTOR1 upstream of mTORC1. Nature 536:229–233
94. Petit CS, Roczniak-Ferguson A, Ferguson SM (2013) Recruitment of folliculin to lysosomes supports the amino acid–dependent activation of Rag GTPases. J Cell Biol 202:1107–1122
95. Martina JA, Chen Y, Gucek M et al (2012) MTORC1 functions as a transcriptional regulator of autophagy by preventing nuclear transport of TFEB. Autophagy 8:903–914
96. Sardiello M, Palmieri M, Di Ronza A et al (2009) A gene network regulating lysosomal biogenesis and function. Science 325:473–477

97. Hsu PP, Kang SA, Rameseder J et al (2011) The mTOR-regulated phosphoproteome reveals a mechanism of mTORC1-mediated inhibition of growth factor signaling. Science 332:1317–1322
98. Plas DR, Thompson CB (2003) Akt activation promotes degradation of tuberin and FOXO3a via the proteasome. J Biol Chem 278:12361–12366
99. Yang G, Murashige DS, Humphrey SJ et al (2015) A positive feedback loop between Akt and mTORC2 via SIN1 phosphorylation. Cell Rep 12:937–943
100. Decuypere J-P, Bultynck G, Parys JB (2011) A dual role for Ca2+ in autophagy regulation. Cell Calcium 50:242–250
101. Bootman MD, Chehab T, Bultynck G et al (2018) The regulation of autophagy by calcium signals: do we have a consensus? Cell Calcium 70:32–46
102. Raiborg C, Wenzel EM, Stenmark H (2015) ER–endosome contact sites: molecular compositions and functions. EMBO J 34:1848–1858
103. Kondratskyi A, Kondratska K, Skryma R et al (2018) Ion channels in the regulation of autophagy. Autophagy 14:3–21
104. Leipnitz G, Mohsen A-W, Karunanidhi A et al (2018) Evaluation of mitochondrial bioenergetics, dynamics, endoplasmic reticulum-mitochondria crosstalk, and reactive oxygen species in fibroblasts from patients with complex I deficiency. Sci Rep 8:1–14
105. Raffaello A, Mammucari C, Gherardi G et al (2016) Calcium at the center of cell signaling: interplay between endoplasmic reticulum, mitochondria, and lysosomes. Trends Biochem Sci 41:1035–1049
106. Medina DL, Di Paola S, Peluso I et al (2015) Lysosomal calcium signalling regulates autophagy through calcineurin and TFEB. Nat Cell Biol 17:288–299
107. Fliniaux I, Germain E, Farfariello V et al (2018) TRPs and Ca2+ in cell death and survival. Cell Calcium 69:4–18
108. East DA, Campanella M (2013) Ca2+ in quality control: an unresolved riddle critical to autophagy and mitophagy. Autophagy 9:1710–1719
109. Williams A, Sarkar S, Cuddon P et al (2008) Novel targets for Huntington's disease in an mTOR-independent autophagy pathway. Nat Chem Biol 4:295–305
110. Valladares D, Utreras-Mendoza Y, Campos C et al (2018) IP3 receptor blockade restores autophagy and mitochondrial function in skeletal muscle fibers of dystrophic mice. Biochim Biophys Acta Mol Basis Dis 1864:3685–3695
111. Yousefi S, Perozzo R, Schmid I et al (2006) Calpain-mediated cleavage of Atg5 switches autophagy to apoptosis. Nat Cell Biol 8:1124–1132
112. Russo R, Berliocchi L, Adornetto A et al (2011) Calpain-mediated cleavage of Beclin-1 and autophagy deregulation following retinal ischemic injury in vivo. Cell Death Dis 2:e144–e144
113. Decuypere J-P, Welkenhuyzen K, Luyten T et al (2011) Ins (1, 4, 5) P 3 receptor-mediated Ca2+ signaling and autophagy induction are interrelated. Autophagy 7:1472–1489
114. Decuypere J-P, Monaco G, Bultynck G et al (2011) The IP3 receptor–mitochondria connection in apoptosis and autophagy. Biochim Biophys Acta Mol Cell Res 1813:1003–1013
115. Geisler S, Holmström KM, Skujat D et al (2010) PINK1/Parkin-mediated mitophagy is dependent on VDAC1 and p62/SQSTM1. Nat Cell Biol 12:119–131
116. Yuan J, Zhang Y, Sheng Y et al (2015) MYBL2 guides autophagy suppressor VDAC2 in the developing ovary to inhibit autophagy through a complex of VDAC2-BECN1-BCL2L1 in mammals. Autophagy 11:1081–1098
117. Vicencio Bustamante JM (2009) The inositol-1, 4, 5-trisphosphate receptor regulates autophagy through its interaction with Beclin 1. Cell Death Differ 16:1006–1017
118. Chung KM, Jeong E-J, Park H et al (2016) Mediation of autophagic cell death by type 3 ryanodine receptor (RyR3) in adult hippocampal neural stem cells. Front Cell Neurosci 10:116
119. Vervliet T, Pintelon I, Welkenhuyzen K et al (2017) Basal ryanodine receptor activity suppresses autophagic flux. Biochem Pharmacol 132:133–142

120. Høyer-Hansen M, Bastholm L, Szyniarowski P et al (2007) Control of macroautophagy by calcium, calmodulin-dependent kinase kinase-β, and Bcl-2. Mol Cell 25:193–205
121. Szalai P, Parys JB, Bultynck G et al (2018) Nonlinear relationship between ER Ca2+ depletion versus induction of the unfolded protein response, autophagy inhibition, and cell death. Cell Calcium 76:48–61
122. Sehgal P, Szalai P, Olesen C et al (2017) Inhibition of the sarco/endoplasmic reticulum (ER) Ca2+-ATPase by thapsigargin analogs induces cell death via ER Ca2+ depletion and the unfolded protein response. J Biol Chem 292:19656–19673
123. Senft D, Ze'ev AR (2015) UPR, autophagy, and mitochondria crosstalk underlies the ER stress response. Trends Biochem Sci 40:141–148
124. Sano R, Reed JC (2013) ER stress-induced cell death mechanisms. Biochim Biophys Acta Mol Cell Res 1833:3460–3470
125. Kishino A, Hayashi K, Hidai C et al (2017) XBP1-FoxO1 interaction regulates ER stress-induced autophagy in auditory cells. Sci Rep 7:1–15
126. Margariti A, Li H, Chen T et al (2013) XBP1 mRNA splicing triggers an autophagic response in endothelial cells through BECLIN-1 transcriptional activation. J Biol Chem 288:859–872
127. Wang P, Li J, Tao J et al (2018) The luminal domain of the ER stress sensor protein PERK binds misfolded proteins and thereby triggers PERK oligomerization. J Biol Chem 293:4110–4121
128. B'chir W, Maurin A-C, Carraro V et al (2013) The eIF2α/ATF4 pathway is essential for stress-induced autophagy gene expression. Nucleic Acids Res 41:7683–7699
129. Gerasimenko JV, Sherwood M, Tepikin AV et al (2006) NAADP, cADPR and IP3 all release Ca2+ from the endoplasmic reticulum and an acidic store in the secretory granule area. J Cell Sci 119:226–238
130. Ezeani M (2019) TRP channels mediated pathological Ca2+-handling and spontaneous ectopy. Front Cardiovasc Med 6:83
131. Prakriya M, Feske S, Gwack Y et al (2006) Orai1 is an essential pore subunit of the CRAC channel. Nature 443:230–233
132. Zhang SL, Yu Y, Roos J et al (2005) STIM1 is a Ca2+ sensor that activates CRAC channels and migrates from the Ca2+ store to the plasma membrane. Nature 437:902–905
133. Prakriya M, Lewis RS (2015) Store-operated calcium channels. Physiol Rev 95:1383–1436
134. Cheng KT, Liu X, Ong HL et al (2008) Functional requirement for Orai1 in store-operated TRPC1-STIM1 channels. J Biol Chem 283:12935–12940
135. Huang GN, Zeng W, Kim JY et al (2006) STIM1 carboxyl-terminus activates native SOC, I crac and TRPC1 channels. Nat Cell Biol 8:1003–1010
136. López J, Salido GM, Pariente JA et al (2006) Interaction of STIM1 with endogenously expressed human canonical TRP1 upon depletion of intracellular Ca2+ stores. J Biol Chem 281:28254–28264
137. Yuan JP, Zeng W, Huang GN et al (2007) STIM1 heteromultimerizes TRPC channels to determine their function as store-operated channels. Nat Cell Biol 9:636–645
138. Cuajungco MP, Silva J, Habibi A et al (2016) The mucolipin-2 (TRPML2) ion channel: a tissue-specific protein crucial to normal cell function. Pflügers Archiv Eur J Physiol 468:177–192
139. Dong X-P, Shen D, Wang X et al (2010) PI (3, 5) P 2 controls membrane trafficking by direct activation of mucolipin Ca2+ release channels in the endolysosome. Nat Commun 1:1–11
140. Zhang X, Li X, Xu H (2012) Phosphoinositide isoforms determine compartment-specific ion channel activity. Proc Natl Acad Sci 109:11384–11389
141. Di Paola S, Scotto-Rosato A, Medina DL (2018) TRPML1: the Ca(2+) retaker of the lysosome. Cell Calcium 69:112–121
142. Vergarajauregui S, Martina JA, Puertollano R (2011) LAPTMs regulate lysosomal function and interact with mucolipin 1: new clues for understanding mucolipidosis type IV. J Cell Sci 124:459–468

143. Vergarajauregui S, Martina JA, Puertollano R (2009) Identification of the penta-EF-hand protein ALG-2 as a Ca2+-dependent interactor of mucolipin-1. J Biol Chem 284:36357–36366
144. Venugopal B, Mesires NT, Kennedy JC et al (2009) Chaperone-mediated autophagy is defective in mucolipidosis type IV. J Cell Physiol 219:344–353
145. Tedeschi V, Petrozziello T, Sisalli MJ et al (2019) The activation of Mucolipin TRP channel 1 (TRPML1) protects motor neurons from L-BMAA neurotoxicity by promoting autophagic clearance. Sci Rep 9:1–11
146. Li X, Rydzewski N, Hider A et al (2016) A molecular mechanism to regulate lysosome motility for lysosome positioning and tubulation. Nat Cell Biol 18:404–417
147. Wang W, Gao Q, Yang M et al (2015) Up-regulation of lysosomal TRPML1 channels is essential for lysosomal adaptation to nutrient starvation. Proc Natl Acad Sci 112:E1373–E1381
148. Coblentz J, St. Croix C, Kiselyov K (2014) Loss of TRPML1 promotes production of reactive oxygen species: is oxidative damage a factor in mucolipidosis type IV? Biochem J 457:361–368
149. Zhang X, Cheng X, Yu L et al (2016) MCOLN1 is a ROS sensor in lysosomes that regulates autophagy. Nat Commun 7:1–12
150. Kim HJ, Soyombo AA, Tjon-Kon-Sang S et al (2009) The Ca2+ channel TRPML3 regulates membrane trafficking and autophagy. Traffic 10:1157–1167
151. Choi S, Kim HJ (2014) The Ca2+ channel TRPML3 specifically interacts with the mammalian ATG8 homologue GATE16 to regulate autophagy. Biochem Biophys Res Commun 443:56–61
152. Wong C-O, Li R, Montell C et al (2012) Drosophila TRPML is required for TORC1 activation. Curr Biol 22:1616–1621
153. Li R-J, Xu J, Fu C et al (2016) Regulation of mTORC1 by lysosomal calcium and calmodulin. elife 5:e19360
154. Onyenwoke RU, Sexton JZ, Yan F et al (2015) The mucolipidosis IV Ca2+ channel TRPML1 (MCOLN1) is regulated by the TOR kinase. Biochem J 470:331–342
155. Dibble CC, Cantley LC (2015) Regulation of mTORC1 by PI3K signaling. Trends Cell Biol 25:545–555
156. Efeyan A, Sabatini DM (2013) Nutrients and growth factors in mTORC1 activation. Portland Press Ltd., London
157. Sancak Y, Peterson TR, Shaul YD et al (2008) The Rag GTPases bind raptor and mediate amino acid signaling to mTORC1. Science 320:1496–1501
158. Gulati P, Gaspers LD, Dann SG et al (2008) Amino acids activate mTOR complex 1 via Ca2+/CaM signaling to hVps34. Cell Metab 7:456–465
159. Levy C, Khaled M, Iliopoulos D et al (2010) Intronic miR-211 assumes the tumor suppressive function of its host gene in melanoma. Mol Cell 40:841–849
160. Mazar J, Deyoung K, Khaitan D et al (2010) The regulation of miRNA-211 expression and its role in melanoma cell invasiveness. PLoS One 5:e13779
161. Ozturk DG, Kocak M, Akcay A et al (2019) MITF-MIR211 axis is a novel autophagy amplifier system during cellular stress. Autophagy 15:375–390
162. Yu Y, Zhao J (2019) Modulated autophagy by microRNAs in osteoarthritis chondrocytes. Biomed Res Int 2019:1484152
163. Wan G, Xie W, Liu Z et al (2014) Hypoxia-induced MIR155 is a potent autophagy inducer by targeting multiple players in the MTOR pathway. Autophagy 10:70–79
164. Korkmaz G, Le Sage C, Tekirdag KA et al (2012) miR-376b controls starvation and mTOR inhibition-related autophagy by targeting ATG4C and BECN1. Autophagy 8:165–176
165. Chen K, Shi W (2016) Autophagy regulates resistance of non-small cell lung cancer cells to paclitaxel. Tumor Biol 37:10539–10544
166. Chen Z, Gao S, Wang D et al (2016) Colorectal cancer cells are resistant to anti-EGFR monoclonal antibody through adapted autophagy. Am J Transl Res 8:1190

167. Sun Y, Xing X, Liu Q et al (2015) Hypoxia-induced autophagy reduces radiosensitivity by the HIF-1α/miR-210/Bcl-2 pathway in colon cancer cells. Int J Oncol 46:750–756
168. Huangfu L, Liang H, Wang G et al (2016) miR-183 regulates autophagy and apoptosis in colorectal cancer through targeting of UVRAG. Oncotarget 7:4735
169. He J, Yu J-J, Xu Q et al (2015) Downregulation of ATG14 by EGR1-MIR152 sensitizes ovarian cancer cells to cisplatin-induced apoptosis by inhibiting cyto-protective autophagy. Autophagy 11:373–384
170. Chen Y, Liersch R, Detmar M (2012) The miR-290-295 cluster suppresses autophagic cell death of melanoma cells. Sci Rep 2:1–10
171. Monkkonen T, Debnath J (2018) Inflammatory signaling cascades and autophagy in cancer. Autophagy 14:190–198
172. Filomeni G, De Zio D, Cecconi F (2015) Oxidative stress and autophagy: the clash between damage and metabolic needs. Cell Death Differ 22:377–388
173. Li L, Tan J, Miao Y et al (2015) ROS and autophagy: interactions and molecular regulatory mechanisms. Cell Mol Neurobiol 35:615–621
174. Scherz-Shouval R, Elazar Z (2011) Regulation of autophagy by ROS: physiology and pathology. Trends Biochem Sci 36:30–38
175. Gurusamy N, Das DK (2009) Autophagy, redox signaling, and ventricular remodeling. Antioxid Redox Signal 11:1975–1988
176. Wang P, Long M, Zhang S et al (2017) Hypoxia inducible factor-1α regulates autophagy via the p27-E2F1 signaling pathway. Mol Med Rep 16:2107–2112
177. Semenza GL (2011) Hypoxia-inducible factor 1: regulator of mitochondrial metabolism and mediator of ischemic preconditioning. Biochim Biophys Acta Mol Cell Res 1813:1263–1268
178. Verzi MP, Shin H, Ho L-L et al (2011) Essential and redundant functions of caudal family proteins in activating adult intestinal genes. Mol Cell Biol 31:2026–2039
179. Hawley SA, Ross FA, Chevtzoff C et al (2010) Use of cells expressing γ subunit variants to identify diverse mechanisms of AMPK activation. Cell Metab 11:554–565
180. Shao D, Oka S-I, Liu T et al (2014) A redox-dependent mechanism for regulation of AMPK activation by Thioredoxin1 during energy starvation. Cell Metab 19:232–245
181. Jang M, Park R, Kim H et al (2018) AMPK contributes to autophagosome maturation and lysosomal fusion. Sci Rep 8:1–10
182. Zhao Y, Li X, Ma K et al (2013) The axis of MAPK1/3-XBP1u-FOXO1 controls autophagic dynamics in cancer cells. Autophagy 9:794–796
183. Kops GJ, Dansen TB, Polderman PE et al (2002) Forkhead transcription factor FOXO3a protects quiescent cells from oxidative stress. Nature 419:316–321

Chapter 2
Acute and Chronic Exercise on Autophagy

Cenyi Wang, Michael Kirberger, and Ning Chen

1 Introduction

Because of modern technology, human beings are becoming increasingly sedentary and expending less energy due to physical inactivity. At present, the average sitting and lying time of adults in Europe is as much as 5 h/day [1], while office workers in the UK may have sitting and lying time averaging as much as 10.6 h on workdays [2]. A prolonged sedentary lifestyle can lead to poor circulation, poor breathing, and slow gastrointestinal motility, and can significantly increase the risks of cardiopulmonary diseases, metabolic diseases including type 2 diabetes mellitus (T2DM) and obesity, cancers, and other diseases as the time extension of physical inactivity [3–5]. Regular and appropriate exercise can strengthen immune system functions, enhance cellular homeostasis, and stimulate the production and secretion of exerkines, thus effectively preventing the occurrence and development of chronic diseases. Previous studies have reported that exercise at an appropriate intensity can maintain physiological stability and significantly improve energy metabolism, neurological function, and cardiopulmonary function, in addition to reducing stress [6, 7]. However, it should be noted that exercise is effective in enhancing cardiorespiratory endurance level and skeletal muscle mass, promoting mitochondrial synthesis, and improving insulin sensitivity, and it might also lead to increased oxidative stress and imbalance in energy metabolism during exercise training period, which is

C. Wang
School of Exercise Science, Soochow University, Suzhou, China

M. Kirberger
School of Science and Technology, Georgia Gwinnett College, Lawrenceville, GA, USA

N. Chen (✉)
Tianjiu Research and Development Center for Exercise Nutrition and Foods, Hubei Key Laboratory of Exercise Training and Monitoring, College of Health Science, Wuhan Sports University, Wuhan, China

N. Chen (ed.), *Exercise, Autophagy and Chronic Diseases*,
https://doi.org/10.1007/978-981-16-4525-9_2

beneficial for the establishment of new balance upon regular exercise training at an appropriate exercise intensity [8–11].

Recent studies have demonstrated that appropriate exercise can activate autophagy as an essential and evolutionarily conserved intracellular dynamic mechanism for maintaining metabolic and energetic stability in skeletal muscle cells, and plays a critical role in the timely removal and recycling of aggregated metabolic wastes from the cytoplasm or aging or damaged cellular organelles [12]. Appropriate exercise can regulate autophagic activity through multiple signal transduction pathways, and autophagic activity changes under different exercise modalities and conditions. Moderate exercise training can promote cellular autophagy, whereas excessive exercise may lead to the disruption of cellular autophagy, and even cause programmed cellular lesions or death, in severe cases.

Based on the duration and frequency, exercise can be divided into two types: acute or chronic exercise. Acute exercise can be defined as one-time or limited exercise, while chronic exercise can be defined as the repeated exercise of a certain amount for a short or long period of time [13]. The influence of the two types of exercise on autophagy will change the level of autophagy due to differences in the duration of exercise and the method of exercise intervention. Therefore, in this chapter, we will review and compare the activation mechanisms of autophagy in muscle cells, induced by acute exercise and chronic exercise, to regulate muscle metabolism and homeostasis levels, achieve exercise adaptation, improve exercise performance, and realize health promotion, as well as delay or rescue a series of chronic diseases resulted from physical inactivity.

2 Exercise and Autophagy

Muscle is the most abundant tissue and the main reservoir of amino acids in the human body. The physiological state of muscle is critical to human health. Excessive or deficient cellular autophagy or impaired autophagy flux is detrimental to muscle health and function and may even lead to muscle atrophy, while optimal autophagy activation potential or capacity can maintain muscle integrity. The underlying mechanisms of exercise effects on myocyte autophagy mainly include the effects on skeletal muscle autophagy and myocardial autophagy, but can also have important effects on liver cells and brain (Fig. 2.1).

2.1 *Effect of Exercise on Skeletal Muscle Autophagy*

Skeletal muscle mass accounts for approximately 40% of body weight. Existing studies have shown that moderate autophagy plays a vital role in the maintenance of skeletal muscle mass and functional regulation. During the process of aging, autophagy in skeletal muscle is at a relatively low level, while cell apoptosis is in

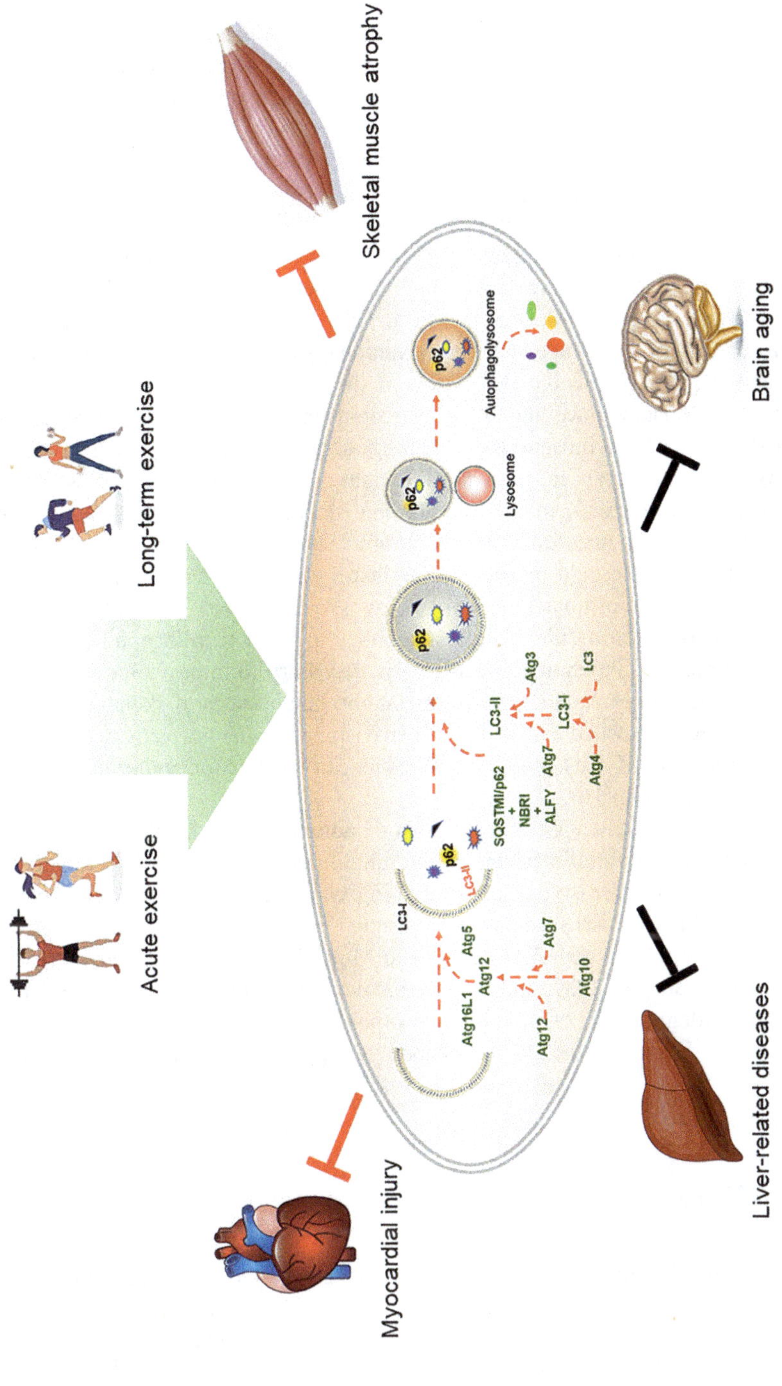

Fig. 2.1 Acute and chronic exercise-induced autophagy is beneficial for health promotion of multiple organs

a relatively vigorous state. High-intensity exercise or exhaustive exercise may lead to skeletal muscle atrophy or muscle weakness due to excessive degradation of skeletal muscle fibers. Conversely, when excessive autophagy in skeletal muscle is taken place, it may lead to autophagic atrophy or weaken mass maintenance of skeletal muscle. Therefore, both exercise modality and exercise intensity are particularly significant for the maintenance of autophagy regulation, the stability and balance of the internal environment of skeletal muscle, and the improvement of skeletal muscle's exercise adaptation and exercise capacity.

The role of cellular autophagy in skeletal muscle mass control is crucial, and its degradation is mainly regulated by the ubiquitin–proteasome system and autophagic lysosomal degradation pathways. Although the exact mechanism is not clear, excessive autophagy can lead to skeletal muscle atrophy. The specific molecules identified as being closely related to skeletal muscle autophagy and skeletal muscle atrophy are Atg7 and Atg5, and specific deficiency of these genes will lead to blurring of sarcoplasmic vesicles, swelling of the sarcoplasmic reticulum, and disruption of homeostasis in skeletal muscle, thus affecting locomotor capacity [14]. Conversely, apoptosis and autophagy are mutually inhibited and regulated, and excessive apoptosis leads to a decrease in the function and degradation of cellular autophagy, which interferes with the elimination of intramuscular garbages and wastes, and the production of intramuscular mitochondria, thereby causing skeletal muscle lesions and myasthenia, and ultimately leading to low skeletal muscle motility [15]. Thus, the activation of intramuscular cellular autophagy to promote mitochondrial productivity and improve mitochondrial function is necessary to improve locomotor performance. The energy associated with metabolic homeostasis of skeletal muscle is directly related to elevated expression of peroxisome proliferator-activated receptor-γ coactivator-1α (PGC-1α), which improves motility adaptation and exercise capacity of skeletal muscle [16].

Exercise training increases the activity of autophagy in skeletal muscle cells. Although exercise training accelerates the rate of cell renewal and is conducive to organism stability, excessive exercise training can alter the sensitivity of cellular autophagy, which in turn leads to skeletal muscle mass disorders. The activity of muscle atrophy-related genes Atrogin-1 and MuRF1, and autophagy-related genes such as Atg7, Beclin1, LC3, and FoxO3, increases from excessive exercise, which leads to overdegradation of proteins and organelles in skeletal muscle cells, thus resulting in decreased cellular resistance, atrophy, and weakness of skeletal muscle [17].

2.2 *Effect of Exercise on Myocardial Autophagy*

Cardiomyocyte is a highly differentiated end-stage cell, and autophagy can degrade misfolded proteins and senescent or malfunctioning organelles to supply energy to cardiomyocytes, promote intracellular material circulation and cellular self-renewal, and maintain cardiac function. Cardiac hypertrophy is an adaptive response of the myocardium to various cardiac diseases, mainly manifested by increased myocyte size, increased protein synthesis, and myonuclear remodeling. Cardiac hypertrophy is accompanied by a decrease in autophagy level. The occurrence of cardiac hypertrophy is also often accompanied by a decrease in LC3 protein expression in cardiomyocytes [18], while a decrease in Atg5 and Atg7 protein levels leads to ventricular dilation and systolic dysfunction in the heart [19]. Under normal stress conditions in cardiomyocytes, autophagy can degrade denatured proteins and produce amino acids, which play an essential role in cardiac cell mass, myocardial morphological structure and function, and intracellular environmental homeostasis. Moderate induction of myocardial autophagy can protect cells from myocardial injury caused by hemodynamic overloading, but when myocardial autophagy is over-regulated or inhibited, abnormal cardiac function will occur [20].

Moderate exercise increases cardiomyocyte autophagy levels and degrades intracellular metabolic wastes, thus maintaining cardiomyocyte homeostasis [21]. In addition, it has also been shown that exercise can improve injured cardiac function by altering the levels of autophagy in the myocardium of the body. A period of exercise can significantly increase the autophagosomes in the membrane bilayer of senescent cardiomyocytes and increase the level of autophagy-related proteins, thus improving cardiac function [22]. As aging cardiomyocytes are often prone to reduced autophagy levels, available evidence suggests that long-term aerobic exercise can improve cardiac function by increasing autophagy levels in aging cardiomyocytes, thereby resisting myocardial aging [23]. Exercise can also reduce myocardial cell injury by enhancing myocardial BNIP3, LC3-II/LC3-I key protein expression, decreasing p62 protein expression, enhancing cardiomyocyte autophagy levels, and accelerating the degradation rate of damaged proteins [24]. Under normal conditions, exercise can induce the activation of autophagy in cells from various tissues and organs, and autophagy in cardiomyocytes plays a vital role in maintaining the self-function of the myocardium.

2.3 *Effect of Exercise on the Regulation of Autophagy Level in Hepatocytes*

The liver is an important organ for synthesis, decomposition, and metabolism. Its normal function is closely related to the stable structure and function of mitochondria: the "energy factory" in liver cells. Mitochondria play a vital role in cells and are multifunctional organelles. The primary function of mitochondria is to generate energy through oxidative phosphorylation, which is the primary source of energy supply for cells. It is particularly crucial for some tissues and organs with high energy demand, such as the heart and the central nervous system [25]. In most healthy cells, the division and fusion of mitochondria always maintain a certain balance, and the mitochondria are constantly undergoing division and fusion, and continue to perform short-distance movement. Mitochondrial dynamics are an important factor in maintaining the health of the mitochondria within the cell. Mitochondria exchange DNA and protein through the process of continuous division and fusion, or separate and eliminate harmful components through the process of autophagy. During cell division, mitochondria can be transmitted to daughter cells through division, thus producing healthy mitochondria. However, the membrane potential of severely damaged mitochondria cannot be recovered, and it is difficult to participate in fusion, thus generating more ROS that endanger the cells, and these must be cleared through mitochondrial autophagy [26].

Mitochondria are an essential source of ROS in most mammalian cells, and the level of ROS is generally elevated in tumor cells, which plays a vital role in regulating cell autophagy and apoptosis. Studies have shown that in hepatocellular carcinoma (HCC), increased mitochondrial division leads to the accumulation of ROS, activates Akt, and promotes MDM3-mediated degradation of TP53 and NF-κBIA, and IKK-mediated transcriptional activation of NF-κB in HCC cells, thereby leading to increased autophagy and improved survival of HCC cells. Furthermore, the interaction between TP53 and NF-κB signal pathways is also involved in regulating chondriokinesis related to the viability of HCC cells. In addition to the regulation of ROS, mitochondrial division can also affect the viability of cells by regulating intracellular Ca^{2+} levels [27]. Exercise can affect the level of mitochondrial division. Studies have found that aerobic exercise affects the occurrence and development of hepatocellular carcinogenesis, the mechanism of which may involve reducing mitochondrial division and affecting the autophagy of cancerous liver cells through the PINK1–Parkin pathway, and increasing the level of apoptosis of cancerous liver cells. Other studies have found that 8 weeks of swimming training can effectively repair damage to the liver leptin-AMPK-ACC signaling pathway; significantly increase liver glycogen; decrease triglyceride (TG) and free fatty acid (FFA); further reduce blood lipids and blood glucose; and improve the sensitivity of leptin and insulin. Furthermore, the leptin-AMPK-ACC signaling pathway in liver cells is still activated after acute exercise, which can significantly improve insulin sensitivity, but cannot reverse dysfunction of glucose and lipid metabolism in liver [28–30]. Therefore, both one-time exercise and long-term exercise training can reduce the risk of liver lipid metabolism damage, but different exercise modes have different effects on liver cells.

2.4 *Effect of Exercise on Brain Function*

In recent years, studies in the field of exercise science have shown that appropriate exercise can promote brain function. Exercise training can enhance the autogenesis capacity of newborn neurons and increase the complexity of neuronal dendrites and synaptic plasticity [31, 32]. In addition, non-neuronal components such as the cerebrovascular system may also play a vital role in the process of improving brain function through exercise training. Imaging studies have demonstrated that physical exercise may increase cerebral blood volume and cerebral blood flow

[31]. Exercise can also facilitate the synthesis and release of many neurotransmitters, including acetylcholine (ACH), dopamine (DA), and nitric oxide (NO) [33]. In recent years, there have been many studies on the effects of ACH and DA on learning and memory capacity of neurodegenerative diseases, while studies on NO have gradually increased in the field of exercise science [33, 34]. By exploring the effects of early exercise intervention on the electrical activity of dopaminergic neurons in the substantia nigra of rats with Parkinson's disease, some researchers have speculated that the underlying mechanism of early exercise intervention-induced changes in the excitability of substantia nigra DA neurons might be related to the protection of DA neurons [35, 36]. Another study has reported that 4 weeks of aerobic exercise can enhance DA metabolism in the hippocampus, striatum, and prefrontal cortex of rats, thereby enhancing the learning and memory capacity of rats [37]. Additionally, after long-term aerobic treadmill exercise, the number and area of neuronal nitric oxide synthase (nNOS)-positive neurons in the hippocampal tissues of aged rats are significantly increased, suggesting that aerobic exercise may enhance nNOS activity and induce long-term effects, thereby enhancing the learning and memory capacity of aged rats [38].

Exercise training not only improves the level of brain function by enhancing the activity of neurotransmitters, but also strengthens the state of brain function by increasing the content of neurotrophic factors. Neurotrophic factors are crucial bioactive substances in the nervous system that can promote and restore the growth, differentiation, and reconstruction of neurons. Multiple studies have confirmed that neurotrophic factors affect neuronal differentiation related to structure, morphology, and function [39]. Brain-derived neurotrophic factor (BDNF) can improve learning and memory capacity [40]. The effect of 28-day treadmill exercise on spatial memory and BDNF expression in epileptic mice has also been explored [41]. The resultant data indicate that the expression of BDNF in the hippocampus of epileptic mice, subjected to exercise intervention, is significantly higher than in epileptic mice without exercise intervention, as accompanied by enhanced learning and memory capacity. Current studies have confirmed that exercise can improve brain function, especially the capacity for learning and memory, which is related to enhancement of the BDNF-TrkB-MEK-ERK signaling pathway [42].

Insulin-like growth factor 1 (IGF-1) can promote an increase in hippocampal neurons, and neuronal differentiation and survival. Exercise-mediated IGF-1 promotes brain function. The underlying mechanism related to this may involve IGF-1-regulated Ca^{2+}/calmodulin-dependent protein kinase II (CAMKII), mitogen-activated protein kinase (MAPK), and synaptophysin-1, as these regulate synaptic plasticity, and thus improve the level of brain function [43]. Previous studies have reported that exercise can improve the learning and memory functions of brain by increasing the level of IGF-1 [44]. Experimental results from a study where rats are subjected to 8 weeks of aerobic treadmill training indicate that appropriate exercise training might improve the structural and functional state of the brain by increasing the expression level of IGF-1 mRNA [42].

3 Acute Exercise-Induced Autophagy

3.1 Underlying Mechanism for the Effect of Acute Exercise on Skeletal Muscle Autophagy

In another study, expression levels of LC3-II, Beclinl, Atg7, and LAMP2 are decreased, and autophagic activity is decreased, in mice subjected to a one-time treadmill training for 50 min at a rate of 12.3 m/min [45]. Conversely, following a one-time training for 90 min (55% VO_{2max}), LC3-II expression is increased significantly, and the LC3-II/LC3-I ratio is also increased [46]. These seemingly contradictory results may be related to differences in energy expenditure caused by the different intensities and durations of exercise. In addition, some studies have found that, after acute exercise, the LC3-II/LC3-I ratio is decreased and p62 remains unchanged at exercise intensity of 55% VO_{2max}, while the LC3-II/LC3-I ratio is increased significantly, and p62 expression is decreased significantly, at exercise intensity of 70% VO_{2max} [47]. These results may indicate that low-intensity exercise causes minor damage to organelles and reduced energy consumption, which results in failure to promote the initiation of autophagic processes. However, some studies have pointed out that LC3-II expression levels can be decreased after one-time moderate- to low-intensity exercise, while the expression levels of MuRF1 protein that promotes protein degradation can be increased, suggesting a compensatory mechanism for the decrease in autophagic activity by the organism to maintain energy balance under conditions of low energy expenditure [45].

Differences in sampling time, after an acute exercise session, can also have an effect on autophagic activity. Based on a study involving a one-time exhaustive treadmill exercise, LC3-II protein expression in mouse skeletal muscle (lateral femoral, muscular soleus, tibialis anterior, and extensor digitorum longus) is found to decrease after 15 min of exercise, followed by gradual increases reported at 30, 50, and 80 min, with stable levels observed at 80 min. Additionally, p62 expression is increased immediately after exercise, and then gradually decreased at 30 min, or 1 h, before increase again after 3 h [48]. Similarly, with the passage of time after exercise, the expression of the Bcl2–Beclinl complex is decreased at 15 min after exercise and is almost undetectable after 30 min [49]. Therefore, changes in autophagic activity after exercise are time-sensitive, and attention should be paid to the sampling time when exploring changes in skeletal muscle autophagy after acute exercise.

Glycogen level in skeletal muscle during the pre-exercise period may also affect the changes in autophagy [50, 51]. The level of glycogen in skeletal muscle before exercise is negatively correlated with the activation status of autophagy. The number of autophagosomes in the soleus, metatarsal, and gastrocnemius muscles, in mice, reveals an increase after fasting for 24 h. On this basis, the LC3-II/LC3-I ratio is decreased after one-time moderate-intensity exercise intervention (12 m/min, 2 h, slope 10°) [52], suggesting that exercise under fasting conditions inhibits autophagic activity in mice. In human experiments, a significant decrease in the LC3-II/LC3-I ratio is also observed in the vastus lateralis immediately after a cycling intervention at exercise intensity of 50% VO_{2max} for 60 min, and preceded by 36 h fasting, with a subsequent increase at 90 min after exercise, which remains significantly lower than before exercise [53]. Autophagy is able to promote catabolism while removing damaged proteins. Exercise can activate autophagy in the absence of muscle glycogen, which may be due to the body's protection of skeletal muscle to prevent excessive degradation of skeletal muscle through apoptosis.

3.2 Underlying Mechanism for the Effect of Acute Exercise on Myocardial Autophagy

The intracellular proteins ATG8 and ATG6 (also identified as LC-3 and Beclin1 in mammals), the mTOR signaling pathway, the ubiquitin-like protein system, and caspases are all involved in regulatory processes related to cellular autophagy [19, 54]. Unlike the ubiquitin–proteasome signaling pathway that degrades short-lived proteins, autophagy is mainly involved in the degradation of the majority of long-lived proteins (LLPs), as well as a portion of cellular organelles. Since various factors such as inflammatory response, apoptosis, mitochondrial dysfunction, protein synthesis, and protein degradation may be involved in the energy metabolism or aging process of skeletal muscle to some extent, autophagy can affect the health and exercise capacity of the organism. During the aging process, the accumulation of a large amount of ROS can cause the release of apoptotic factors and cytochrome C, thus resulting in mitochondrial damage, and eventually causing the death of cardiomyocytes [55, 56].

Apoptosis is programmed cell death related to cellular regulation, and the apoptosis of cardiomyocytes is closely related to the maintenance of cardiac homeostasis and the involvement of cardiac pathological processes [57]. Acute exercise can decrease the expression of Bcl-2 protein, especially when the duration of acute exercise exceeds 1.5 h, where the expression of apoptosis-related Bcl-2 appears to be significantly reduced and reaches a minimal level [58]. It can be inferred that the induction of autophagy by acute exercise also activates the apoptotic signal pathway simultaneously. This may be due to the instantaneous high-intensity exercise stimulation during acute exercise, thereby leading to a large number of apoptotic cells, and the body is unable to rapidly activate the relevant regulatory mechanisms to inhibit apoptosis within a short period of time. Alternatively, it could be because the excessive autophagy may instead stimulate apoptosis. Recent studies have demonstrated that ischemia and hypoxia, oxidative stress, reperfusion injury, and other stimuli may trigger myocardial cell necrosis and apoptosis [59, 60].

Acute exercise can promote the expression of autophagy-related protein Beclin1 to different degrees, and the expression of autophagy-related protein Beclin1 appears to increase significantly and maintain peak levels when the exercise time reaches 1.5 h [61]. Additionally, different modalities of exercise, such as treadmill running and swimming, can differentially increase the levels of cellular autophagy-related protein Beclin1 in myocardial tissues, thus enhancing autophagic activity in tissue cells, and contributing maintenance of intracellular homeostasis, rational scavenging of free radicals and misfolded proteins, and other senescent organelles, and providing energy for cellular metabolism [62, 63]. The results of these studies suggest that treadmill running and swimming are more likely to increase the level of the cellular autophagy-associated protein Beclin1 and enhance cellular autophagic activity. It has also been well documented that cellular autophagy has an essential role in protecting cardiac function and maintaining myocardial viability due to the low differentiation and regeneration capacity of cardiomyocytes, while cellular autophagy can promote myocardial protection against injury and alleviate cell death [64–66]. Thus, moderate-intensity exercise can activate cellular autophagy to prevent and protect the heart, reduce myocardial ischemia and reperfusion injury, and alleviate various cardiac diseases, thereby ensuring normal physiological function of the myocardium, maintaining exercise capacity, and delaying the onset of fatigue [67].

Nuclear factor erythroid 2-related factor 2 (NRF2), belonging to the basic leucine-zipper protein (bZIP) family, plays an essential role in the expression of anti-inflammatory and antioxidant genes mediated by anti-inflammatory response elements (ARE) [68]. Acute exercise significantly upregulates the expression of the antioxidative stress factor NRF2, and exercise-activated NRF2 contributes to the increase in its nuclear-entering capacity, which in turn initiates antioxidant gene expression, contributes to the increase in transcriptional capacity, reduces the generation of oxygen free radicals in the organism, and ultimately improves the antioxidant capacity of the organism [69].

4 Chronic Exercise Induces Autophagy

4.1 Underlying Mechanism for the Effect of Chronic Exercise on Skeletal Muscle Autophagy

Current studies on the effects of chronic exercise on skeletal muscle autophagy have focused on observing adaptive changes in basal autophagy and autophagic flux after exercise interventions. Therefore, samples are taken 24–48 h after the final training session to observe changes in autophagic activity after exercise [17, 70]. The period of exercise intervention and the type of skeletal muscle are the major factors for determining the effect of chronic exercise on skeletal muscle autophagy [71–73]. In another study, after running intervention by mice for five consecutive days (30 m/min, 60 min/day), no changes are observed in the levels of Beclin1, Atg12, Atg4, Atg7, LC3, Atg12-Atg5 complex, or cathepsin L proteins, in soleus muscle [74]. In contrast, some researchers have demonstrated that after 4 weeks of running intervention in mice, the expression of LC3-II, and LC3-II/LC3-I ratio in the soleus muscle of mice, is increased significantly, with a significant decrease in the level of p62 protein, while the protein expression levels of Beclin1, BNIP3, and Atg7 are reportedly increased [71]. Meanwhile, different skeletal muscle fiber types/tissues can produce different responses to the stimulation of autophagy [71]. After 4 weeks of exercise intervention, autophagy-related proteins Atg6 and Atg7, mitochondrial autophagy protein BNIP3, and mitochondrial synthesis proteins PGC-1α and COX-4 in soleus and metatarsal muscles are significantly upregulated. The autophagy-related proteins in soleus muscle are significantly higher than those in metatarsal muscle, suggesting that exercise intervention can significantly promote the autophagy activity of skeletal muscle dominated by chronic skeletal muscle fibers. Moreover, both long-term endurance exercise and resistance exercise can increase the expression levels of autophagy-related proteins and autophagic flux in skeletal muscle [14, 75]. Additionally, both resistance exercise and aerobic exercise for 8 weeks can reportedly increase the expression levels of LC3, Beclin1, and Atg7 in skeletal muscle of rats. However, the changes in autophagy activity in skeletal muscle after resistance exercise are mainly concentrated in flexor digitorum profundus, while the changes in autophagy activity in extensor digitorum longus are more pronounced after aerobic exercise [76, 77].

Although both resistance exercise and aerobic exercise can increase skeletal muscle autophagic activity, prolonged exercise intervention may lead to adaptive changes in skeletal muscle, but no changes in Beclin1 or LC3-II/LC3-I ratios in skeletal muscle, hippocampus, and liver tissues, after 36 weeks of moderate-intensity exercise intervention in rats [78]. No significant changes are observed for protein expression levels of Beclin1, LC3-I, LC3-II, Atg7, or Atg9, in the skeletal muscle of rats that engaged in persistent habitual running exercises. These results indicate that prolonged exercise intervention cycles may lead to the adaptation of skeletal muscle, liver, and hippocampus tissues, to exercise stimuli, with no further changes in autophagic activity [79].

In addition to normal experimental subjects, rats with specific knockout of the relevant genes have also been studied, and data from such studies indicate that when disordered energy metabolism occurs, the autophagic activity of the organism is usually not altered, and chronic exercise training does not improve the locomotor capacity of mice with autophagy deficiency [80]. It has also been found that exercise intervention cannot stimulate Bcl2 phosphorylation in $Bcl2^{AAA}$ mice, thus inhibiting Beclin1 dissociation from the Bcl2–Beclin1 complex. Therefore, neither energy metabolism level nor autophagic activity is altered in $Bcl2^{AAA}$ mice after exercise intervention [21]. The knockout of PGC-1α in mice indicates that exercise intervention can be failed to affect LC3-II and p62 mRNA expression levels in skeletal muscle [81].

4.2 *Underlying Mechanism for the Effect of Chronic Exercise on Myocardial Autophagy*

A number of studies have reported that, through appropriate long-term chronic exercise intervention, intracellular autophagy is activated, and the level of cellular autophagy is upregulated, thus accelerating the degradation of metabolites in cells, and maintaining the stable structure and function of cells. Moderate exercise can stimulate increased levels of cellular autophagy, thereby preventing or alleviating the development of various cardiac diseases [55]. Exercise can effectively induce autophagy, and both acute exercise and long-term aerobic endurance exercise can increase the expression of autophagy-related proteins in normal cells [82]. Long-term endurance exercise is also found to increase cellular autophagy levels and improve the adaptation of the body to exercise training by upregulating the expression of LC3-I, LC3-II, and Beclin1 [83]. Four weeks of continuous treadmill running upregulated the LC3-II/LC3-I ratio through an animal model of myocardial infarction [22]. Another study indicated that 8 weeks of aerobic exercise could increase LC3-II/LC3-I ratio, while reducing the expression of p62 protein and mRNA. Therefore, long-term aerobic exercise can reduce the level of cardiomyocyte damage and improve cardiac function by increasing myocardial autophagic activity, thus accelerating the clearance of damaged organelles and renewing misfolded or denatured proteins [84, 85].

Chronic exercise can stimulate the compensatory activation of signaling pathways, increase the level of autophagy, remove damaged organelles and misfolded proteins from cardiomyocytes, accelerate the degradation of damaged proteins, reduce cardiomyocyte damage, and maintain the metabolic balance of the heart. The biomolecules produced by autophagy after degrading damaged organelles and proteins can also become substrates required by anabolism, thus providing a material basis for maintaining the structure and normal physiological functions of cardiomyocytes [48, 85, 86]. Unlike acute exercise, chronic exercise induces autophagy and inhibits apoptosis, suggesting that chronic exercise can better maintain cellular homeostasis, promote autophagy, and inhibit apoptosis [87, 88]. Stress loading stimulates the autophagy-related protein Beclin1 in the myocardium. However, no changes in cardiac weight and hemodynamic extravasation stress have been observed in the absence of a single dose of myocardial-specific Beclin1 expression, presumably moderating stress-induced myocardial hypertrophy by the inhibition of autophagy [89, 90].

In terms of the effect of exercise training on the mitochondrial quality of cardiomyocytes, chronic exercise can significantly enhance the orderly arrangement of cardiac myofibers, and cardiomyocytes can also significantly contribute to the stabilization of the internal environment of cardiomyocytes, and the quality control of mitochondria, through mitochondrial autophagy [91, 92]. These results indicate that chronic exercise training can improve cardiomyocyte autophagy or mitochondrial plasticity, and promote the renewal of mitochondria and mitochondrial quality in cardiomyocytes, thus delaying apoptotic injury of cardiomyocytes and achieving a protective effect on the myocardium from long-term moderate exercise training. It has been found that after long-term exercise training with different exercise modalities such as treadmill running and swimming, the number and volume of mitochondria in mouse cardiomyocytes exhibit a significant increase, and long-term exercise training with moderate intensity is effective in increasing the quantity and quality of myocardial mitochondria, and reducing the number of senescent or damaged mitochondria [58, 93]. Chronic exercise can significantly improve the quality of mitochondria and maintain the internal environmental stability of cardiomyocytes, and the quality control of mitochondria through cellular autophagy, thereby protecting the myocardium through long-term moderate exercise training. Moderate-intensity long-term exercise training could effectively improve the quantity and quality of myocardial mitochondria, reduce the number of aging or damaged mitochondria, meet the requirements of cardiac output during exercise, establish exercise cardiac remodeling, enhance myocardial function, and help the heart to supply blood and oxygen [67].

5 Conclusion and Prospects

In summary, exercise training has a dual regulatory effect. Appropriate exercise intensity can appropriately upregulate the level of cellular autophagy, thus degrading intracellular residues and damaged organelles, and maintaining intracellular stability. High-intensity exercise may cause excessive activation of cellular autophagy, thus resulting in excessive cytoplasmic and organelle decline, and causing cellular damage and fatigue. Meanwhile, chronic exercise can strengthen the plasticity of cellular autophagy activation and further improve the stability of the intracellular environment and the balance of energy metabolism, which plays an important role in the improvement of muscle exercise adaptation and exercise capacity.

However, the studies on exercise and autophagy are still relatively limited, and further studies are required to investigate the activation mechanism of exercise on autophagy and the reasons for adaptive changes in autophagy during chronic exercise. Future experimental studies can explore the effect of exercise on autophagy from the source of exercise energy metabolic pathways and factor changes, and provide an essential basis for further exploring the health benefits of exercise.

References

1. Bennie JA, Chau JY, Van Der Ploeg HP et al (2013) The prevalence and correlates of sitting in European adults—a comparison of 32 Eurobarometer-participating countries. Int J Behav Nutr Phys Act 10:107
2. Smith L, Hamer M, Ucci M et al (2015) Weekday and weekend patterns of objectively measured sitting, standing, and stepping in a sample of office-based workers: the active buildings study. BMC Public Health 15:9

3. Bowden Davies KA, Sprung VS, Norman JA et al (2019) Physical activity and sedentary time: association with metabolic health and liver fat. Med Sci Sports Exerc 51:1169–1177
4. Booth FW, Roberts CK, Thyfault JP et al (2017) Role of inactivity in chronic diseases: evolutionary insight and pathophysiological mechanisms. Physiol Rev 97:1351–1402
5. Hartman SJ, Marinac CR, Bellettiere J et al (2017) Objectively measured sedentary behavior and quality of life among survivors of early stage breast cancer. Support Care Cancer 25:2495–2503
6. Ziaaldini MM, Marzetti E, Picca A et al (2017) Biochemical pathways of sarcopenia and their modulation by physical exercise: a narrative review. Front Med (Lausanne) 4:167
7. Wang L, Wang J, Cretoiu D et al (2020) Exercise-mediated regulation of autophagy in the cardiovascular system. J Sport Health Sci 9:203–210
8. Lundby C, Jacobs RA (2016) Adaptations of skeletal muscle mitochondria to exercise training. Exp Physiol 101:17–22
9. Bergman BC, Goodpaster BH (2020) Exercise and muscle lipid content, composition, and localization: influence on muscle insulin sensitivity. Diabetes 69:848–858
10. Alves CRR, Neves WD, De Almeida NR et al (2020) Exercise training reverses cancer-induced oxidative stress and decrease in muscle COPS2/TRIP15/ALIEN. Mol Metab 39:101012
11. Diaz-Castro J, Mira-Rufino PJ, Moreno-Fernandez J et al (2020) Ubiquinol supplementation modulates energy metabolism and bone turnover during high intensity exercise. Food Funct 11:7523–7531
12. Angulo J, El Assar M, Lvarez-Bustos A et al (2020) Physical activity and exercise: strategies to manage frailty. Redox Biol 35:101513
13. Sellami M, Gasmi M, Denham J et al (2018) Effects of acute and chronic exercise on immunological parameters in the elderly aged: can physical activity counteract the effects of aging? Front Immunol 9:2187
14. Masiero E, Agatea L, Mammucari C et al (2009) Autophagy is required to maintain muscle mass. Cell Metab 10:507–515
15. Salminen A, Kaarniranta K, Kauppinen A (2013) Crosstalk between oxidative stress and SIRT1: impact on the aging process. Int J Mol Sci 14:3834–3859
16. Preobrazenski N, Islam H, Drouin PJ et al (2020) A novel gravity-induced blood flow restriction model augments ACC phosphorylation and PGC-1α mRNA in human skeletal muscle following aerobic exercise: a randomized crossover study. Appl Physiol Nutr Metab 45:641–649
17. Feng Z, Bai L, Yan J et al (2011) Mitochondrial dynamic remodeling in strenuous exercise-induced muscle and mitochondrial dysfunction: regulatory effects of hydroxytyrosol. Free Radic Biol Med 50:1437–1446
18. Zhou L, Ma B, Han X (2016) The role of autophagy in angiotensin II-induced pathological cardiac hypertrophy. J Mol Endocrinol 57:R143–R152
19. Nakai A, Yamaguchi O, Takeda T et al (2007) The role of autophagy in cardiomyocytes in the basal state and in response to hemodynamic stress. Nat Med 13:619–624
20. Wang ZV, Rothermel BA, Hill JA (2010) Autophagy in hypertensive heart disease. J Biol Chem 285:8509–8514
21. He C, Sumpter R Jr, Levine B (2012) Exercise induces autophagy in peripheral tissues and in the brain. Autophagy 8:1548–1551
22. Chen CY, Hsu HC, Lee BC et al (2010) Exercise training improves cardiac function in infarcted rabbits: involvement of autophagic function and fatty acid utilization. Eur J Heart Fail 12:323–330
23. Silva KS, Leary EV, Olver TD et al (2020) Tissue-specific small heat shock protein 20 activation is not associated with traditional autophagy markers in Ossabaw swine with cardiometabolic heart failure. Am J Physiol Heart Circ Physiol 319:H1036–H1043
24. Jiang L, Shen X, Dun Y et al (2021) Exercise combined with trimetazidine improves anti-fatal stress capacity through enhancing autophagy and heat shock protein 70 of myocardium in mice. Int J Med Sci 18:1680–1686

25. Ikeda Y, Sciarretta S, Nagarajan N et al (2014) New insights into the role of mitochondrial dynamics and autophagy during oxidative stress and aging in the heart. Oxidative Med Cell Longev 2014:210934
26. Dorn GW 2nd, Kitsis RN (2015) The mitochondrial dynamism-mitophagy-cell death interactome: multiple roles performed by members of a mitochondrial molecular ensemble. Circ Res 116:167–182
27. Huang Q, Cao H, Zhan L et al (2017) Mitochondrial fission forms a positive feedback loop with cytosolic calcium signaling pathway to promote autophagy in hepatocellular carcinoma cells. Cancer Lett 403:108–118
28. Langfort J, Viese M, Ploug T et al (2003) Time course of GLUT4 and AMPK protein expression in human skeletal muscle during one month of physical training. Scand J Med Sci Sports 13:169–174
29. Stephens TJ, Chen ZP, Canny BJ et al (2002) Progressive increase in human skeletal muscle AMPKalpha2 activity and ACC phosphorylation during exercise. Am J Physiol Endocrinol Metab 282:E688–E694
30. Joseph LJ, Prigeon RL, Blumenthal JB et al (2011) Weight loss and low-intensity exercise for the treatment of metabolic syndrome in obese postmenopausal women. J Gerontol A Biol Sci Med Sci 66:1022–1029
31. Stranahan AM, Khalil D, Gould E (2007) Running induces widespread structural alterations in the hippocampus and entorhinal cortex. Hippocampus 17:1017–1022
32. Bettio L, Thacker JS, Hutton C et al (2019) Modulation of synaptic plasticity by exercise. Int Rev Neurobiol 147:295–322
33. Lessmann V, Stroh-Kaffei S, Steinbrecher V et al (2011) The expression mechanism of the residual LTP in the CA1 region of BDNF k.o. mice is insensitive to NO synthase inhibition. Brain Res 1391:14–23
34. Cheng XR, Zhou WX, Zhang YX (2014) The behavioral, pathological and therapeutic features of the senescence-accelerated mouse prone 8 strain as an Alzheimer's disease animal model. Ageing Res Rev 13:13–37
35. Rezaee Z, Marandi SM, Alaei H et al (2019) Effects of preventive treadmill exercise on the recovery of metabolic and mitochondrial factors in the 6-hydroxydopamine rat model of Parkinson's disease. Neurotox Res 35:908–917
36. Oliveira LOD, Da Silva PIC, Filho RPR et al (2020) Prior exercise protects against oxidative stress and motor deficit in a rat model of Parkinson's disease. Metab Brain Dis 35:175–181
37. Jiangbo N, Liyun Z (2018) Effect of donepezil hydrochloride & aerobic exercise training on learning and memory and its mechanism of action in an Alzheimer's disease rat model. Pak J Pharm Sci 31:2897–2901
38. Hashemiaghdam A, Mroczek M (2020) Microglia heterogeneity and neurodegeneration: the emerging paradigm of the role of immunity in Alzheimer's disease. J Neuroimmunol 341:577185
39. Schliebs R, Arendt T (2011) The cholinergic system in aging and neuronal degeneration. Behav Brain Res 221:555–563
40. Zhen YF, Zhang J, Liu XY et al (2013) Low BDNF is associated with cognitive deficits in patients with type 2 diabetes. Psychopharmacology (Berl) 227:93–100
41. Sartori CR, Pelágio FC, Teixeira SA et al (2009) Effects of voluntary running on spatial memory and mature brain-derived neurotrophic factor expression in mice hippocampus after status epilepticus. Behav Brain Res 203:165–172
42. Cassilhas RC, Lee KS, Fernandes J et al (2012) Spatial memory is improved by aerobic and resistance exercise through divergent molecular mechanisms. Neuroscience 202:309–317
43. Yau SY, Lau BW, Zhang ED et al (2012) Effects of voluntary running on plasma levels of neurotrophins, hippocampal cell proliferation and learning and memory in stressed rats. Neuroscience 222:289–301

44. Cetinkaya C, Sisman AR, Kiray M et al (2013) Positive effects of aerobic exercise on learning and memory functioning, which correlate with hippocampal IGF-1 increase in adolescent rats. Neurosci Lett 549:177–181
45. Kim YA, Kim YS, Song W (2012) Autophagic response to a single bout of moderate exercise in murine skeletal muscle. J Physiol Biochem 68:229–235
46. Levine B, Kroemer G (2008) Autophagy in the pathogenesis of disease. Cell 132:27–42
47. Schwalm C, Jamart C, Benoit N et al (2015) Activation of autophagy in human skeletal muscle is dependent on exercise intensity and AMPK activation. FASEB J 29:3515–3526
48. He C, Bassik MC, Moresi V et al (2012) Exercise-induced BCL2-regulated autophagy is required for muscle glucose homeostasis. Nature 481:511–515
49. Kim SH, Asaka M, Higashida K et al (2013) β-Adrenergic stimulation does not activate p38 MAP kinase or induce PGC-1α in skeletal muscle. Am J Physiol Endocrinol Metab 304:E844–E852
50. Derave W, Ai H, Ihlemann J et al (2000) Dissociation of AMP-activated protein kinase activation and glucose transport in contracting slow-twitch muscle. Diabetes 49:1281–1287
51. Wright DC, Geiger PC, Holloszy JO et al (2005) Contraction- and hypoxia-stimulated glucose transport is mediated by a Ca2+-dependent mechanism in slow-twitch rat soleus muscle. Am J Physiol Endocrinol Metab 288:E1062–E1066
52. Zheng DM, Bian Z, Furuya N et al (2015) A treadmill exercise reactivates the signaling of the mammalian target of rapamycin (mTor) in the skeletal muscles of starved mice. Biochem Biophys Res Commun 456:519–526
53. Møller AB, Vendelbo MH, Christensen B et al (2015) Physical exercise increases autophagic signaling through ULK1 in human skeletal muscle. J Appl Physiol (1985) 118:971–979
54. Cao DJ, Gillette TG, Hill JA (2009) Cardiomyocyte autophagy: remodeling, repairing, and reconstructing the heart. Curr Hypertens Rep 11:406–411
55. Golbidi S, Laher I (2011) Molecular mechanisms in exercise-induced cardioprotection. Cardiol Res Pract 2011:972807
56. Parry TL, Starnes JW, O'Neal SK et al (2018) Untargeted metabolomics analysis of ischemia-reperfusion-injured hearts ex vivo from sedentary and exercise-trained rats. Metabolomics 14:8
57. Kiyuna LA, Albuquerque RPE, Chen CH et al (2018) Targeting mitochondrial dysfunction and oxidative stress in heart failure: challenges and opportunities. Free Radic Biol Med 129:155–168
58. Yuan Y (2017) The role of autophagy functional status in exercise adaptation of cardiac muscle and skeletal muscle. Wuhan Sports University, Wuhan, pp 18–20
59. Dang L (2018) Effects of different aerobic exercise time on myocardial cell apoptosis in mice. Wei Sheng Yan Jiu 47:794–797
60. Lai CC, Tang CY, Chiang SC et al (2015) Ischemic preconditioning activates prosurvival kinases and reduces myocardial apoptosis. J Chin Med Assoc 78:460–468
61. Brandt N, Gunnarsson TP, Bangsbo J et al (2018) Exercise and exercise training-induced increase in autophagy markers in human skeletal muscle. Physiol Rep 6:e13651
62. Li Y, Sun D, Zheng Y et al (2020) Swimming exercise activates aortic autophagy and limits atherosclerosis in ApoE(−/−) mice. Obes Res Clin Pract 14:264–270
63. Pan G, Jin L, Shen W et al (2020) Treadmill exercise improves neurological function by inhibiting autophagy and the binding of HMGB1 to Beclin1 in MCAO juvenile rats. Life Sci 243:117279
64. Li JY, Pan SS, Wang JY et al (2019) Changes in autophagy levels in rat myocardium during exercise preconditioning-initiated cardioprotective effects. Int Heart J 60:419–428
65. Krylatov AV, Maslov LN, Voronkov NS et al (2018) Reactive oxygen species as intracellular signaling molecules in the cardiovascular system. Curr Cardiol Rev 14:290–300
66. Yuan JQ, Yuan Y, Pan SS et al (2020) Altered expression levels of autophagy-associated proteins during exercise preconditioning indicate the involvement of autophagy in cardioprotection against exercise-induced myocardial injury. J Physiol Sci 70:10

67. Jugdutt BI (2010) Aging and heart failure: changing demographics and implications for therapy in the elderly. Heart Fail Rev 15:401–405
68. Lu MC, Ji JA, Jiang ZY et al (2016) The Keap1-Nrf2-ARE pathway as a potential preventive and therapeutic target: an update. Med Res Rev 36:924–963
69. Done AJ, Traustadóttir T (2016) Nrf2 mediates redox adaptations to exercise. Redox Biol 10:191–199
70. Casuso RA, Al-Fazazi S, Hidalgo-Gutierrez A et al (2019) Hydroxytyrosol influences exercise-induced mitochondrial respiratory complex assembly into supercomplexes in rats. Free Radic Biol Med 134:304–310
71. Lira VA, Okutsu M, Zhang M et al (2013) Autophagy is required for exercise training-induced skeletal muscle adaptation and improvement of physical performance. FASEB J 27:4184–4193
72. Luo L, Lu AM, Wang Y et al (2013) Chronic resistance training activates autophagy and reduces apoptosis of muscle cells by modulating IGF-1 and its receptors, Akt/mTOR and Akt/FOXO3a signaling in aged rats. Exp Gerontol 48:427–436
73. Hu F, Liu F (2014) Targeting tissue-specific metabolic signaling pathways in aging: the promise and limitations. Protein Cell 5:21–35
74. Smuder AJ, Kavazis AN, Min K et al (2011) Exercise protects against doxorubicin-induced markers of autophagy signaling in skeletal muscle. J Appl Physiol (1985) 111:1190–1198
75. Tong JF, Yan X, Zhu MJ et al (2009) AMP-activated protein kinase enhances the expression of muscle-specific ubiquitin ligases despite its activation of IGF-1/Akt signaling in C2C12 myotubes. J Cell Biochem 108:458–468
76. Kwon I, Jang Y, Cho JY et al (2018) Long-term resistance exercise-induced muscular hypertrophy is associated with autophagy modulation in rats. J Physiol Sci 68:269–280
77. Kim YA, Kim YS, Oh SL et al (2013) Autophagic response to exercise training in skeletal muscle with age. J Physiol Biochem 69:697–705
78. Bayod S, Del Valle J, Pelegri C et al (2014) Macroautophagic process was differentially modulated by long-term moderate exercise in rat brain and peripheral tissues. J Physiol Pharmacol 65:229–239
79. Wohlgemuth SE, Seo AY, Marzetti E et al (2010) Skeletal muscle autophagy and apoptosis during aging: effects of calorie restriction and life-long exercise. Exp Gerontol 45:138–148
80. Ju JS, Jeon SI, Park JY et al (2016) Autophagy plays a role in skeletal muscle mitochondrial biogenesis in an endurance exercise-trained condition. J Physiol Sci 66:417–430
81. Vainshtein A, Tryon LD, Pauly M et al (2015) Role of PGC-1α during acute exercise-induced autophagy and mitophagy in skeletal muscle. Am J Physiol Cell Physiol 308:C710–C719
82. Klionsky DJ (2005) The molecular machinery of autophagy: unanswered questions. J Cell Sci 118:7–18
83. Sun M, Shen W, Zhong M et al (2013) Nandrolone attenuates aortic adaptation to exercise in rats. Cardiovasc Res 97:686–695
84. Chen H (2016) Effects of long-term aerobic exercise on changes of myocardial basal autophagy in SHR rats and its mechanism. Shaanxi Normal University, pp 25–27
85. Fiuza-Luces C, Delmiro A, Soares-Miranda L et al (2014) Exercise training can induce cardiac autophagy at end-stage chronic conditions: insights from a graft-versus-host-disease mouse model. Brain Behav Immun 39:56–60
86. Takemura G, Kanamori H, Okada H et al (2018) Anti-apoptosis in nonmyocytes and pro-autophagy in cardiomyocytes: two strategies against postinfarction heart failure through regulation of cell death/degeneration. Heart Fail Rev 23:759–772
87. Vescovo G, Volterrani M, Zennaro R et al (2000) Apoptosis in the skeletal muscle of patients with heart failure: investigation of clinical and biochemical changes. Heart 84:431–437
88. Mcmillan EM, Paré MF, Baechler BL et al (2015) Autophagic signaling and proteolytic enzyme activity in cardiac and skeletal muscle of spontaneously hypertensive rats following chronic aerobic exercise. PLoS One 10:e0119382
89. Zhu H, Tannous P, Johnstone JL et al (2007) Cardiac autophagy is a maladaptive response to hemodynamic stress. J Clin Invest 117:1782–1793

90. Cao DJ, Wang ZV, Battiprolu PK et al (2011) Histone deacetylase (HDAC) inhibitors attenuate cardiac hypertrophy by suppressing autophagy. Proc Natl Acad Sci U S A 108:4123–4128
91. Guichard JL, Rogowski M, Agnetti G et al (2017) Desmin loss and mitochondrial damage precede left ventricular systolic failure in volume overload heart failure. Am J Physiol Heart Circ Physiol 313:H32–H45
92. Yancey DM, Guichard JL, Ahmed MI et al (2015) Cardiomyocyte mitochondrial oxidative stress and cytoskeletal breakdown in the heart with a primary volume overload. Am J Physiol Heart Circ Physiol 308:H651–H663
93. Liu W, Xia Y, Kuang H et al (2019) Proteomic profile of carbonylated proteins screen the regulation of calmodulin-dependent protein kinases-AMPK-Beclin1 in aerobic exercise-induced autophagy in middle-aged rat hippocampus. Gerontology 65:620–633

Chapter 3
The Beneficial Roles of Exercise-Mediated Autophagy in T2DM

Shuaiwei Qian and Ning Chen

1 The Background of T2DM

The occurrence of type 2 diabetes mellitus (T2DM) has reached up to the pandemic level all over the world. Its inducers include obesity, poor diet, physical inactivity, increasing age, family history, and ethnicity. T2DM is characterized by reduced insulin sensitivity and increased insulin resistance combined with impaired insulin secretion. Serious complications of T2DM include cardiovascular diseases, stroke, chronic kidney failure, nephropathy, foot ulcers, neuropathy, and eye damage [1]. The pathological mechanisms of T2DM include chronic systemic inflammation, oxidative stress, mitochondrial homeostasis disorder, endoplasmic reticulum stress, excessive apoptosis and deficient autophagy, or abnormal functional status of autophagy. However, available therapeutic modalities for T2DM have achieved limited success. Honestly, metformin, a popular pharmacological treatment, has been confirmed to reveal the low efficacy for preventing the onset of T2DM than lifestyle interventions directed at diet and exercise [2].

Autophagy is a self-degradative process that targets cell constituents (damaged or aging mitochondria and sarcoplasmic reticulum, unfolded or misfolded proteins, and intracellular pathogens) to lysosomes for digestion or degradation. Under basal conditions, autophagy occurs at a lower level in most tissues contributing to the routine turnover of cellular trash and the maintenance of cellular homeostasis. However, the dysfunction of autophagy (such as autophagy inhibition or overactivation) leads to many acute and chronic diseases, such as obesity, T2DM, nonalcoholic fatty liver disease (NAFLD), cancer, neurodegenerative diseases including Parkinson's disease (PD), Alzheimer's disease (AD), Huntington's

S. Qian · N. Chen (✉)
Tianjiu Research and Development Center for Exercise Nutrition and Foods, Hubei Key Laboratory of Exercise Training and Monitoring, College of Health Science, Wuhan Sports University, Wuhan, China

N. Chen (ed.), *Exercise, Autophagy and Chronic Diseases*,
https://doi.org/10.1007/978-981-16-4525-9_3

disease (HD), and amyotrophic lateral sclerosis (ALS), cardiovascular diseases, infection, and aging [3]. Recently, more and more studies have demonstrated that autophagic dysfunction is closely involved in the pathogenesis of insulin resistance and T2DM [4–6], and exercise-mediated autophagy can exert a curative effect on insulin resistance and T2DM [7–9].

Herein, we summarize the aberrant changes in autophagy resulting in insulin resistance and T2DM in liver, skeletal muscle, adipose tissue, and pancreatic β-cells that are more sensitive to insulin originally, and the rescuing of abnormal functional status of autophagy upon exercise intervention for the prevention and treatment of insulin resistance and T2DM via improving insulin sensitivity, sustaining mitochondrial quality control, and maintaining skeletal muscle mass and function.

2 Autophagy in Insulin Target Tissues of T2DM

2.1 Hepatic Autophagy in T2DM

The liver serves as the target for metabolic diseases and is essential for the maintenance of metabolic homeostasis and organismal energetic balance. Autophagy plays a critical role in the regulation of energy metabolism homeostasis in liver, which is in charge of degrading lipid droplets (namely lipophagy), and damaged or aged organelles including mitochondria, endoplasmic reticulum, and ribosomes, to produce free fatty acids or maintain multiple quality controls. Dysregulated autophagy in liver is common in many metabolic diseases such as obesity, T2DM, and NAFLD.

High-fat diet (HFD) and streptozocin (STZ)-induced T2DM display impaired glucose tolerance, reduced insulin sensitivity, dysglycemia, and dyslipidemia, aggravated hepatic steatosis and inflammation, and suppressed hepatic autophagy. However, rapamycin and metformin can ameliorate insulin resistance, hepatic steatosis, and inflammation, and eventually improve hepatic glycolipid metabolism via the reactivation of autophagy [8]. In addition, hepatic Rab24 can effectively promote the recovery of hepatic steatosis and control glucose homeostasis through increasing autophagic flux and improving mitochondrial plasticity [10]. Moreover, the expression of human carnitine palmitoyltransferase 1A (hCPT1AM) can enhance hepatic fatty acid oxidation and autophagy, alleviate liver steatosis, maintain glucose homeostasis, and eventually repress HFD-related derangements [11]. Furthermore, human umbilical cord MSC-derived exosomes (HucMDEs) can improve hepatic glycolipid metabolism in T2DM via adenosine 5′-monophosphate (AMP)-activated protein kinase (AMPK)-induced autophagy, which also provides a novel evidence to develop HucMDE potentials for the treatment of T2DM in clinical practice [12]. X-box-binding protein 1 (sXBP1) transcription factor EB (TFEB) signaling-mediated autophagy is also of great importance to insulin resistance and hepatic steatosis [13]. Deletion of hepatic Xbp1 (Xbp1 LKO) can execute the suppression on the transcription of TFEB, a major regulator of autophagy and lysosomal biogenesis, whereas the overexpression of hepatic spliced XBP1 can enhance TFEB transcription and autophagy–lysosomal pathway. Moreover, the overexpression of TFEB in liver tissues of the Xbp1 LKO mice ameliorates glucose intolerance and hepatic steatosis in HFD-fed mice, while the deletion of TFEB and the disturbance of autophagy–lysosomal pathway impair the protective effects of sXBP1 on hepatic steatosis and glucose intolerance. These results provide a specific therapeutic option for metabolic diseases such as obesity, insulin resistance, T2DM, and NAFLD associated with dysfunctional autophagy.

As we all know, excessive accumulation of lipid droplets often coincides with insulin resistance and disturbed endoplasmic reticulum (ER) proteostasis in steatotic livers on account of the inhibition of autophagy. However, the reactivation of autophagy plays a critical role in alleviating insulin resistance via improving ER stress and degrading accumulated lipids. For instance, the suppression of hepatic Atg7 can lead to increased lipid droplets, defective insulin signaling, and elevated ER stress in liver, while the restoration of hepatic Atg7 can result in dampened ER stress, and improved hepatic insulin sensitivity and glucose tolerance [14]. Meanwhile, fibroblast growth factor 1 (FGF1) has emerged as the potentially safe candidate for stabilizing euglycemia and controlling T2DM, without inducing overt adverse effects. FGF1 protects against diabetes-induced liver steatosis, fibrosis, and apoptosis by reducing oxidative stress and ER stress, and restoring defective hepatic autophagy [15], suggesting that autophagy is an important regulator of hepatic insulin resistance and glycolipid metabolism homeostasis.

Hepatic lipid catabolism is also thought to be completed through adipose triglyceride lipase (ATGL) or lipophagy. ATGL-mediated signaling by SIRT1 promotes autophagy/lipophagy as an intrinsic mechanism to accelerate hepatic lipid catabolism and fatty acid oxidation [16]. Fibroblast growth factor 21 (FGF21) signaling can also activate hepatic autophagy/lipophagy-mediated lipid degradation via Jumonji-D3 (JMJD3/KDM6B) histone demethylase [17].

2.2 *Adipose Autophagy in T2DM*

Autophagy in adipose tissue has also been demonstrated in connection to obesity, insulin resistance, NAFLD, and T2DM. Brown adipose tissue (BAT) and white adipose tissue (WAT) can profoundly affect lipid storage and overall energy homeostasis. The metabolic disorders of BAT and WAT induced by autophagy/mitophagy dysfunction contribute to insulin resistance and T2DM.

However, the deletion of essential autophagy genes in adipocytes paradoxically leads to a unique anti-obesity and insulin sensitization effect. Atg7, an autophagy-related gene, plays an important role in normal adipogenesis, and the ablation of autophagy by deleting adipocyte-specific Atg7 significantly affects the differentiation of WAT, thereby leading to a lean phenotype [18]. Consistently, the knockdown of Atg7, Atg5, or pharmacological inhibition of autophagy in 3T3-L1 preadipocytes results in the obvious inhibition of lipid accumulation and adipocyte differentiation [19]. Similarly, adipocyte-specific Atg7 KO mice show the decreased WAT mass, increased BAT mass, and enhanced insulin sensitivity, as well as a lean body mass [19]. It is suggested that the deletion of adipocyte-specific autophagy genes may lead to an improved insulin resistance through compensatorily increasing BAT mass.

In contrast, the deletion of other autophagy-related genes or mitophagy-specific genes in adipocytes can lead to the inconsistent effects on diet-induced obesity and insulin resistance. For instance, the depletion of Atg3 or Atg16L1 induces inflammation and mitochondrial dysfunction, which eventually promotes insulin resistance on account of reduced autophagic flux in WAT and BAT [20], suggesting the important role of inflammation and mitochondrial function in the development of insulin resistance. The deletion of Pink1 in brown adipocytes can increase NLRP3 to induce white-like adipocytes from brown adipocyte precursors, thus leading to BAT dysfunction and obesity-related insulin resistance and glycolipid metabolism disorder [21]. Conversely, adipose-specific Atg7 KO mice present significantly lower expression of NLRP3 in the BAT, therefore leading to a lean phenotype [11]. The ablation of mitophagy receptor FUNDC1 leads to the accumulation of dysfunctional mitochondria in WAT. The ROS outburst from the damaged mitochondria can lead to mitogen-activated protein kinase (MAPK) activation and WAT remodeling, and subsequent insulin resistance in WAT [22]. These results indicate that the deletion of mitophagy-specific genes in adipocytes can induce the dysfunction of mitochondrial quality control and thereby lead to obesity, insulin resistance, and even T2DM.

On the whole, autophagy and mitophagy are required for regulating lipid accumulation via controlling adipocyte differentiation and determining the metabolic balance and energy homeostasis between BAT and WAT.

2.3 *Skeletal Muscle Autophagy in T2DM*

Skeletal muscle is a highly metabolic organ comprising approximately 40% of total body weight and approximately 80% of total insulin-stimulated glucose uptake under normal circumstances. The dysfunction of skeletal muscle can lead to metabolic disorders, especially insulin resistance and T2DM. The development of insulin resistance across skeletal muscle is certainly a crucial factor in the peripheral insulin resistance at the onset of T2DM. Autophagy is highly required for maintaining glucose homeostasis, improving skeletal muscle mass and function, and keeping metabolic balance in skeletal muscle.

The autophagy-deficient mice (termed BCL2AAA mice) containing knock-in mutations of BCL2 phosphorylation sites (Thr69Ala, Ser70Ala, and Ser84Ala) reveal decreased endurance exercise capacity and disordered glucose metabolism in skeletal muscle during acute exercise, as well as impaired chronic exercise-mediated protection against diet-induced glucose intolerance and metabolic disorders [7]. However, the reactivation of autophagy by dihydromyricetin [23] or exercise [24] can increase glucose uptake and ameliorate insulin resistance in skeletal muscle.

It is generally accepted that T2DM induces skeletal muscle atrophy due to its mass and function loss, which is partly attributed to the deficiency or impaired functional status of autophagy. Skeletal muscle-specific knockout of Atg7 has shown to result in skeletal muscle mass loss or atrophy, and aging-related loss-of-force production with the accumulation of defective mitochondria, excessive oxidative stress, sarcoplasmic reticulum distension, and sarcomere disorder, suggesting that autophagy is of great importance to maintain skeletal muscle mass and myofiber integrity [25]. Inconsistently, skeletal muscle-specific Atg7 knockout mice have shown the enhanced energy expenditure, less skeletal muscle and fat mass, greater glucose tolerance, increased fatty acid oxidation, and browning of WAT, thereby protecting against diet-induced obesity and insulin resistance [26]. This is probably attributed to the induction of FGF21 via ATF4 activation on account of autophagy-induced mitochondrial dysfunction [26].

Aging is generally related to increased insulin resistance and reduced autophagic activity. The expression of skeletal muscle autophagy-related genes (ATG14, RB1CC1/FIP200, GABARAPL1, SQSTM1/p62, and WIPI1) and proteins (LC3BII, SQSTM1/p62, and ATG5) is significantly decreased in aged T2DM patients, and autophagic activity in skeletal muscle is correspondingly weakened [27]. In addition, insulin resistance, apoptosis, and serious cellular senescence are confirmed after repressing autophagy with palmitate in C2C12 myotubes and can be alleviated after increasing autophagic flux by resveratrol treatment [28]. Another study has also found that autophagy can regulate the mitochondrial ADP sensitivity, redox balance, glucose tolerance, and insulin sensitivity in aged skeletal muscle [29], suggesting that the declining autophagic flux may lead to a significant increase in the risk of insulin resistance and senescence during the aging process of skeletal muscle. Therefore, it is concluded that autophagy is of great importance to glucose homeostasis, muscle mass and function, and cellular senescence in skeletal muscle.

2.4 *Pancreatic β-Cell Autophagy in T2DM*

Pancreatic β-cells play a significant role in whole-body glucose homeostasis, and the very important reason is that β-cells are the major components of the only organ responsible for insulin production and secretion. Recently, increasing reports have demonstrated that autophagy is of essential importance for regulating the architecture, mass, and function of β-cells, and is eventually involved in insulin secretion and glucose homeostasis.

On the other hand, β-cell-specific Atg7 knockout mice exhibit the increased apoptosis and decreased proliferation of β-cells with resultant reduction in β-cell mass, thereby leading to a defective insulin secretion, impaired glucose tolerance, and subsequent hyperglycemia regardless of basal or high-glucose stimulation [30], further suggesting that the degradation of unnecessary cellular components by constitutive autophagy is essential for the maintenance of pancreatic β-cell architecture, mass, and function. The depletion of Atg7 in β-cells in mice can also result in decreased β-cell mass, impaired glucose tolerance, defective insulin secretion, and excessive apoptosis when combined with high-fat and high-glucose diet, suggesting that autophagy may serve as an intrinsic protective mechanism of β-cells during a high-calorie stress [31]. However, the treatment with exendin-4, a glucagon-like peptide-1 receptor agonist, can efficiently improve glucose tolerance mainly by reducing apoptotic cell death and stimulating cell proliferation and insulin secretion in Atg7-deficient beta cells [32]. These evidences fully support the protective role of autophagy in the maintenance of pancreatic β-cell architecture, mass, and function.

T2DM develops as a consequence of combined insulin resistance and relative insulin deficiency. Generally, mitochondrial dysfunction and excessive ER stress response have been considered as the potential causes [33]. Autophagy plays a crucial role in improving insulin resistance via the degradation of damaged or aged organelles such as mitochondria and endoplasmic reticulum, which are closely related to the pathogenesis of T2DM. As reported, the deficiency of Atg7 in β-cells can suppress the expression of mitochondrial respiratory complexes, such as complexes I and II, and eventually contribute to the accumulation of damaged mitochondria and mitochondrial dysfunction [34]. β-cell-specific Atg7-null mice, when bred with ob/ob mice, can develop as severe diabetes, suggesting that autophagy-deficient β-cells can handle basal metabolic stress such as lipid-induced ER stress rather than severe metabolic stress [35]. Intermittent hypoxia activates autophagy via the ER stress-related PERK/eIF2α/ATF4 signaling pathway, which is considered to be a protective mechanism to intermittent hypoxia (IH)-induced pancreatic β-cell apoptosis [36].

In sum, autophagy can maintain β-cell architecture, mass, and function, and promote insulin production and secretion through improving mitochondrial function and ER stress.

3 Exercise-Mediated Autophagy in T2DM

3.1 Exercise-Mediated Autophagy Improves Insulin Sensitivity in T2DM

As we all know, regular exercise training exerts a positive effect on the prevention or treatment of T2DM, thus leading to increased glycolipid uptake and utilization, enhanced insulin sensitivity, reduced oxidative stress and inflammatory response, and improved skeletal muscle mass and myofiber integrity. However, the cellular and molecular mechanisms underlying these effects are largely unrevealed. Autophagy, as an important intracellular recycling system, can regulate glucose homeostasis in skeletal muscle and contribute to the reduction in insulin resistance in response to exercise.

Exercise-induced autophagy involves the disruption of the Bcl-2–Beclin1 complex, an important mediator of autophagy signaling, indicating that long-term exercise-induced autophagy may be crucial for promoting GLUT4 plasma membrane localization and improving glucose intolerance in HFD-induced obesity. However, these beneficial effects can be blocked by a mutation in BCL2 that prevents its release from an inhibitory interaction with Beclin1 [7], thereby highlighting an essential role of autophagy in the beneficial metabolic effect of exercise in insulin-resistant state. The malfunction of autophagy has been implicated with impaired glucose metabolism and progression of T2DM from prediabetes. In addition, exercise induces autophagy immediately postexercise and can be recovered after exercise for 4 h in control participants, but not in participants with prediabetes [37]. Exercise-mediated short-chain fatty acid upregulation may ameliorate insulin resistance and T2DM via increasing skeletal muscle autophagy upon binding to G protein-coupled receptor 43 (GPR43), which provides a theoretical basis for targeting gut bacterial metabolites to prevent T2DM [24]. Exercise can also stimulate AMPK and the expression of Sestrins in skeletal muscle, thus leading to increased interaction between Sestrins and AMPK. Subsequently, AMPK rapidly stimulates ULK1 activation either by suppressing mTOR, or by directly phosphorylating ULK1, and eventually increases insulin sensitivity in skeletal muscle [38]. A single bout of exercise can induce skeletal muscle autophagy in wild-type mice rather than in AMPKα$2^{-/-}$ mice. The discrepancy is related to increased amounts of Sestrin2/3 combined with AMPKα2. Chronic exercise can also significantly increase the basal autophagy flux and protein expression of Sestrin2/3 in skeletal muscle of both normal chow and HFD mice, suggesting that exercise-induced AMPK and Sestrin interaction may be involved in the beneficial metabolic effects of exercise by activating autophagy [39]. Another report has documented that acute aerobic exercise recovers the accumulation of Sestrin2, increases ULK1 phosphorylation, and promotes autophagy in skeletal muscle of aging mice, thus leading to the improvement of insulin sensitivity [9]. The transient receptor potential canonical channel 1 (TRPC1) is a Ca^{2+}-permeable channel in metabolic tissues, such as hypothalamus, adipose tissue, and skeletal muscle. TRPC1 can lead to insulin resistance and diabetes via the abnormal function status of autophagy and apoptosis, and inhibit the positive effect of exercise on T2DM risk under a HFD environment [40].

AMPK/SIRT1 is regarded as an energy center in response to nutrient deprivation or exercise, and is inhibited under an excessive energy environment. Moderate-intensity swimming training can improve lipid droplet metabolism disorder and enhance insulin sensitivity in liver through the activation of AMPK/SIRT1-mediated lipophagy [41]. Endurance exercise training counteracts glucose intolerance, insulin resistance, dyslipidemia, and hepatic steatosis in HFD mice, and exercise also can activate AMPK-mediated phosphorylation of acetyl-CoA carboxylase (ACC) and ULK1, thus leading to increased fatty acid oxidation and improved autophagy in the kidney of HFD mice [42]. TFEB is a master regulator of lysosomal biogenesis and autophagy for the adaptive response to exercise. TFEB nuclear translocation and subsequent autophagy–lysosomal pathway induced by exercise can act as a central coordinator of insulin sensitivity, glucose homeostasis, lipid oxidation, and mitochondrial function in skeletal muscle in a PGC-1α-independent manner [43].

Chronic intermittent hypoxia-associated obstructive sleep apnea is a major risk factor for insulin resistance, characterized by impaired insulin sensitivity and lower AKT phosphorylation in adipose tissue and liver [44]. Autophagy is involved in the pathophysiology of insulin resistance, and exercise has increasingly emerged as a potential therapy. However, high-intensity training improves intermittent hypoxia-induced insulin resistance without involvement of basal autophagy [44].

3.2 *Exercise-Mediated Autophagy Maintains Mitochondrial Quality Control in T2DM*

As the critical organelles responsible for energy metabolism, mitochondria can form complex and highly dynamic networks via mitochondrial biogenesis, mitochondrial dynamics (fusion and fission), and mitophagy. These dynamic coordination processes are integrated into mitochondrial quality control, which is important in maintaining the normal structure and function of mitochondrial homeostasis network [45]. However, many internal or external factors, such as high-glucose and high-fat diet, oxidative stress injury, physical inactivity, or drug treatment, may lead to the disorder of mitochondrial quality control and subsequent abnormal mitochondrial homeostasis. Recently, more and more emerging evidence supports a causal relationship between defective mitochondrial quality control and the development of insulin resistance and T2DM [45, 46]. Mitochondrial quality control machinery and mitochondrial homeostasis can be improved by exercise training through mitochondrial biogenesis, mitochondrial fusion and fission, and removal of damaged/dysfunctional mitochondria via autophagy/mitophagy, and thus repress the development of insulin resistance and T2DM.

Insulin resistance and T2DM are associated with the reduced mitochondrial mass or oxidative function and defective overall mitochondrial activity. Thus, the activation of mitochondrial biogenesis via exercise training may be a potential therapeutic strategy to treat or relieve insulin resistance and T2DM. PGC-1α is a critical transcriptional co-activator of mitochondrial biogenesis in response to exercise and can be considered as a critical strategy for preventing and reversing insulin resistance and obesity [47, 48]. The upregulation of PGC-1α mediated by exercise activates nuclear respiratory factors 1 and 2 (NRF1 and NRF2), which in turn promotes the transcription of mitochondrial transcription factor A (TFAM). TFAM directly interacts with mitochondrial DNA (mtDNA), in concert with mitochondrial transcription specificity factors TFB1M and TFB2M, and thus promotes mtDNA replication and transcription [48]. However, PGC-1α-mediated mitochondrial biogenesis in response to exercise can be positively controlled by many critical signaling pathways such as AMP/ATP-AMPK, Ca^{2+}/CaMK, ROS/p38 MAPK, and NAD^{+}/NADH/SIRT1 signaling pathways [49], which are eventually reported to improve mitochondrial mass or oxidative function, and overall mitochondrial activity under an insulin resistance or T2DM state. TFEB also can induce the expression of genes involved in mitochondrial biogenesis, glucose homeostasis, fatty acid oxidation, and oxidative phosphorylation in the adaptive metabolic response to exercise [43]. TFEB drives PGC-1α expression to protect against diet-induced obesity, insulin resistance, and metabolic dysfunction [50].

The mitochondria undergo fusion and fission to maintain mitochondrial homeostasis. Previous studies have reported that decreased mitochondrial dynamics is closely associated with insulin resistance, metabolic changes, and even T2DM [51–53]. However, moderate aerobic exercise can mitigate insulin resistance by improving mitochondrial function, and reverse mitochondrial structural damage by improving mitochondrial dynamics and mitophagy, suggesting that exercise may play a therapeutic role in protecting against mitochondrial impairments and insulin resistance [54]. Western diet reduces mitochondrial content (COX IV and cytochrome C proteins) and *PGC-1a* mRNA content in the liver, decreases mitochondrial fusion and fission markers, and eventually disturbs mitochondrial quality control and mitochondrial homeostasis. Moderate exercise appears to enhance basal autophagy, alleviates the disorder in mitochondrial quality control, and thus protects against hepatic fat accumulation and insulin resistance [55]. The regulation of autophagy/mitophagy and mitochondrial dynamics by exercise plays a significant role in maintaining cellular and mitochondrial homeostasis. Exercise and/or recovery can increase protein content of ATG7, p62/SQSTM1, FOXO3A, MFN2, and $DRP1^{Ser616}$, whereas can decrease LC3B-II content immediately after exercise in T2DM skeletal muscle. Exercise increases ULK1 phosphorylation at Ser555 site and reduces its Ser757 phosphorylation, suggesting that autophagic initiation and mitochondrial dynamics are improved by exercise in T2DM skeletal muscle [56]. Mitochondrial content and function are reduced in T2DM skeletal muscle. Three-month endurance exercise can significantly increase oxidative phosphorylation (OXPHOS) subunits in skeletal muscle, especially mitochondrial complex II, but p-FoxO3a, mitochondrial ubiquitin ligase (MUL1), BNIP3, and LC3-II/LC3-I ratio does not reveal significant difference. It means that exercise-induced expression of OXPHOS subunits is not accompanied by changes in mitophagy-associated proteins in T2DM [57]. However, endurance exercise has been confirmed to restore mitochondrial capacity in T2DM. Since the deficiency of mitochondrial dynamics and mitophagy may contribute to T2DM, elevating mitochondrial fission may result in mitochondrial dysfunction and insulin resistance. Correspondingly, a prescription with 12-week aerobic exercise has decreased DRP1 phosphorylation at Ser616, upregulated the gene expression of DRP1 and OPA1, and stimulated a growing trend toward MFN1/2 expression, suggesting that aerobic exercise may realize the improvement of insulin sensitivity and fat oxidation in insulin resistance through DRP1 inhibition [58]. As previously mentioned, Western diet-induced obesity can impair mitochondrial biogenesis resulted from the failure of PGC-1α induction and mitochondrial fusion due to reduced expression of MFN2 and OPA1, but it does not exert an obvious impact on mitochondrial fission, suggesting a shift in balance of dynamic regulation favoring fission. Western diet also blocks mitophagy through reducing BNIP3 expression, and moderate exercise appears to increase mitochondrial content through enhancing mitochondrial biogenesis and promoting health mitochondrial network by stimulating mitochondrial fusion and mitophagy [59]. Meanwhile, moderate-intensity exercise can improve mitochondrial function, enhance mitochondrial β-oxidation, attenuate systemic insulin resistance and glucose intolerance, promote autophagy influx, and mitigate hepatic steatosis and fibrosis in the liver of HFD-fed mice, thereby accomplishing beneficial contributions to insulin resistance and diabetes [60]. The underlying mechanism associated with exercise-induced mitophagy is that AMPK activation through phosphorylation at Thr172 is required for exercise-induced phosphorylation of ULK at Ser555 and lysosomal biogenesis in skeletal muscle [61].

3.3 *Exercise-Mediated Autophagy Maintains Muscle Mass and Function in T2DM*

Skeletal muscle mass is controlled by the dynamic balance between protein synthesis and protein degradation. Protein anabolism in skeletal muscle requires the activation of insulin signaling cascade such as insulin/Akt/mTOR signal pathway. Conversely, protein degradation in skeletal muscle mainly requires the mechanisms of ubiquitin–proteasome pathway and autophagy–lysosomal protein degradation. As we all know, hyperglycemia, insulin resistance, and T2DM can decrease skeletal muscle mass and induce skeletal muscle atrophy and dysfunction, which therefore contributes to poor physical fitness and increased risks of disability. Although it has already been demonstrated that these adverse effects can be alleviated by exercise-mediated autophagy, the underlying mechanisms are largely unknown yet.

Autophagy is required to maintain skeletal muscle mass and protect against skeletal muscle atrophy. Energy surplus such as high-fat and/or high-sucrose diet can lead to dysfunctional muscle and decreased grip strength. However, exercise in combination with restriction of sugar-sweetened beverages improves glucose tolerance and muscle function, reduces p62 levels, and increases GLUT4 and PGC-1α protein contents, suggesting that it is an effective way to restore glucose tolerance and muscle function via the reactivation of autophagy/mitophagy [62]. Insulin deficiency and uncontrolled diabetes can lead to the atrophy and declined strength of skeletal muscle. FoxO family members, such as FoxO1/3/4, are the critical regulators of diabetes-induced protein degradation and skeletal muscle atrophy, which can be regarded as the potential therapeutic targets for the prevention and treatment of skeletal muscle atrophy of patients with diabetes [63]. T2DM-mediated overactivation of p300 contributes to skeletal muscle atrophy due to the suppressed autophagy, and the repression of p300 can partially reduce the mass loss of skeletal muscle and rescue the atrophy of skeletal muscle in db/db mice due to the autophagic reactivation [64]. Paradoxically, another study has demonstrated that the overactivation of autophagy in diabetes could contribute to the atrophy of skeletal muscle through hypercatabolic metabolism, but chronic aerobic exercise can modify or reverse diabetic skeletal muscle atrophy or wasting by suppressing the overactivation of autophagy [65], suggesting that it may be highly necessary for controlling the functional status of autophagy to maintain the cellular homeostasis, such as autophagic activation or normal autophagic flux. FGF21 is a secreting factor that can be released in the bloodstream by peripheral organs such as skeletal muscle, liver, heart, WAT, and BAT in response to exercise. FGF21 can control skeletal muscle mass through the regulation of the anabolic/catabolic balance. FGF21 is required for fasting-induced skeletal muscle atrophy and weakness. The contribution of FGF21 to skeletal muscle atrophy has been supported by FGF21 overexpression, which is sufficient to induce autophagy and skeletal muscle loss, together with a reduction in Bnip3-mediated mitophagy flux [66]. Skeletal muscle of db/db diabetic mice displays decreased AKT phosphorylation, mTORC1 activity, and FoxO phosphorylation. In addition, the expression level of Atrogin-1 and MuRF1 is enhanced, LC3B-II is increased, and p62 is decreased in skeletal muscle of db/db diabetic mice [67]. Similarly, long-term voluntary weightlifting training also can significantly increase the mass and power of skeletal muscle through activating PI3K-AKT-mTOR signaling pathway with the overexpressed Raptor, 4E-BP-1, and p70S6K to promote protein synthesis. Importantly, weightlifting training can improve glucose clearance in whole body and insulin sensitivity in skeletal muscle along with enhanced autophagy, suggesting that resistance training promotes skeletal muscle adaptation and insulin sensitivity through enhancing autophagy [68]. Fructose ingestion can impair the expression of PGC-1α, FNDC5, NR4A3, GLUT4, Atg9, Lamp2, Ctsl, MuRF1, and MAFBx/Atrogin-1 in skeletal muscle of both sedentary and exercised animals, while the expression of ERRα and PPARδ is impaired only in the exercise group, suggesting that fructose ingestion impairs the expression of genes involved in biological processes relevant to exercise-induced skeletal muscle remodeling [69] (Fig. 3.1).

4 Conclusion

Autophagy is a self-degradative process that targets cell constituents to lysosomes for digestion or degradation. The dysfunction of autophagy leads to the etiology of many diseases, including obesity, T2DM, NAFLD, cancer, neurodegenerative diseases, cardiovascular diseases, infection disease, and aging. Autophagy is also a significant endogenous regulatory mechanism for the homeostasis of cells and tissues, especially insulin-sensitive tissues, such as skeletal muscle, liver, adipose tissue, and pancreatic β-cells. The regulatory mechanism controlled by exercise-mediated autophagy to treat and prevent T2DM is highly diverse, and the final effect on exercise-mediated autophagy is tissue-specific and metabolic context-dependent. Generally, through improving insulin sensitivity, and sustaining mitochondrial quality control and skeletal muscle mass, exercise-mediated autophagy not only can help to alleviate or depress T2DM by protecting against already existing damage, but also can attenuate the production or promote the elimination of potentially harmful intermediates of glycolipid metabolism. Hence, the possibility of regulating

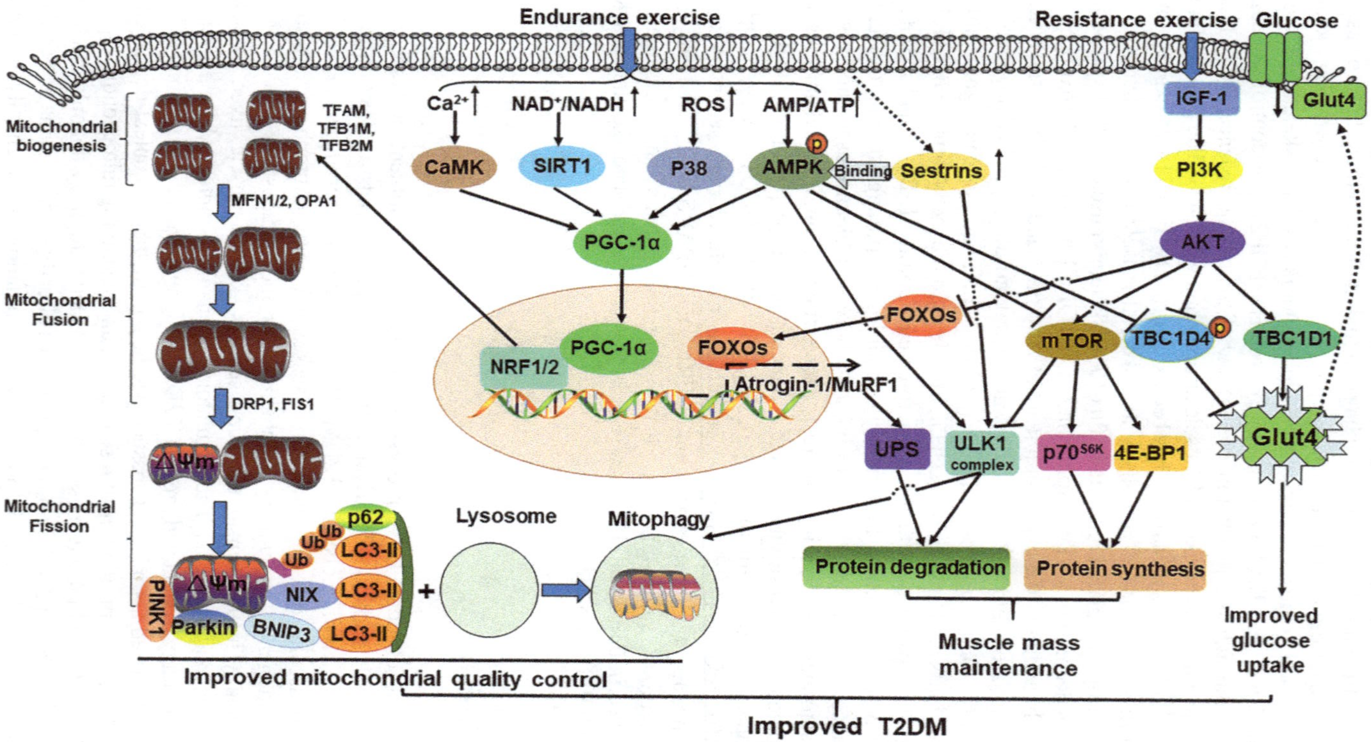

Fig. 3.1 Exercise-mediated autophagy in T2DM. (1) Endurance exercise upregulates PGC-1α via activating AMP/ATP-AMPK, Ca^{2+}/CaMK, ROS/p38 MAPK, and NAD^+/NADH/SIRT1, and then, PGC-1α enters the nucleus and activates NRF1/2, which in turn promotes the transcription of TFAM, and thus promotes mitochondrial biogenesis. Subsequently, mitochondria undergo dynamically mitochondrial fission and fusion, and the former is controlled by DRP1

and FIS1, while the latter is promoted by OPA1 and MFN1/2, all of which are regulated by PGC-1α. Finally, the damaged or aged mitochondria can be selectively degraded by autophagy/mitophagy, which is regulated by PINK/Parkin and NIX/BNIP3, thereby improving mitochondrial quality control and preventing T2DM. (2) Resistance exercise promotes protein synthesis via IGF1-PI3K-AKT-mTOR-$p70^{S6K1}$/4E-BP1, while endurance exercise reasonably increases protein degradation via AMPK/Sestrin-mediated autophagy, and finally maintains skeletal muscle mass and prevents T2DM. (3) Endurance exercise can promote GLUT4 membrane translocation via AMPK/TBC1D1, while resistance exercise can achieve the same effect via IGF1-PI3K-AKT-TBC1D1/TBC1D4, and improve glucose uptake and prevent T2DM autophagy through exercise intervention is a promising nondrug interventional strategy in T2DM treatment.

References

1. Yang JS, Lu CC, Kuo SC et al (2017) Autophagy and its link to type II diabetes mellitus. Biomedicine (Taipei) 7:8
2. Knowler WC, Barrett-Connor E, Fowler SE et al (2002) Reduction in the incidence of type 2 diabetes with lifestyle intervention or metformin. N Engl J Med 346:393–403
3. Choi AM, Ryter SW, Levine B (2013) Autophagy in human health and disease. N Engl J Med 2013(368):651–662
4. Lim YM, Lim H, Hur KY et al (2014) Systemic autophagy insufficiency compromises adaptation to metabolic stress and facilitates progression from obesity to diabetes. Nat Commun 5:4934
5. Ma LY, Lv YL, Huo K et al (2017) Autophagy-lysosome dysfunction is involved in Abeta deposition in STZ-induced diabetic rats. Behav Brain Res 320:484–493
6. Zheng HJ, Zhang X, Guo J et al (2020) Lysosomal dysfunction-induced autophagic stress in diabetic kidney disease. J Cell Mol Med 24:8276–8290
7. He C, Bassik MC, Moresi V et al (2012) Exercise-induced BCL2-regulated autophagy is required for muscle glucose homeostasis. Nature 481:511–515
8. Zhou W, Ye S (2018) Rapamycin improves insulin resistance and hepatic steatosis in type 2 diabetes rats through activation of autophagy. Cell Biol Int 42:1282–1291
9. Lenhare L, Crisol BM, Silva V et al (2017) Physical exercise increases Sestrin 2 protein levels and induces autophagy in the skeletal muscle of old mice. Exp Gerontol 97:17–21
10. Seitz S, Kwon Y, Hartleben G et al (2019) Hepatic Rab24 controls blood glucose homeostasis via improving mitochondrial plasticity. Nat Metab 1:1009–1026
11. Weber M, Mera P, Casas J et al (2020) Liver CPT1A gene therapy reduces diet-induced hepatic steatosis in mice and highlights potential lipid biomarkers for human NAFLD. FASEB J 34:11816–11817
12. He Q, Wang L, Zhao R et al (2020) Mesenchymal stem cell-derived exosomes exert ameliorative effects in type 2 diabetes by improving hepatic glucose and lipid metabolism via enhancing autophagy. Stem Cell Res Ther 11:223
13. Zhang Z, Qian Q, Li M et al (2021) The unfolded protein response regulates hepatic autophagy by sXBP1-mediated activation of TFEB. Autophagy 17:1841–1855
14. Yang L, Li P, Fu S et al (2010) Defective hepatic autophagy in obesity promotes ER stress and causes insulin resistance. Cell Metab 11:467–478
15. Xu Z, Wu Y, Wang F et al (2020) Fibroblast growth factor 1 ameliorates diabetes-induced liver injury by reducing cellular stress and restoring autophagy. Front Pharmacol 11:52
16. Sathyanarayan A, Mashek MT, Mashek DG (2017) ATGL promotes autophagy/lipophagy via SIRT1 to control hepatic lipid droplet catabolism. Cell Rep 19:1–9
17. Byun S, Seok S, Kim YC et al (2020) Fasting-induced FGF21 signaling activates hepatic autophagy and lipid degradation via JMJD3 histone demethylase. Nat Commun 11:807
18. Zhang Y, Goldman S, Baerga R et al (2009) Adipose-specific deletion of autophagy-related gene 7 (atg7) in mice reveals a role in adipogenesis. Proc Natl Acad Sci U S A 106:19860–19865
19. Singh R, Xiang Y, Wang Y et al (2009) Autophagy regulates adipose mass and differentiation in mice. J Clin Invest 119:3329–3339
20. Cai J, Pires KM, Ferhat M et al (2018) Autophagy ablation in adipocytes induces insulin resistance and reveals roles for lipid peroxide and Nrf2 signaling in adipose-liver crosstalk. Cell Rep 25:1708–1717

21. Ko MS, Yun JY, Baek IJ et al (2020) Mitophagy deficiency increases NLRP3 to induce brown fat dysfunction in mice. Autophagy 17:1205–1227
22. Wu H, Wang Y, Li W et al (2019) Deficiency of mitophagy receptor FUNDC1 impairs mitochondrial quality and aggravates dietary-induced obesity and metabolic syndrome. Autophagy 15:1882–1898
23. Shi L, Zhang T, Zhou Y et al (2015) Dihydromyricetin improves skeletal muscle insulin sensitivity by inducing autophagy via the AMPK-PGC-1alpha-Sirt3 signaling pathway. Endocrine 50:378–389
24. Yang L, Lin H, Lin W et al (2020) Exercise ameliorates insulin resistance of type 2 diabetes through motivating short-chain fatty acid-mediated skeletal muscle cell autophagy. Biology (Basel) 9:203
25. Masiero E, Agatea L, Mammucari C et al (2009) Autophagy is required to maintain muscle mass. Cell Metab 10:507–515
26. Kim KH, Jeong YT, Oh H et al (2013) Autophagy deficiency leads to protection from obesity and insulin resistance by inducing Fgf21 as a mitokine. Nat Med 19:83–92
27. Moller AB, Kampmann U, Hedegaard J et al (2017) Altered gene expression and repressed markers of autophagy in skeletal muscle of insulin resistant patients with type 2 diabetes. Sci Rep 7:43775
28. Chang YC, Liu HW, Chen YT et al (2018) Resveratrol protects muscle cells against palmitate-induced cellular senescence and insulin resistance through ameliorating autophagic flux. J Food Drug Anal 26:1066–1074
29. Buch BT, Halling JF, Ringholm S et al (2020) Colchicine treatment impairs skeletal muscle mitochondrial function and insulin sensitivity in an age-specific manner. FASEB J 34:8653–8670
30. Jung HS, Chung KW, Won KJ et al (2008) Loss of autophagy diminishes pancreatic beta cell mass and function with resultant hyperglycemia. Cell Metab 8:318–324
31. Sheng Q, Xiao X, Prasadan K et al (2017) Autophagy protects pancreatic beta cell mass and function in the setting of a high-fat and high-glucose diet. Sci Rep 7:16348
32. Abe H, Uchida T, Hara A et al (2013) Exendin-4 improves beta-cell function in autophagy-deficient beta-cells. Endocrinology 154:4512–4524
33. Rocha M, Diaz-Morales N, Rovira-Llopis S et al (2016) Mitochondrial dysfunction and endoplasmic reticulum stress in diabetes. Curr Pharm Des 22:2640–2649
34. Kim MJ, Choi OK, Chae KS et al (2015) Mitochondrial complexes I and II are more susceptible to autophagy deficiency in mouse beta-cells. Endocrinol Metab (Seoul) 30:65–70
35. Quan W, Jung HS, Lee MS (2013) Role of autophagy in the progression from obesity to diabetes and in the control of energy balance. Arch Pharm Res 36:223–229
36. Song S, Tan J, Miao Y et al (2018) Intermittent-hypoxia-induced autophagy activation through the ER-stress-related PERK/eIF2alpha/ATF4 pathway is a protective response to pancreatic beta-cell apoptosis. Cell Physiol Biochem 51:2955–2971
37. McCormick JJ, King KE, Dokladny K et al (2019) Effect of acute aerobic exercise and rapamycin treatment on autophagy in peripheral blood mononuclear cells of adults with prediabetes. Can J Diabetes 43:457–463
38. Dagon Y, Mantzoros C, Kim YB (2015) Exercising insulin sensitivity: AMPK turns on autophagy! Metabolism 64:655–657
39. Liu X, Niu Y, Yuan H et al (2015) AMPK binds to Sestrins and mediates the effect of exercise to increase insulin-sensitivity through autophagy. Metabolism 64:658–665
40. Krout D, Schaar A, Sun Y et al (2017) The TRPC1 Ca(2+)-permeable channel inhibits exercise-induced protection against high-fat diet-induced obesity and type II diabetes. J Biol Chem 292:20799–20807
41. Li H, Dun Y, Zhang W et al (2021) Exercise improves lipid droplet metabolism disorder through activation of AMPK-mediated lipophagy in NAFLD. Life Sci 273:119314

42. Juszczak F, Vlassembrouck M, Botton O et al (2020) Delayed exercise training improves obesity-induced chronic kidney disease by activating AMPK pathway in high-fat diet-fed mice. Int J Mol Sci 22:350
43. Mansueto G, Armani A, Viscomi C et al (2017) Transcription factor EB controls metabolic flexibility during exercise. Cell Metab 25:182–196
44. Pauly M, Assense A, Rondon A et al (2017) High intensity aerobic exercise training improves chronic intermittent hypoxia-induced insulin resistance without basal autophagy modulation. Sci Rep 7:43663
45. Gundersen AE, Kugler BA, McDonald PM et al (2020) Altered mitochondrial network morphology and regulatory proteins in mitochondrial quality control in myotubes from severely obese humans with or without type 2 diabetes. Appl Physiol Nutr Metab 45:283–293
46. Jheng HF, Tsai PJ, Guo SM et al (2012) Mitochondrial fission contributes to mitochondrial dysfunction and insulin resistance in skeletal muscle. Mol Cell Biol 32:309–319
47. McCarty MF (2005) Up-regulation of PPARgamma coactivator-1alpha as a strategy for preventing and reversing insulin resistance and obesity. Med Hypotheses 64:399–407
48. Sergi D, Naumovski N, Heilbronn LK et al (2019) Mitochondrial (dys)function and insulin resistance: from pathophysiological molecular mechanisms to the impact of diet. Front Physiol 10:532
49. Hood DA, Tryon LD, Carter HN et al (2016) Unravelling the mechanisms regulating muscle mitochondrial biogenesis. Biochem J 473:2295–2314
50. Evans TD, Zhang X, Jeong SJ et al (2019) TFEB drives PGC-1alpha expression in adipocytes to protect against diet-induced metabolic dysfunction. Sci Signal 12:eaau2281
51. Zorzano A, Liesa M, Palacin M (2009) Mitochondrial dynamics as a bridge between mitochondrial dysfunction and insulin resistance. Arch Physiol Biochem 115:1–12
52. Leduc-Gaudet JP, Reynaud O, Chabot F et al (2018) The impact of a short-term high-fat diet on mitochondrial respiration, reactive oxygen species production, and dynamics in oxidative and glycolytic skeletal muscles of young rats. Physiol Rep 6:e13548
53. Dube JJ, Collyer ML, Trant S et al (2020) Decreased mitochondrial dynamics is associated with insulin resistance, metabolic rate, and fitness in African Americans. J Clin Endocrinol Metab 105:1210–1220
54. Heo JW, No MH, Cho J et al (2021) Moderate aerobic exercise training ameliorates impairment of mitochondrial function and dynamics in skeletal muscle of high-fat diet-induced obese mice. FASEB J 35:e21340
55. Rosa-Caldwell ME, Lee DE, Brown JL et al (2017) Moderate physical activity promotes basal hepatic autophagy in diet-induced obese mice. Appl Physiol Nutr Metab 42:148–156
56. Kruse R, Pedersen AJ, Kristensen JM et al (2017) Intact initiation of autophagy and mitochondrial fission by acute exercise in skeletal muscle of patients with type 2 diabetes. Clin Sci (Lond) 131:37–47
57. Brinkmann C, Przyklenk A, Metten A et al (2017) Influence of endurance training on skeletal muscle mitophagy regulatory proteins in type 2 diabetic men. Endocr Res 42:325–330
58. Fealy CE, Mulya A, Lai N et al (2014) Exercise training decreases activation of the mitochondrial fission protein dynamin-related protein-1 in insulin-resistant human skeletal muscle. J Appl Physiol (1985) 117:239–245
59. Greene NP, Lee DE, Brown JL et al (2015) Mitochondrial quality control, promoted by PGC-1alpha, is dysregulated by Western diet-induced obesity and partially restored by moderate physical activity in mice. Physiol Rep 3:e12470
60. Wang B, Zeng J, Gu Q (2017) Exercise restores bioavailability of hydrogen sulfide and promotes autophagy influx in livers of mice fed with high-fat diet. Can J Physiol Pharmacol 95:667–674
61. Laker RC, Drake JC, Wilson RJ et al (2017) Ampk phosphorylation of Ulk1 is required for targeting of mitochondria to lysosomes in exercise-induced mitophagy. Nat Commun 8:548

62. Zhang D, Lee JH, Shin HE et al (2021) The effects of exercise and restriction of sugar-sweetened beverages on muscle function and autophagy regulation in high-fat high-sucrose-fed obesity mice. Diabetes Metab J. Online ahead of print. https://doi.org/10.4093/dmj.2020.0157
63. O'Neill BT, Bhardwaj G, Penniman CM et al (2019) FoxO transcription factors are critical regulators of diabetes-related muscle atrophy. Diabetes 68:556–570
64. Fan Z, Wu J, Chen QN et al (2020) Type 2 diabetes-induced overactivation of P300 contributes to skeletal muscle atrophy by inhibiting autophagic flux. Life Sci 258:118243
65. Lee Y, Kim JH, Hong Y et al (2012) Prophylactic effects of swimming exercise on autophagy-induced muscle atrophy in diabetic rats. Lab Anim Res 28:171–179
66. Oost LJ, Kustermann M, Armani A et al (2019) Fibroblast growth factor 21 controls mitophagy and muscle mass. J Cachexia Sarcopenia Muscle 10:630–642
67. Kim KW, Baek MO, Choi JY et al (2019) Analysis of the molecular signaling signatures of muscle protein wasting between the intercostal muscles and the gastrocnemius muscles in db/db mice. Int J Mol Sci 20:6062
68. Cui D, Drake JC, Wilson RJ et al (2020) A novel voluntary weightlifting model in mice promotes muscle adaptation and insulin sensitivity with simultaneous enhancement of autophagy and mTOR pathway. FASEB J 34:7330–7344
69. Goncalves NG, Cavaletti SH, Pasqualucci CA et al (2017) Fructose ingestion impairs expression of genes involved in skeletal muscle's adaptive response to aerobic exercise. Genes Nutr 12:33

Chapter 4
Exercise-Induced Autophagy and Obesity

Yanju Guo and Ning Chen

1 Introduction

Modern lifestyle characterized by energy-rich diets and physical inactivity due to the fast-paced rhythm of life has led to the global epidemic of obesity [1]. The imbalance between energy intake and expenditure of the body is due to the increase in the size and number of adipocytes, in contributing to steatosis, which is closely related to insulin resistance, mitochondrial dysfunction, endoplasmic reticulum (ER) stress, inflammation, and so on [2–4]. Autophagy is a lysosomal-dependent degradation pathway widely found in eukaryotic cells. In this process, long-lived or damaged proteins or organelles are degraded and recycled to produce ATP and new organelles [5]. Autophagy is a fundamental process that regulates cellular metabolism and energy balance. Exercise has always been considered as the best healthy and effective way to lose body weight, but it must be combined with calorie restriction, so-called tube shut up and move around, to achieve significant effects. Previous studies have confirmed that exercise or calorie restriction can activate autophagy in peripheral and central tissues, which is an important molecular mechanism of exercise to promote human health [6, 7]. Therefore, exercise-induced autophagy may be an effective nonpharmacological intervention to prevent and treat obesity and its complications. Herein, we will summarize the previous studies, briefly describe the characteristics and underlying mechanisms of autophagy in regulating obesity, discuss the relationship between autophagy and obesity, and how exercise-induced autophagy affects obesity, and then finally put forward exercise intervention measures for targeting autophagy in the prevention and rehabilitation of obesity.

Y. Guo · N. Chen (✉)
Tianjiu Research and Development Center for Exercise Nutrition and Foods, Hubei Key Laboratory of Exercise Training and Monitoring, College of Health Science, Wuhan Sports University, Wuhan, China

N. Chen (ed.), *Exercise, Autophagy and Chronic Diseases*,
https://doi.org/10.1007/978-981-16-4525-9_4

2 General Characteristics and Mechanism of Autophagy

Autophagy is a general term for the process of lysosome degradation of intracellular material components unique to eukaryotic cells and is mainly responsible for the degradation and utilization of long-lived proteins and some organelles [4, 8]. According to the different ways in which the intracellular substrate is transported to the lysosome cavity, autophagy can be divided into macroautophagy (hereafter referred to as autophagy), microautophagy, and chaperone-mediated autophagy (CMA) [4]. In addition, according to the different selectivity of substrate proteins, autophagy can be further divided into mitochondrial autophagy (mitophagy), protein aggregation autophagy (aggrephagy), peroxisomal autophagy (pexophagy), lipid droplet autophagy (lipophagy), endoplasmic reticulum autophagy (reticulophagy), xenophagy, and so on [9].

Autophagy is a response of cells to the changes in internal and external environmental stimuli. These predisposing factors come from outside the cell, such as external nutrients, ischemia, hypoxia, and the deprivation of growth factors, and also come from intracellular stimulation, such as metabolic stress, aging or damaged organelles, protein folding errors, or protein aggregation. Due to the long-term existence of these factors, the cells maintain a very low, basal autophagy activity to maintain their own homeostasis. During starvation, autophagy provides an internal source of nutrients for energy production and cell survival, whereas elevated autophagy level is often accompanied by caloric restriction. In addition, hormones, especially insulin and glucagon, are key autophagy regulators. Insulin inhibits autophagy, while glucagon stimulates autophagy [10]. Insulin resistance is a key indicator of obesity [11].

The formation of autophagosome is one of the critical steps in the process of autophagy activation, which are strictly regulated by 18 different autophagy-related genes (Atg). The formation of autophagosomes includes three major steps: initiation, nucleation, and elongation/closure. The initiation of autophagosomes is controlled by the ULK1-ATG13-FIP200 complex. The nucleation stage requires Beclin1 class III phosphatidylinositol 3-kinase (PI3K) complex, which includes Beclin1, Vps34 (class III PI3K), Vps15, Atg14L (Barkor), and Ambra1. Two conjugate systems are involved in the extension and elongation processes. One is the Atg12-Atg5 coupling reaction mediated by Atg7 and Atg10 ligases. Atg5 also binds to Atg16 to form the Atg12-Atg5-Atg16 complex. The second way is to cleave LC3 (Atg8) by Atg4 to form a soluble form of LC3-I and then couple with phosphatidylethanolamine (PE) with the participation of Atg7 and Atg3. This lipid combination forms the LC3-II protein associated with the autophagy double membrane, which closes the autophagic vacuole. The appearance of LC3-II indicates the formation of autophagosomes.

As emphasized by the 2016 Nobel Prize in Physiology or Medicine, autophagy is now considered to be the major mechanism by which cells eliminate unnecessary or toxic cellular constituents [12]. Basal autophagy protects the health of cells and tissues through replacing expired or damaged cellular components with new ones.

Notably, abnormal functional status of autophagy (excessive activation or deficiency) can lead to the occurrence of many human diseases, such as cancer, cardiovascular diseases, obesity, diabetes, and aging [4, 13]. Recent studies have reported its role and regulation in obesity-related complications.

3 Relationship Between Autophagy and Obesity

Due to the modern high-calorie food intake and lack of exercise lifestyle, calorie storage in the body is far greater than its expenditure, thus resulting in the accumulation of lipids and leading to overweight or obesity [5, 14]. Obesity is often associated with multiple comorbidities such as diabetes, hypertension, dyslipidemia, cardiovascular diseases, and cancer [5, 15], and has become a major risk factor affecting global human health in the twenty-first century. Autophagy is highly sensitive to changes in the nutritional environment, especially high-fat and/or high-calorie diets [5]. Obesity is often due to the accumulation of adipose tissue in the body, which triggers metabolic stress by increasing inflammation and the levels of fatty acids, triglycerides, and low-density lipoproteins, thus leading to a series of complications such as insulin resistance, dyslipidemia, diabetes, and hypertension [16]. Autophagy plays a protective role against these stress damages. For example, the removal of lipid droplets through the autophagy signal pathway (called lipophagy) provides a mechanism to reduce adipose content and thereby normalizes lipid metabolism in the tissues of obese individuals. Autophagy can also limit the production of reactive oxygen species (ROS) under obesity-related pathological conditions by removing dysfunctional mitochondria (called mitophagy). During ER stress, autophagy can eliminate damaged or redundant parts of the ER (called reticulophagy), thus restoring the dynamic balance of ER [2]. In turn, the altered autophagy, especially the defect of autophagy, may also promote the occurrence of metabolic disorders and obesity, so autophagy has a dual function in maintaining human health.

3.1 Lipophagy and Obesity

Adipose tissue is mainly composed of adipose cells. According to the structure and function of adipocytes, adipose tissue in the body can be divided into white adipose tissue (WAT) and brown adipose tissue (BAT). WAT is composed of highly differentiated cells that can store excessive energy in the body, in the form of neutral fat, and release it when the body is in urgent need [17], and with the average diameter of 30–230μm [18]. BAT contains a large number of mitochondria and can emit lipids and glucose in the form of heat through a reaction mediated by the specifically expressed mitochondrial uncoupling protein UCP1 [19, 20]. Obesity is a direct result of the accumulation of WAT [21]. In adipose tissue, autophagy is a key regulator of

WAT and BAT adipogenesis, and dysregulation of autophagy will impair fat accumulation both in vivo and in vitro [19, 22]. A study has found that 3-methyladenine (3-MA), an autophagy inhibitor, can significantly increase triglyceride (TG) content in mouse hepatocytes [22]. When knocked out Atg5 in mouse hepatocytes, TG content is also significantly increased, and the amounts and the size of TG are significantly increased [22], indicating that the inhibited autophagy could result in the significant increase in TG content, meanwhile indirectly indicating that the enhancement of autophagy activity could promote lipid metabolism [22].

Lipophagy, a special autophagy first discovered in the study of liver cells, is the process by which parts or entire lipid droplets in an organism form autophagy into the lysosome for degradation, between individually and together with other cytobacterial components [23]. Under this particular autophagy, triglycerides and cholesterol bind to lysosomes through autophagosomes and then degraded to generate free fatty acids (FFAs) for oxidative phosphorylation. Lipophagy could regulate intracellular fat reserve, FFA contents, and energy metabolism balance. Obesity or excessive calorie intake can lead to defective regulation of autophagy in adipose tissue [24]. Recent studies have demonstrated that the development and functional maintenance of adipose tissue depend on the quality and differentiation of adipocytes, among which autophagy plays an important role [4, 25]. Autophagy is upregulated in adipose tissue of obese individuals, and is more significant in visceral adipose tissue (VAT) than in subcutaneous adipose (SAT), especially when accompanied by insulin resistance and/or type 2 diabetes mellitus (T2DM) [26, 27]. In order to explore the role of autophagy in the differentiation of white adipose tissue, an adipocyte model has been established through deleting autophagy-related gene 5 (*Atg5*) in mouse primary embryonic fibroblasts (MEFs) to prevent fat production in vitro. Then, the mice carrying *flox*-flanked *Atg7* gene are crossbred with mice carrying aP2-Cre to evaluate the effect of adipose tissue specific autophagy deficiency on fat formation in offspring mice. The results show that the fat deposition mass of mutant mice is much smaller than the wild-type mice, and mutant adipocytes exhibit unusual characteristics, including multilocular lipid droplets and the substantial increase in the number of mitochondria. In addition, under the same conditions of food and water consumption, mutant mice are significantly slimmer than wild-type mice and show higher levels of basic physical activity. It is noted that mutant mice are resistant to HFD-induced obesity and can significantly increase the sensitivity to insulin [21, 28].

3.2 Mitophagy and Obesity

Except mature red blood cells, mitochondria are present in almost all human cells. Mitochondria in adipocytes mainly reflect the following four functions: (1) Providing ATP for adipocytes to use [29]: Mitochondria decompose FFA into acetyl-CoA and propionyl-CoA through β-oxidation, and these coenzymes are further oxidized to produce ATP by tricarboxylic acid cycle (TCA cycle) and oxidative phosphorylation. In particular, the abundant mitochondria in BAT and the specific expression of uncoupling protein 1 (UCP1) can enable BAT to quickly radiate energy in the form of heat to maintain body temperature. (2) Promoting lipogenesis: In this process, acetyl-CoA from various sources such as glucose metabolism is used to synthesize fatty acids, which are then esterified with glycerides to form TG. Lipogenesis occurs primarily in hepatocytes and also in adipocytes [30]. Mitochondria play an important role in adipogenesis by providing key intermediates. Most of acetyl-CoA involved in fatty acid synthesis is produced by pyruvate dehydrogenase in mitochondria. (3) Promoting lipolysis: Mitochondria are the only organelles in mammals that further oxidize short- and medium-chain aliphatic groups, and are also important organelles that further oxidize long-chain aliphatic groups. In BAT, most of FFAs produced by adipose decomposition are degraded through β-oxidation in mitochondria. Acetyl-CoA, the final product of β-oxidation in mitochondria, is further oxidized through the TCA cycle and electron transport chain, and is the energy source of heat dissipation under the action of UCP1, while catecholamine can activate the expression of UCP1 in BAT, suggesting a direct link between lipolysis and mitochondrial function [31]. WAT has fewer mitochondria, almost no expression of UCP1, and lower level of β-oxidation. However, previous studies have found that the increase in cAMP level in white adipocytes leads to acute decoupling of mitochondria and increased oxygen consumption, indicating that short-term and effective mitochondrial oxidation of lipids has been activated in response to lipolysis signals [20, 32]. (4) Affecting adipokine secretion: Mitochondrial function is critical for the synthesis of adiponectin, one of the major antihyperglycemic adipokines produced by adipose tissue [33]. Impaired mitochondrial function may reduce adiponectin production by activating c-Jun N-terminal kinase (JNK) and then inducing the activating transcription factor 3 (ATF3).

Adipose tissue is the key to regulating the entire body metabolism, while mitochondria play an important role in the function of adipocytes [20, 34]. Mitochondrial damage is a direct cause of many human diseases, and mitophagy selectively removing mitochondria is an important factor in regulating mitochondrial content and enhancing mitochondrial quality control [9]. Since the substrate of mitophagy is usually damaged mitochondria, when depolarization damage occurs, the corresponding ligand proteins on the outer mitochondrial membrane will accumulate and then recruit and bind to the corresponding receptor proteins. Under traction, it binds to Atg8 or LC3 proteins to complete the package of autophagosomes. In eukaryotes, the major receptor proteins for mitophagy are Atg32 (yeast), BNIP3 (Bcl-2/adenovirus E1B 19-kDa-interacting protein 3), NIX/BNIP3L, and FUNDC1.

Mitophagy has been emphasized as an important process of white fat formation. Increasing evidence suggests that adipogenic differentiation is related to an increase in the number of autophagosomes containing mitochondria [35]. After the consumption of high-fat diet, adipocytes may undergo significant metabolic changes before WAT macrophages infiltrate, including the loss of mitochondrial biosynthesis and the downregulation of mitochondrial proteins [36]. In addition to its role in white adipogenesis, the regulation of mitochondrial content and quality is essential for browning of WAT and bleaching of beige and brown adipocytes [37, 38]. Similarly, Parkin-mediated downregulation of mitochondrial autophagy during WAT browning process is also observed [37]. In contrast, the restoration of beige fat cells to the nonthermogenic white fat cell-like state depends on the removal of autophagic mitochondria [39]. In animal models, deficient autophagy due to the deletion of Atg5 or Atg12 or drug inhibition can promote the retention of UCP1 and maintain the characteristics of beige adipocytes. Taken together, these effects are associated with the reduced susceptibility to diet-induced obesity and insulin resistance [39].

HFD intake is correlated with BAT albinism and enhanced expression of mitophagy biomarkers [40]. Autophagy flux, the expression of Parkin and Pink1, and the number of mitochondria close to autophagosomes have been reported to increase after the consumption of high-fat diets, suggesting that the response of WAT to high-fat diets is at least partly dependent on mitochondrial remodeling [36]. Interestingly, a recent study has demonstrated that mice lacking the mitochondrial autophagy receptor FUNDC1 have an impaired response to high-fat diets, thus showing greater obesity and insulin resistance, as well as more infiltration of pro-inflammatory macrophages [41]. At least in part, mitochondrial autophagy is also necessary to maintain normal mitochondrial function and avoid cellular dysfunction [41]. Therefore, in addition to the function of brown/beige adipocytes, the changes that white adipocytes undergo are also closely related to mitochondrial homeostasis, especially to mitochondrial autophagy in order to adapt to high-fat diets.

3.3 *Reticulophagy and Obesity*

Endoplasmic reticulum (ER) is the main protein "assembly factory" in cells and acts as a reservoir of calcium ions that facilitate the folding of newly synthesized proteins. Unfolded or misfolded protein accumulation, glucose deficiency, or oxidative stress results in unfold protein response (UPR) activation. UPR promotes protein folding, degradation, and endoplasmic reticulum homeostasis. In mammals, UPR can be activated by three independent transmembrane sensors, thereby leading to the transcription of a number of genes, including protein kinase R (PKR)-like endoplasmic reticulum kinase (PERK), activated transcription factor 6 (ATF6), and inositol required kinase 1 (IRE1) [42]. However, continuous activation of UPR can induce ER stress.

ER stress can be caused by obesity and lipotoxicity, and then promote the process of reticulophagy [2, 43]. Obesity itself is a powerful inducer of liver ER stress, and obesity-related ER stress can aggravate fat accumulation, insulin resistance, and liver damage. Therefore, the induction of autophagy may be a defense mechanism against ER stress-induced damage. In hematopoietic cells, ER stress induced by tunicamycin alters cell metabolism, including glucose uptake and utilization, followed by mitochondrial activation, which increases cellular oxygen consumption and overall ATP synthesis [44].

In yeast, Atg39 and Atg40 are the receptors of reticulophagy, and other receptors for reticulophagy in mammals include FAM134B [45], SEC62 [46], reticulon 3 (RTN3) [47], and cell-cycle progression gene 1 (CCPG1) [48]. These ER receptor proteins have the similar function with autophagy receptors and can interact with LC3 proteins. Reticulophagy plays an important role in the recovery of ER stress. Insulin stimulation in Atg7 gene-deficient cells can decrease insulin receptor β-subunit and protein kinase B (PKB or Akt) phosphorylation, thereby leading to insulin resistance. If the Atg7 gene is silenced with short hairpin RNA (shRNA), the protein expression levels of the PERK and the C/EBP homologous protein (CHOP) are increased, suggesting increased ER stress. The restoration of Atg7 expression in hepatocytes can alleviate ER stress, mitigate glucose tolerance, and promote insulin sensitivity. Therefore, hyperinsulinemia induced by insulin resistance can inhibit liver autophagy, and the suppressed autophagy in liver can aggravate ER and insulin resistance in a vicious cycle [49]. At present, studies on reticulophagy in other metabolic diseases are not much involved.

4 The Effect of Exercise-Induced Autophagy on Obesity

Exercise is a new predisposing factor of autophagy in the body, and it is also an effective means to prevent and treat obesity. Several studies have shown that treadmill exercise can optimize body weight and visceral fat mass, and reduce insulin resistance and blood lipid level in HFD-induced rats, thereby effectively treating obesity [50, 51]. Beth Levine, a professor of Southwestern Medical Center at the University of Texas, once said, "I've always known exercise is good for you, but when we found that it increases autophagy, I finally got a treadmill." [7]. It is sufficient to illustrate the critical role of autophagy in body regulation and the importance of exercise-mediated autophagy in health promotion. Exercise-induced autophagy occurs in multiple organs involved in metabolism, such as skeletal muscle, liver, pancreas, and adipose tissue [7, 52]. Different exercise modes have different effects on autophagy flux in different tissues and organs of the body. Therefore, understanding how endurance and resistance exercise affect autophagy in various tissues and organs of the body is important for formulating precise exercise prescriptions for the prevention and treatment of metabolic diseases.

4.1 Endurance Exercise

Exercise has long been considered to be the best way to lose body weight, and endurance exercise in particular has a significant effect on reducing systemic and visceral fat and alleviating insulin resistance by promoting glucose metabolism. After C57BL/6J mice are fed with a Western diet (WD) for 8 weeks, regardless of obesity, the autophagy biomarker LC3-II/LC3-I ratio reveals an increase by 50% and p62 protein content exhibits a decrease by 40% in voluntary wheel running (VWR) group, but the decrease in the content of BNIP3 protein, a typical molecular biomarker for measuring mitochondrial content, does not present the significant difference from that in the WD group [53]. These results indicate that the disruption of mitochondrial quality control caused by diet-induced obesity may precede the onset of liver inflammation, and that moderate physical activity may enhance basal autophagy in liver, thereby protecting the liver from lipid accumulation [53]. However, for HFD-induced SD rats, 8-week treadmill training has no significant effect on the content of autophagy biomarkers Beclin1 and p62 in epididymal white adipose tissue (eVAT), but it can effectively attenuate the apoptosis-related signals in HFD-fed rats, such as increased expression of delta-like homologue 1 (*DLK1*)/*pre*adipocyte factor 1 (*PREF1*) and Bcl-2 proteins, and decreased activity of Bax and Caspase-9, Caspase-8, and Caspase-3 [50]. Another study has also shown that treadmill exercise can optimize body weight, body fat, and obesity-related biochemical indicators in HFD-induced obese SD rats, but has no effect on autophagy in the soleus [51]. Compared to autophagy and mitophagy markers in skeletal muscle of long-term endurance runners and sedentary nonobese people before and after high-fat meal (HFM) challenge, it is found that the content of total LC3 protein in skeletal muscle of endurance runners is significantly higher than that of sedentary nonobese people from the control group, and the critical biomarkers reflecting mitochondrial autophagy are also significantly higher than that in the control group, which is not affected by HFM diet [54]. Due to the different types and intensity of exercise, the time point of analysis, and the conditions and tissues of the experimental animals used, could result in the differences in autophagy-related molecular markers (Table 4.1). As such, to reveal the potential therapeutic effect of exercise intervention on obesity-related diseases from the perspective of autophagy, further studies on corresponding signal transduction pathways are needed.

4.2 Resistance Exercise or Combined Endurance Exercise

Most studies on the improvement of obesity and obesity-related metabolic diseases have focused on aerobic endurance exercises, such as treadmill training. Although autophagy changes have been observed in skeletal muscle after resistance exercise, but this exercise mode mainly focuses on the treatment and improvement of geriatric degenerative sarcopenia [56, 57] and skeletal muscle strength rehabilitation. Previous studies have also reported that long-term chloroquine (CQ) intervention can cause sporadic inclusion body myositis (sIBM)-related pathological features in rats, characterized by the decrease in skeletal muscle strength, whereas resistance exercise (RE) can rescue abnormal autophagy in soleus muscle of rats from the CQ group, but the expression of autophagy-related proteins in flexor hallucis longus (FHL) muscle has no significant difference among sham, CQ control, and CQ-RE groups, so the improvement of this abnormal autophagy shows tissue-type specificity [57]. The chronic effects of protein synthesis in skeletal muscle of obese elderly people

Table 4.1 Impact of exercise on autophagy markers in different tissues of HFD-induced obesity

Reference	Model	Intervention	Tissue	Autophagy biomarker
Rosa-Caldwell et al. (2017) [53]	8-week-old C57BL/6J mice	4-week voluntary wheel running	Liver	↑ LC3-II/LC3-I ↓ p62
Tarpey et al. (2017) [54]	18- to 45-year-old male endurance runners (BMI 18–30)	HFM challenge and endurance training	Skeletal muscle (VL)	↑ Total LC3 ↑ LC3-I, without change in LC3-II, p62, Bcl-2, Beclin1, ULK1 ↑ FoxO3a, p-FoxO3a ↑ p-Pink1^{Thr257}, without change in total Pink1 ↓ Total Parkin ↑ p-Parkin^{Ser65}
Cho et al. (2017) [51]	8-week-old SD rats	10-week treadmill exercise	Skeletal muscle (Sol)	No change in Beclin1, p62, LC3, Lamp2
Rocha-Rodrigues et al. (2018) [50]	17-week-old SD rats	8-week treadmill exercise	eWAT	↑ Beclin1, p62
Yang et al. (2021) [55]	7-week-old C57BL/6J mice	7-month DHA + treadmill exercise	Liver	↑ Atg5, Atg7, LC3-II/LC3-I, ↓ p62

Abbreviations: *Sol* soleus muscle, *eWAT* epididymal white adipose tissue, *HFM* high-fat meal, *VL* vastus lateralis, *DHA* docosahexenoic acid, *C group* control group

through body weight management and different exercise methods have also been evaluated [58]. The results show that the chronic expression of autophagy regulators such as Atg101, Lamp2, and Atg17 is decreased in the lateral femoral muscle tissue of the subjects in the aerobic exercise combined with resistance exercise group, heat shock protein 70 (HSP70), a kind of protein chaperones, reveals the more significant decrease than the simple aerobic exercise group [58]. However, further studies are highly needed to enhance the expression of molecular signal pathways involved in muscle protein synthesis by combinatorial exercise interventions.

5 Exercise Interventions for Obese Patients Targeting Autophagy

According to the 2017 Global Nutrition Report, 2 billion adults worldwide are either overweight (BMI 25–29.9 kg/m^2) or obese (BMI ≥30 kg/m^2), and 41 million children are overweight [59]. Obesity and overweight have become an international public health problem, and also affect the quality of life of the individuals, thereby increasing the risks of various chronic metabolic diseases [60]. Current interventions for obesity mainly focus on diet, drugs, surgical treatment, and physical activity. Changes in dietary structure include the currently popular modes, such as ketogenic diet [61] and the Mediterranean diet [62], while any improvement of dietary structure and habit must aim at lifelong behavior change. Long-term drug treatment for obesity can result in many side effects. Surgical treatment is more obvious for the body weight loss of severely obese people, but a variety of complications often occur. Therefore, physical activity intervention as an important part of daily energy expenditure is currently considered to be the most effective means to resist and reverse obesity. A review of lifestyle, dietary changes, medications, and the implementation of bariatric surgery for weight loss [60] will not be elaborated, but physical activity interventions will be focus on here.

The current international guidelines recommend that obese people should perform aerobic exercise at least 300 min at the continuous moderate intensity per week and be supplied with resistance training under the condition permit [63, 64]. Physical activity is a complex intervention and whose effectiveness depends on the duration and intensity of exercise. Different exercise training programs, such as strength training, endurance training, or a combination of both, may have different effects. In addition, due to the differences in age and health status of obese people, the formulation of exercise prescriptions should also be based on the reference of health intervention guidelines according to individual characteristics (Fig. 4.1). For example, for obese children and adolescents, in addition to strictly controlling the intake of high-calorie foods, high-quality protein foods such as fish, meat, eggs, milk, and soybean products as well as foods rich in minerals such as calcium, magnesium, and zinc should be supplemented. Physical activity is performed in the sunlight for at least 1 h every day, meanwhile adequate sleeping should also be ensured. Severely obese individuals are recommended to participate in the body weight loss training camp for targeted exercise training. Overweight or obesity in adults is mainly due to sedentary, late night, and physical inactivity. In order to achieve better effect on body weight loss, physical exercise should be guaranteed for 3–5 times/week, more than 30 min each time, and the heart rate reaches up to 120–150 times/min. The major exercise modes should include aerobic exercise, or combined with resistance exercise, because obese elderly people suffer from progressive decline in skeletal muscle mass, increased prevalence of cardiovascular diseases, and declined exercise performance. Therefore, when making exercise prescription for body weight loss, subjects should first have a comprehensive health medical examination and physical assessment, and then make corresponding exercise plan for body weight loss according to the assessment results. During the process of exercise interventions, the exercise intensity should not be too large, and the exercise time should be dispersed to several times a day, with aerobic exercise supplemented by simple resistance exercise. Obese elderly people should strengthen medical supervision during physical exercise, such as monitoring and recording heart rate, blood pressure, and health status to prevent excessive fatigue and avoid exercise injuries.

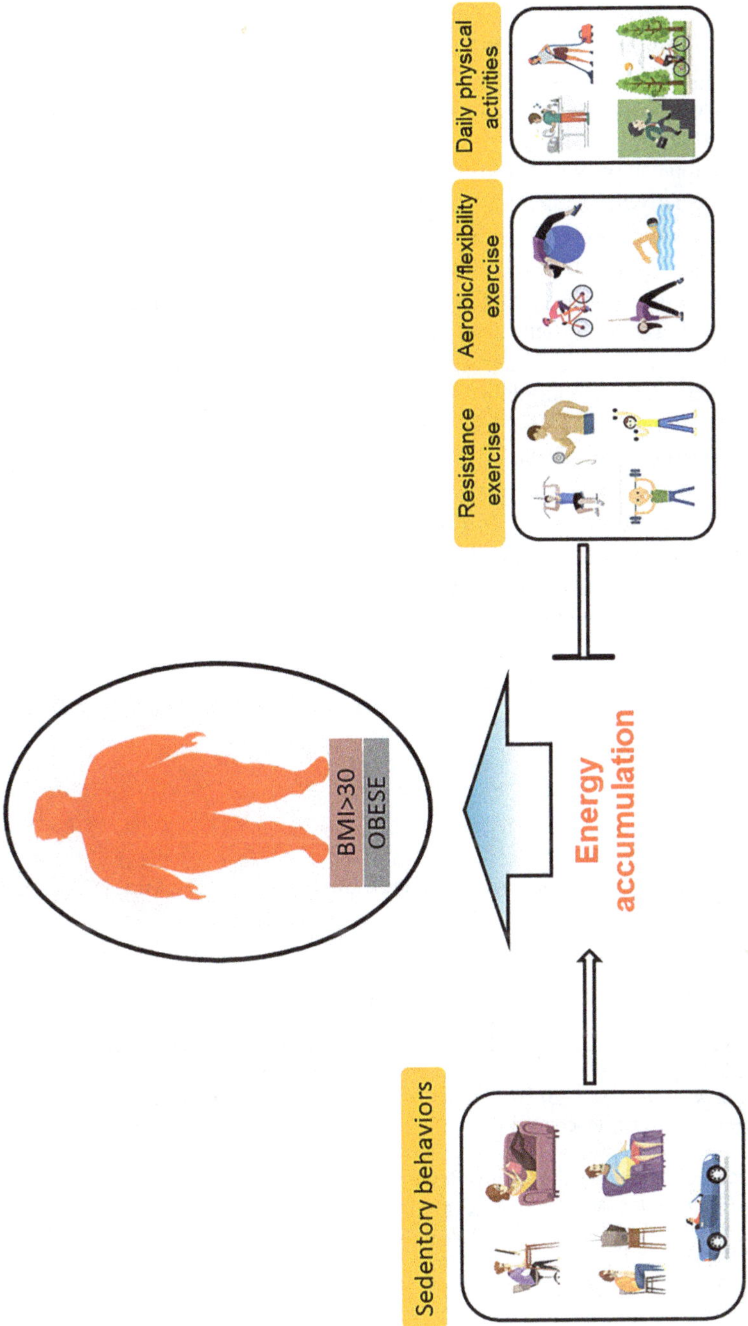

Fig. 4.1 Physical activity pyramid recommends healthy lifestyle for different obese people [65]

6 Conclusion

Autophagy, a subcellular process that promotes the recycling of energy substances and protects cells from damage, plays a complex and important role in the regulation of tissue homeostasis. The close correlation between autophagy and obesity contributes the important role of autophagy in the initiation and development of obesity. Calorie restriction can prolong lifespan, which may be partly attributed to autophagy activation, especially lipophagy and mitophagy can lead to reduced oxidative stress and lipid content, thus preventing and delaying the development of obesity-related complications. ER stress-induced adipocyte autophagy can enhance insulin sensitivity and plays an important role in the alleviation of obesity-induced insulin resistance and T2DM. Physical activity combined with caloric restriction is an important nonpharmacological intervention strategy for body weight loss, and both interventions can activate autophagy to improve metabolic flexibility and insulin sensitivity. However, there are still many questions to be addressed before autophagy-targeting therapy can be applied in the clinical treatment of obesity, including the specific role of autophagy in the insulin signaling pathway, the change in functional status of autophagy in obese people, the underlying mechanisms of autophagy in insulin resistance, and the regulation of autophagy on the formation of lipid droplets. Further studies should be focused on better exploring the potential mechanisms of exercise-mediated autophagy in the initiation and progression of obesity-related diseases, so as to facilitate the implementation of precise body weight loss interventions at the aspect of autophagy.

References

1. Frühbeck G, Yumuk V (2014) Obesity: a gateway disease with a rising prevalence. Obes Facts 7 (Suppl 2):33–36
2. Namkoong S, Cho CS, Semple I et al (2018) Autophagy dysregulation and obesity-associated pathologies. Mol Cells 41:3–10
3. Kosacka J, Kern M, Klöting N et al (2015) Autophagy in adipose tissue of patients with obesity and type 2 diabetes. Mol Cell Endocrinol 409:21–32
4. Lavallard VJ, Meijer AJ, Codogno P et al (2012) Autophagy, signaling and obesity. Pharmacol Res 66:513–525
5. Zhang Y, Sowers JR, Ren J (2018) Targeting autophagy in obesity: from pathophysiology to management. Nat Rev Endocrinol 14:356–376
6. He C, Bassik MC, Moresi V et al (2012) Exercise-induced BCL2-regulated autophagy is required for muscle glucose homeostasis. Nature 481:511–515
7. Garber K (2012) Autophagy. Explaining exercise. Science 335:281
8. Yorimitsu T, Klionsky DJ (2005) Autophagy: molecular machinery for self-eating. Cell Death Differ 12(Suppl 2):1542–1552
9. Clemente-Postigo M, Tinahones A, El Bekay R et al (2020) The role of autophagy in white adipose tissue function: implications for metabolic health. Meta 10:179
10. Stienstra R, Haim Y, Riahi Y et al (2014) Autophagy in adipose tissue and the beta cell: implications for obesity and diabetes. Diabetologia 57:1505–1516

11. Kahn SE, Hull RL, Utzschneider KM (2006) Mechanisms linking obesity to insulin resistance and type 2 diabetes. Nature 444:840–846
12. Tooze SA, Dikic I (2016) Autophagy captures the Nobel prize. Cell 167:1433–1435
13. Rabinowitz JD, White E (2010) Autophagy and metabolism. Science 330:1344–1348
14. Bastien M, Poirier P, Lemieux I et al (2014) Overview of epidemiology and contribution of obesity to cardiovascular disease. Prog Cardiovasc Dis 56:369–381
15. Flegal KM, Kruszon-Moran D, Carroll MD et al (2016) Trends in obesity among adults in the United States, 2005 to 2014. JAMA 315:2284–2291
16. Choi AM, Ryter SW, Levine B (2013) Autophagy in human health and disease. N Engl J Med 368:651–662
17. Ro SH, Jang Y, Bae J et al (2019) Autophagy in adipocyte browning: emerging drug target for intervention in obesity. Front Physiol 10:22
18. Singh R, Xiang Y, Wang Y et al (2009) Autophagy regulates adipose mass and differentiation in mice. J Clin Invest 119:3329–3339
19. Ferhat M, Funai K, Boudina S (2019) Autophagy in adipose tissue physiology and pathophysiology. Antioxid Redox Signal 31:487–501
20. Zhang Y, Zeng X, Jin S (2012) Autophagy in adipose tissue biology. Pharmacol Res 66:505–512
21. Goldman S, Zhang Y, Jin S (2010) Autophagy and adipogenesis: implications in obesity and type II diabetes. Autophagy 6:179–181
22. Singh R, Kaushik S, Wang Y et al (2009) Autophagy regulates lipid metabolism. Nature 458:1131–1135
23. Gatica D, Lahiri V, Klionsky DJ (2018) Cargo recognition and degradation by selective autophagy. Nat Cell Biol 20:233–242
24. Nuñez CE, Rodrigues VS, Gomes FS et al (2013) Defective regulation of adipose tissue autophagy in obesity. Int J Obes 37:1473–1480
25. Baerga R, Zhang Y, Chen PH et al (2009) Targeted deletion of autophagy-related 5 (atg5) impairs adipogenesis in a cellular model and in mice. Autophagy 5:1118–1130
26. Maixner N, Bechor S, Vershinin Z et al (2016) Transcriptional dysregulation of adipose tissue autophagy in obesity. Physiology (Bethesda) 31:270–282
27. Kovsan J, Blüher M, Tarnovscki T et al (2011) Altered autophagy in human adipose tissues in obesity. J Clin Endocrinol Metab 96:E268–E277
28. Zhang Y, Goldman S, Baerga R et al (2009) Adipose-specific deletion of autophagy-related gene 7 (atg7) in mice reveals a role in adipogenesis. Proc Natl Acad Sci USA 106:19860–19865
29. Rong JX, Qiu Y, Hansen MK et al (2007) Adipose mitochondrial biogenesis is suppressed in db/db and high-fat diet-fed mice and improved by rosiglitazone. Diabetes 56:1751–1760
30. Large V, Peroni O, Letexier D et al (2004) Metabolism of lipids in human white adipocyte. Diabetes Metab 30:294–309
31. Lowell BB, Spiegelman BM (2000) Towards a molecular understanding of adaptive thermogenesis. Nature 404:652–660
32. Yehuda-Shnaidman E, Buehrer B, Pi J et al (2010) Acute stimulation of white adipocyte respiration by PKA-induced lipolysis. Diabetes 59:2474–2483
33. Koh EH, Park JY, Park HS et al (2007) Essential role of mitochondrial function in adiponectin synthesis in adipocytes. Diabetes 56:2973–2981
34. Tseng YH, Cypess AM, Kahn CR (2010) Cellular bioenergetics as a target for obesity therapy. Nat Rev Drug Discov 9:465–482
35. Goldman SJ, Zhang Y, Jin S (2011) Autophagic degradation of mitochondria in white adipose tissue differentiation. Antioxid Redox Signal 14:1971–1978
36. Cummins TD, Holden CR, Sansbury BE et al (2014) Metabolic remodeling of white adipose tissue in obesity. Am J Physiol Endocrinol Metab 307:E262–E277
37. Taylor D, Gottlieb RA (2017) Parkin-mediated mitophagy is downregulated in browning of white adipose tissue. Obesity (Silver Spring) 25:704–712

38. Cairó M, Villarroya J (2020) The role of autophagy in brown and beige adipose tissue plasticity. J Physiol Biochem 76:213–226
39. Altshuler-Keylin S, Shinoda K, Hasegawa Y et al (2016) Beige adipocyte maintenance is regulated by autophagy-induced mitochondrial clearance. Cell Metab 24:402–419
40. Shimizu I, Aprahamian T, Kikuchi R et al (2014) Vascular rarefaction mediates whitening of brown fat in obesity. J Clin Invest 124:2099–2112
41. Wu H, Wang Y, Li W et al (2019) Deficiency of mitophagy receptor FUNDC1 impairs mitochondrial quality and aggravates dietary-induced obesity and metabolic syndrome. Autophagy 15:1882–1898
42. Abdrakhmanov A, Gogvadze V, Zhivotovsky B (2020) To eat or to die: deciphering selective forms of autophagy. Trends Biochem Sci 45:347–364
43. Li L, Xu J, Chen L et al (2016) Receptor-mediated reticulophagy: a novel promising therapy target for diseases. Acta Biochim Biophys Sin 48:774–776
44. Wang X, Eno CO, Altman BJ et al (2011) ER stress modulates cellular metabolism. Biochem J 435:285–296
45. Kurth I, Pamminger T, Hennings JC et al (2009) Mutations in FAM134B, encoding a newly identified Golgi protein, cause severe sensory and autonomic neuropathy. Nat Genet 41:1179–1181
46. Bergmann TJ, Fumagalli F, Loi M et al (2017) Role of SEC62 in ER maintenance: a link with ER stress tolerance in SEC62-overexpressing tumors? Mol Cell Oncol 4:e1264351
47. Shi Q, Ge Y, Sharoar MG et al (2014) Impact of RTN3 deficiency on expression of BACE1 and amyloid deposition. J Neurosci 34:13954–13962
48. Smith MD, Harley ME, Kemp AJ et al (2018) CCPG1 is a non-canonical autophagy cargo receptor essential for ER-phagy and pancreatic ER proteostasis. Dev Cell 44:217–232.e211
49. Yang L, Li P, Fu S et al (2010) Defective hepatic autophagy in obesity promotes ER stress and causes insulin resistance. Cell Metab 11:467–478
50. Rocha-Rodrigues S, Gonçalves IO, Beleza J et al (2018) Effects of endurance training on autophagy and apoptotic signaling in visceral adipose tissue of prolonged high fat diet-fed rats. Eur J Nutr 57:2237–2247
51. Cho DK, Choi DH, Cho JY (2017) Effect of treadmill exercise on skeletal muscle autophagy in rats with obesity induced by a high-fat diet. J Exerc Nutr Biochem 21:26–34
52. He C, Sumpter R Jr, Levine B (2012) Exercise induces autophagy in peripheral tissues and in the brain. Autophagy 8:1548–1551
53. Rosa-Caldwell ME, Lee DE, Brown JL et al (2017) Moderate physical activity promotes basal hepatic autophagy in diet-induced obese mice. Appl Physiol Nutr Metab 42:148–156
54. Tarpey MD, Davy KP, Mcmillan RP et al (2017) Skeletal muscle autophagy and mitophagy in endurance-trained runners before and after a high-fat meal. Mol Metab 6:1597–1609
55. Yang J, Sáinz N, Félix-Soriano E et al (2021) Effects of long-term DHA supplementation and physical exercise on non-alcoholic fatty liver development in obese aged female mice. Nutrients 13:501
56. Mejías-Peña Y, Estébanez B, Rodriguez-Miguelez P et al (2017) Impact of resistance training on the autophagy-inflammation-apoptosis crosstalk in elderly subjects. Aging (Albany, NY) 9:408–418
57. Jeong JH, Yang DS, Koo JH et al (2017) Effect of resistance exercise on muscle metabolism and autophagy in sIBM. Med Sci Sports Exerc 49:1562–1571
58. Colleluori G, Aguirre L, Phadnis U et al (2019) Aerobic plus resistance exercise in obese older adults improves muscle protein synthesis and preserves myocellular quality despite weight loss. Cell Metab 30:261–273.e266
59. Endalifer ML, Diress G (2020) Epidemiology, predisposing factors, biomarkers, and prevention mechanism of obesity: a systematic review. J Obes 2020:6134362
60. Bray GA, Frühbeck G, Ryan DH et al (2016) Management of obesity. Lancet 387:1947–1956
61. Roberts MN, Wallace MA, Tomilov AA et al (2017) A ketogenic diet extends longevity and healthspan in adult mice. Cell Metab 26:539–546.e535

62. Estruch R, Ros E (2020) The role of the Mediterranean diet on weight loss and obesity-related diseases. Rev Endocr Metab Disord 21:315–327
63. Medicine ACOS (2018) ACSM's guidelines for exercise testing and prescription, 10th edn. Wolters Kluwer Health, Philadelphia
64. Yumuk V, Tsigos C, Fried M et al (2015) European guidelines for obesity management in adults. Obes Facts 8:402–424
65. Williams G (ed) (2009) Obesity science to practice. John Wiley & Sons, Hoboken, NJ

Chapter 5
Exercise-Mediated Autophagy and Nonalcoholic Fatty Liver Disease

Fengxing Li, Kai Zou, and Ning Chen

1 Introduction

1.1 Epidemiology and Diagnosis of Nonalcoholic Fatty Liver Disease (NAFLD)

Nonalcoholic fatty liver disease (NAFLD) is becoming one of the major clinical burdens of liver diseases worldwide [1]. In many developed countries and regions, the incidence of NAFLD is much higher than that of various infectious liver diseases, and it affects 20–40% of the global population [2]. The strong correlation between NAFLD and type II diabetes mellitus (T2DM) makes it complicated for the prevention and treatment of these diseases. A meta-analysis including 49,419 cases of T2DM patients with NAFLD describes the global prevalence rate of 55.5% for NAFLD [3]. The even more worrying is the pandemic of NAFLD in children, which is closely related to the development of adult cirrhosis, T2DM, and cardiovascular diseases [4]. Despite the high prevalence of chronic diseases in the world, the pathogenesis of NAFLD is complicated and remains unknown. Moreover, there are series of problems including side effects of current drugs. Taken together, there is an urgent need to develop new treatment strategies that are safe and feasible. Therefore, comprehensive investigation of the pathogenesis of NAFLD can be beneficial for its early diagnosis and treatment.

F. Li · N. Chen (✉)
Tianjiu Research and Development Center for Exercise Nutrition and Foods, Hubei Key Laboratory of Exercise Training and Monitoring, College of Health Science, Wuhan Sports University, Wuhan, China

K. Zou
Department of Exercise and Health Sciences, University of Massachusetts Boston, Boston, MA, USA

N. Chen (ed.), *Exercise, Autophagy and Chronic Diseases*,
https://doi.org/10.1007/978-981-16-4525-9_5

NAFLD is usually defined as liver steatosis that requires radiological or histological confirmation. In the absence of excessive alcohol consumption, liver steatosis exceeds 5% [5]. In addition, systemic diseases, drug abuse, and alcohol drinking more than 140 g/week for men and 70 g/week for women in the past 12 months should be excluded [6]. Due to the close relationship with metabolic syndromes, NAFLD is often considered as a unique manifestation of systemic metabolic syndrome in liver.

NAFLD is the most common cause of chronic liver diseases in Western countries. A large proportion of patients eventually become liver cirrhosis or even liver cancer [7, 8]. The clinical burden is not simply limited to the consideration on incidence rate and mortality rate of liver diseases, and it should consider larger impacts of NAFLD on overall health because it is a multisystem disease affecting extrahepatic organs through regulating corresponding signal pathways. Most of the modalities with NAFLD could result in cardiovascular diseases (CVD) [9, 10]. Therefore, metabolic diseases have gained tremendous attentions, and now, there is an urgent demand for the prevention and treatments of NAFLD with the simultaneous intervention for the overall health of human beings.

1.2 Pathological Process of NAFLD

The most significant characteristic of NAFLD is the elevated intrahepatic triglycerides (IHTGs), accompanied by inflammation and fibrosis at varying degrees [11–13]. NAFLD is initially considered as a benign disease. However, from the recent evidence [14–17], its pathological process integrates insulin resistance, excessive lipid deposition, low-grade inflammatory activation, and fibroblast proliferation. The suppressed insulin sensitivity and excessive lipid toxicity at the early stage of NAFLD have been studied in depth [18, 19], but at the later stage, the corresponding process becomes complex and shows great individual differences.

Nonalcoholic steatohepatitis (NASH), as one of the most common inflammatory subtypes of NAFLD, is an important landmark in the development and progression of NAFLD. In addition to steatosis, NAFLD is also accompanied by balloon-like injury and inflammation of hepatocytes. Currently, no drugs can effectively reverse the inflammation associated with NAFLD [20].

The physiological and pathological processes such as inflammation, apoptosis, cell proliferation, differentiation, and metabolism are involved in the pathogenesis of most chronic diseases. NAFLD is a multifactorial disease with the involvement of multiple genetic factors. Exploring and identifying these mutant genes are helpful for scientists and clinical physicians to understand more clearly the specific process of NAFLD. Patatin-like phospholipase domain-containing 3 (PNPLA3) is one of the most studied genes involved in the formation and decomposition of lipid droplets, which is the most directly related to the development and progression of NAFLD.

Apart from the specific population with high expression of susceptible genes, unhealthy lifestyles such as sedentary lifestyle [21, 22] and overnutrition are the major risk factors for NAFLD. As NAFLD is defined as excessive lipid accumulation, reducing lipid intake and increasing lipid oxidation are important prevention and control measures of NAFLD. Of course, the further exploration and clarification of specific molecular mechanisms for the initiation and progression of NAFLD and the development of its prevention and treatments are highly desired. The classic multiple hypotheses hold that obesity, insulin resistance, imbalanced intestinal flora, oxidative stress, and inflammatory cytokine increase are involved in the onset and progression of NAFLD [23].

1.3 Autophagy in NAFLD

Autophagy is a highly conserved process for cells to achieve their own metabolic function through a selective self-recycling manner, thereby maintaining cellular homeostasis. In various chronic diseases, the functional status of autophagy may be responsible for the initiation, development, and progression of chronic diseases. As a highly conserved intracellular process, autophagy can not only recover and decompose various damaged organelles through the degradation in lysosomes, but also act as an important material source for the generation of new organelles. Due to the dynamic regulation by autophagy-related genes (Atg), autophagy can be activated or blocked by its regulators. Accurate monitoring and measuring functional status of autophagy is essential to understand the functional importance and regulatory role of autophagy under physiological and pathological conditions [24].

1.3.1 Autophagy Regulates Lipid Storage in Hepatocytes

Autophagy is a cascade of reactions strictly regulated by a variety of autophagy-related genes. Its basic function is to degrade damaged cell components for providing new raw materials under the condition of limited nutrition, so as to maintain the normal function of cells [25].

Lipid metabolism is involved in lipid generation and decomposition in hepatocytes. Emerging evidence has demonstrated the regulatory roles of autophagy in lipid metabolism [26–28]. For example, autophagy deficiency caused by high-fat diet (HFD) can promote lipid accumulation, which further suppresses the functional status of autophagy and accelerates lipid accumulation in a closed vicious circle. In hepatocytes, lipid clearance can be accomplished by autophagy through encapsulation by autophagosomes, delivery to lysosomes, and decomposition in acidic lysosomes, thereby producing new fatty acids. In NAFLD models, the reduced acidity of the lysosome could result in the impaired autophagy flux and the suppressed autophagy. To some extent, the impaired autophagy in NAFLD is caused by the disordered lysosomal acidification [29].

In an obese mouse model, autophagy-related genes especially Atg7 in liver are seriously downregulated [30]. Similarly, by comparing the rates of triglyceride synthesis and free fatty acid β-oxidation in liver tissues of the mice with silenced Atg5 gene and the control subjects, it is confirmed that autophagy has a direct regulatory effect on triglyceride decomposition, and excessive exogenous fat intake also can destroy the autophagy-mediated lipid decomposition [27]. Therefore, targeting autophagy for the prevention and treatment of NAFLD could be a promising interventional strategy because autophagy is associated with oxidative stress, obesity, insulin resistance, and other metabolic dysfunctions. Further studies are needed to understand how autophagy-mediated metabolic changes affect the development of progression of human diseases including NAFLD.

1.3.2 Autophagy Regulates the Differentiation and Production of Adipocytes

The deficiency of autophagy is always the common feature of many metabolic diseases [31–33]. In fact, autophagy can be activated in various tissues or organs of human and animals, and even present an opposite effect in different organs from the same disease sometimes. Therefore, more and more people are beginning to pay attention to the metabolic changes in nonliver tissues in NAFLD, such as adipose tissue and skeletal muscle. The most common occurrence in liver cells of NAFLD is the accumulation of lipids.

For a long time, it is believed that brown adipocytes only exist in infants and young children, and then gradually degenerate and disappear with the extension of age. In contrast, with the development of nuclear medicine, the study has found that this kind of adipose tissue with high metabolic activity still exists in adults. Brown adipose tissue is the opposite of traditional adipose tissue composed of white adipocytes and usually characterized by energy storage, abundant mitochondria, and fast energy supply [34]. In recent years, a study has documented that autophagy is closely correlated with adipocyte differentiation [35]. Essentially, both white adipocytes and brown adipocytes are differentiated from pre-adipocytes. Therefore, both mature cells retain the potential to differentiate into another type of cells. For example, there are two opposite cellular processes including white adipocyte browning and brown adipocyte whitening, under the regulation by specific genes during the maturation and differentiation of adipocytes. In addition, autophagy, especially mitochondrial autophagy, shows an important maintenance effect on the function of adipocyte tissues. Brown adipocyte precursors (BAPs) in the mice with the knockout of mitochondrial autophagy-related gene *Pink1* could result in the accumulation of large lipid droplets and an obvious downregulation of the brown adipose-related genes *Ppargc1a* and *Prdm16*, and a clear trend of brown adipocyte whitening. In addition, genetically defective mice with global or brown adipocyte-specific deletion of *Pink1* could result in the disorder to maintain body temperature, suggesting the intolerance to cold exposure and dysfunctional brown adipose tissues [36].

Uncoupling protein 1 (*UCP1*) gene is one of the most critical genes for heat production in brown adipocytes [37]. The mouse primary adipocytes are harvested to establish autophagy activation models through serum starvation and rapamycin administration. After administering autophagy inhibitor 3-methyladenine (3-MA), the thermogenic gene *UCP1* presents an obvious activation, thereby significantly reduced accumulation of lipid droplets, and smaller lipid droplets are observed. This result indirectly indicates that the function of primary adipocytes with autophagy deficiency is close to that of brown adipocytes [38]. After specifically knocking out autophagy-related gene 7 (*Atg7*) in mouse brown adipocytes, the excessive clearance of mitochondria in the cell is alleviated and more mitochondria in these cells than normal brown adipocytes are observed. Due to the higher mitochondrial content, insulin sensitivity is also improved, and energy metabolism is more smooth [39]. The knockout of *Atg7* gene can result in a large number of small lipid droplets in mouse white adipose tissue instead of single eye lipid droplets, suggesting the characteristics of brown adipose tissue, with increased intracellular mitochondria, and the improved lipid metabolism. However, the expression of *UCP1* and other brown adipocyte-specific genes has not changed, indicating that even if adipocytes are not completely transformed, the changes in the intracellular organelles can affect lipid metabolism in the body [40]. The autophagy activation caused by the knockout of the specific Rptor gene in mouse adipose tissue also can greatly degrade intracellular lipid droplets by increasing autophagy isolation and lipolysis of lipid droplets. The degradation of a large number of lipid droplets directly leads to fat development. When the *Rptor* and *Atg7* genes are knocked out at the same time, the autophagy defect reverses the immaturity of fat development caused by the knockout of the *Rptor* gene alone, suggesting the important role of autophagy in the development and maturation of adipocytes [41]. Previous studies have also shown that autophagy can affect the ratio of two adipocytes by regulating the differentiation of pre-adipocytes [42], thereby reducing lipid storage and maintaining the balance of systemic metabolism, curbing obesity, NAFLD, and other metabolic abnormalities [43].

In short, the regulation of autophagy and the intervention of adipocytes through exercise are expected to become a new target for the intervention of NAFLD. During exercise, skeletal muscle is always regarded as the core variable, and the effect of exercise on adipose tissue is often ignored. Therefore, there is still no definite answer to the effect of exercise on adipose tissue, and further studies are highly needed. The optimal exercise protocols with defined exercise methods and exercise durations should be further explored and developed. Therefore, whether there is an optimal exercise prescription for regulating the functional status of autophagy in other tissues while inhibiting the level of autophagy in adipose tissue as a specific issue is worth exploring.

2 Effect of Exercise Intervention in NAFLD and Underlying Mechanisms

2.1 *Exercise Increases Lipid Metabolism Via Regulation of Autophagy*

Exercise is effective to reverse cell defects in NAFLD, suggesting it may be a viable approach for preventing and treating NAFLD [44]. The activation of autophagy, especially the basal autophagy in hepatocytes, is critical to reverse cell defects. However, some special styles of autophagy processes in cells, such as mitochondrial autophagy and lipophagy, maintain the balance of metabolism by selectively decomposing damaged organelles and excessive lipid droplets [45]. The damaged organelles such as mitochondria could lead to energy disorders in cells. Interestingly, 12-week swimming intervention in HFD-fed C57BL/6 mice can result in the downregulation of fatty acid-binding protein 1 (FABP1) signal transduction pathway. In addition, long-term exercise training can improve lipid homeostasis of NAFLD and alleviate hepatic steatosis [44]. These benefits are validated to be correlated with the improved autophagy lysosomal pathway. Similarly, another animal study has demonstrated that 8-week moderate intensity (60% of maximum aerobic speed) treadmill running offers the reduced progression of steatosis and inflammation in liver tissue by upregulating AMPK phosphorylation at Thr172 site and peroxisome proliferation-activated receptor alpha (PPARα) protein expression [46].

2.2 *Exercise Alleviates NAFLD by Reducing Mitochondrial Oxidative Stress*

High-energy food consumption and increasing convenient transportation have allowed people to achieve a gradual decrease in daily activity. People generally consume more energy, but their daily consumption continues to decrease. Cells, while undergoing oxidative metabolism, also generate an array of by-products such as reactive oxygen species (ROS), which include small molecule superoxide radicals ($O_2^{-\bullet}$) in the presence of unpaired electrons and hydrogen peroxide (H_2O_2) [47]. Excessive energy intake and reduced energy consumption can render the storage of excessive energy without full utilization in the body due to the incapability of clearing these oxygen free radicals as the metabolic by-products in time, thereby causing the direct consequences with the loss of mitochondrial integrity, massive accumulation of free radicals, and even DNA fragments, as well as more serious cell death [48]. In the physiological state, the generation and elimination of ROS are always in a dynamic equilibrium, and free ROS can also act as the receptor and regulator in many pathways to respond to the challenge of the external harsh environment [49].

As NAFLD is characterized by massive lipid accumulation within hepatocytes, excessive fatty acids also impose a tremendous lipid pressure on hepatocytes. Initially, mitochondria can cope with increasing biosynthesis of ROS, but as a metabolic progress, the continuous accumulation of ROS, which are by-products of metabolism, begins to poison mitochondria. The function of mitochondria is disrupted by oxidative stress-induced ROS and inflammation as an important mechanism, thereby leading to hepatocyte death and tissue damage [50].

Increasing evidence has also proven that diet-induced NAFLD presents the dysfunctional mitochondrial quality control [51]. During the development and progression of NAFLD, the structure and function of mitochondria are damaged to some extent, as shown in swollen and irregular shape. At the early stage of mitochondrial damage, in order to cope with excessive lipid pressure, the oxidation rate may remain unchanged or even upregulated. However, the excessive lipid accumulation may lead to the overload of ROS in mitochondria and eventually the damaged mitochondrial function. As an important organelle, mitochondria are in charge of cellular function and metabolism. Once energy circulation in mitochondria is blocked, the vitality of the whole cell will be affected. Moderate physical activity seems to enhance the basal phagocytic function of cells in liver, and the induced autophagy can prevent lipid accumulation in liver by maintaining the integrity and permeability of mitochondria [52].

The potential of exercise training in attenuating the levels of hepatic oxidative stress and inflammation in patients with NAFLD is substantial, as shown in a modest effect on oxidative stress in a variety of metabolic diseases during exercise intervention [53–56]. The combinatorial treatment including voluntary exercise and electrically stimulated muscle contraction, a novel exercise training originally designed to prevent skeletal muscle atrophy during space flights, is recently shown to attenuate the level of oxidative stress and improve insulin sensitivity of the liver in patients with NAFLD [57]. A study of endurance exercise provides the evidence that endurance exercise protects hepatocytes from oxidative stress triggered by hypercholesterolemia by enhancing the activity of the antioxidant system and enhancing the activity of hepatic glutathione peroxidases through enhanced expression of hepatic superoxide dismutase 2 (SOD-2) [58]. Similarly, a study has demonstrated that chronic aerobic exercise in both high-fat fed and normal fed rats enhances the expression of nuclear factor erythroid 2-related factor 2 (Nrf2), an important regulator of antioxidant processes, activates the antioxidant enzyme superoxide dismutase 1 (SOD-1), and attenuates oxidative damage of liver of the rats [59]. Therefore, exercise training has shown great potential in relieving hepatocytes from oxidative damage. Gratifyingly, the increasing exercise training methods also provide people with more options in addition to drugs for the prevention and treatment of NAFLD.

2.3 Exercise Regulates microRNA-Mediated Autophagy in NAFLD

microRNA is a class of noncoding single-stranded RNA molecules with the length of approximately 22 nucleotides, with the function of regulating posttranscriptional gene expression. microRNAs are involved in the development and progression of a series of diseases including NAFLD [60–63].

The upregulation of miR-212 is detected in the liver of mice following HFD feeding, while 16-week treadmill exercise training downregulates the expression of miR-212, thereby protecting liver damage from HFD-induced steatosis [64]. In order to explore the underlying mechanism, fibroblast growth factor 21 (FGF-21) is elucidated as the target gene of miR-212 involved in lipid formation. It is reasonable to speculate that the downregulation of miR-212 exerts the protective effect on NAFLD by targeting FGF-21 upon exercise intervention [65]. In addition, miR-188 is upregulated in liver of obese mice, and liver-specific overexpression of miR-188 exacerbates liver steatosis and insulin resistance by targeting autophagy-related gene 12 (Atg12) [66], suggesting that microRNA-mediated autophagic signal pathway may be a therapeutic target. Similarly, in NAFLD mice, miR-130b-5p is upregulated, while insulin-like growth factor-binding protein-2 (IGFBP2) is downregulated. The miR-130b-5p can target IGFBP2 to downregulate its expression and to induce the inactivation of Akt signaling pathway, thereby resulting in metabolic disorders. In contrast, the downregulated miR-130b-5p can prevent lipid accumulation and insulin resistance in NAFLD [67]. Some upstream targets of microRNA-regulated autophagy flux are also screened and identified; for example, liver X receptor-α (LXR-α) promotes the accumulation of lipid droplets through suppressing autophagy and lipophagy due to the dysregulation of Atg4b and Ras-related protein Rab-8b [68] responsible for the formation of autophagosomes and autolysosomes, during inducing miR-34a. Therefore, during the activation of LXR-α, autophagy activity is suppressed and lipid phagocytosis is weakened. At the same time, mitochondrial oxidation capacity is also reduced, which leads to the increase in lipid level and the deterioration of liver steatosis in NAFLD [69]. In vitro experiment, using human HepG2 cell line, the endogenous expression level of miR-26a is significantly decreased after treatment with free fatty acids, so the accumulation of lipids is further reduced after the overexpression of miR-26a, and the mRNA expression levels of inflammatory biomarkers such as IL-6 and fibrosis markers including TGF-β1 and TGF-β2 are decreased. What is more, through the detection of autophagy biomarkers, it is found that miR-26a can also induce autophagy to alleviate NAFLD [70], which is similar to the previous study on the regulatory role of miR-26a in alcoholic fatty liver disease [71].

Aerobic exercise has been proved to play a lipid-lowering role through activating autophagy signal pathway in a miR-33-dependent manner with increasing AMPK phosphorylation and inhibiting mTOR phosphorylation [72]. Similarly, swimming exercise can upregulate miR-451 to suppress macrophage migration inhibitory factor in liver through inhibiting Akt signaling and inducing autophagy, thereby realizing the prevention and alleviation of NAFLD. In recent years, high-intensity interval training (HIIT) is often compared with aerobic endurance exercise during the intervention of chronic diseases. Current studies have documented that HIIT at intensity of 85–90% VO_{2max} for 2 min and at intensity of 30–40% VO_{2max} for 2 min with five repetitions each time reveals the more upregulation of miR-122 than aerobic training at intensity of 60–65% VO_{2max} for 30 min, indicating that HIIT-induced miR-122 results in more obviously reduced expression of fatty acid synthase (FAS), acetyl CoA carboxylase (ACC), and sterol regulatory element-binding 1c (SREBP1c) [73] (Fig. 5.1).

2.4 Exercise Promotes Autophagy in NAFLD by Increasing H_2S Activity

Hydrogen sulfide (H_2S) is a kind of gaseous signal molecule with the benefits for cardiovascular homeostasis. H_2S is an effective stimulant of autophagy flux. Similarly, as a signal molecule of liver function, H_2S is involved in the etiology of various liver diseases, such as NASH, liver fibrosis, and liver cancer [74]. Exogenous H_2S can induce pro-inflammatory cytokines, and activate nucleotide-binding oligomerization domain-like receptor protein 3 (NLRP3) inflammasomes and NF-κB signaling in HepG2 cells in a dose-dependent manner, which is an alternative treatment candidate for NAFLD [75].

H_2S can mitigate HFD-induced NAFLD by suppressing apoptosis and inducing autophagy through ROS/PI3K/Akt/mTOR signaling pathway [76]. It is found that the inhibition of AMPK in AMPKα$2^{-/-}$ mice can effectively block autophagy by sodium hydrosulfide (NaHS, a widely used H_2S donor), which further proves that H_2S can activate autophagy in liver through AMPK-mTOR signal pathway to reduce serum triglyceride level and improve NAFLD [77]. In contrast, the injection of NaHS in HFD-induced NAFLD model mice can alleviate NAFLD by reducing liver lipid deposition and improving pathological changes in liver. Using omics approaches such as metabonomics and proteomics [78], several candidate proteins for exercise-mediated regulation of lipolysis in NAFLD are screened and identified. Similarly, some natural foods, such as algae in seawater, *E. prolifera*, have also been proved to have beneficial effects on NAFLD by upregulating cystathionine β-synthase (CBS), a key enzyme of H_2S, to reduce the level of triglyceride in the body [79].

Exercise training restores the bioavailability of H_2S and promotes autophagy flux in liver to alleviate the systemic insulin resistance, glucose intolerance, hepatic steatosis, and fibrosis of HFD-induced mice, and significantly enhance β-oxidation of mitochondria in hepatocytes [80]. Therefore, it indicates that targeting key enzymes involved in H_2S generation may be a novel effective therapeutic strategy of NAFLD.

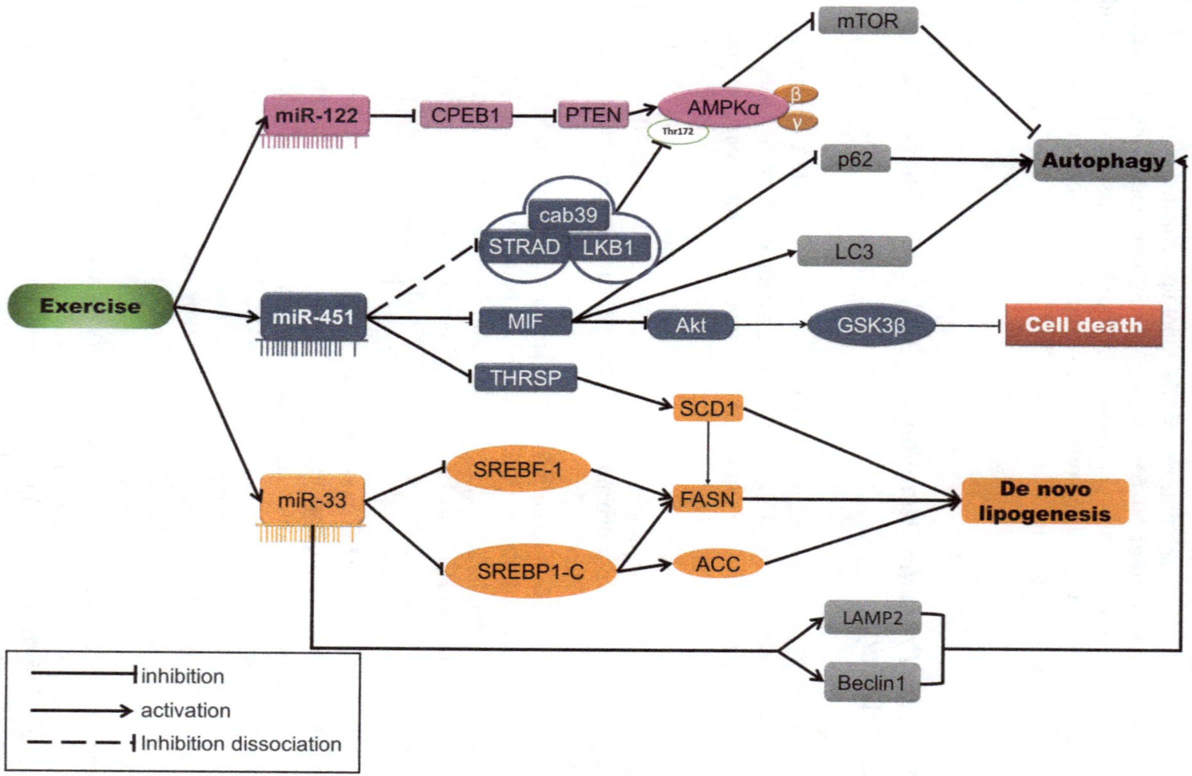

Fig. 5.1 Molecular mechanisms of microRNA-mediated regulation of NAFLD during exercise intervention. Exercise induces miR-122, miR-451, and miR-33 for alleviating NAFLD through inhibiting de novo synthesis of fat, suppressing cell death, and activating autophagy

2.5 *Exercise-Mediated Autophagy Regulates Liver Fibrosis and Late-Stage NAFLD*

Currently, quantitative studies have found that exercise has a beneficial effect on the prognosis of liver cancer. A meta-analysis shows that high-level physical activity reduces the risk of liver cancer by 25% when compared with low physical activity [81]. With the continuous progression and aggravation of NAFLD, the patients themselves usually cannot feel the significant abnormality. Therefore, the development and progression of liver fibrosis or advanced liver diseases are often ignored. When the phenotype of liver steatosis is only initiated in histology, the interventions of exercise training or body weight loss may have an obvious blocking effect on liver steatosis. In contrast, the progression of liver steatosis to severe histological changes may result in irreversible status or develop into fibrosis and canceration. The suppression of autophagy has been shown to have a therapeutic effect on cancers, but it is also controversial because autophagy plays different roles in different scenarios. So, an important question is whether autophagy activity should be upregulated or downregulated, by what way, and the most important regulation stage. Are there differences in different ways of regulating autophagy? The regulatory effect of exercise on autophagy has been widely proved. The clinical experimental conditions are complex, and the relationship between diet and exercise cannot be completely separated. In the animal model of nonobesity, nutritional deficiency, and high-fat diet, physical activity has a positive effect on tumor recovery, and the average number of nodules in liver is reduced. It has preliminarily proved the inhibitory effect of physical activity on hepatocarcinogenesis during the late stage of NAFLD and also validated the inhibitory mechanism of exercise on cancer cells from the activation of AMPK and mTOR/S6K signal pathways [20]. As an important autophagy inhibitor, mTOR is at an inhibited state under the stable environment; however, under the starvation or exercise intervention condition, the suppression of mTOR-mediated autophagy is weakened and autophagy flux is enhanced; therefore, cellular homeostasis is maintained without the progression to the malignant proliferation of cancer cells. However, once excessive proliferation of tumor cells occurs, autophagy can drive the survival of tumor cells by continuously providing nutrition for liver cancer cells, thus aggravating the progression of the disease and threatening its prognosis [82].

3 Weight Loss Is a Feasible Strategy for NAFLD?

Currently, there are no recognized drugs that can slow down hepatic steatosis. One of the major risk factors for NAFLD is the lack of physical activity, especially in developed European and American countries with more convenient transportation and lifestyle. Another major reason for less physical activity is due to the less exercise time and interest from huge pressure of life [83]. Based on the studies on rodent models, exercise is not only an effective means to prevent NAFLD, but also a powerful therapeutic strategy.

The overall health effect of exercise is to reduce body fat and weight [84]. The benefits of body weight loss by exercise are not limited to the liver, and exercise interventions also can effectively promote the improvement of the overall health status to accomplish the synergistic effect on multiple organs and systems. Although the direct organ of exercise is skeletal muscle, the exercise-induced myokines will also be beneficial to many organs and tissues.

The most intuitive response to the benefits of body weight loss is in the metabolic system. Previous studies have documented that subjects with the loss of 10% body weight have higher remission rates of liver fibrosis and inflammation. If the goal of exercise is not only to improve short-term metabolic conditions, but also to achieve significant intervention in chronic metabolic diseases such as NAFLD, the standard for body weight loss should reach at least 7–10% according to previous literatures [85]. However, in actual experiments, many volunteers still cannot meet the body weight loss standards designed before the experiment. This actually reflects the lack of physical activity in the intervention of NAFLD, that is, poor compliance, especially encouraging patients to exercise at home rather than under the supervision of a doctor or exercise physiologist.

Epigenetic studies have shown that modern fast-paced lifestyle is an important inducement of various chronic diseases, and diet and lifestyle are the most important factors for obesity-related chronic diseases. Therefore, regular exercise and healthy diet habits or lifestyle can greatly reduce the risks of chronic diseases. Although the improvement of diet structure can delay chronic diseases to a certain extent, no diet can reverse the development of chronic diseases, so calorie restriction is still the best choice. Of course, the vast majority of doctors' orders and drinking has an obvious damage to hepatocytes, which should be avoided as far as possible [86]. Moreover, experts and scholars in the field of exercise science are surprised to find that epigenetic mechanisms of lifestyle changes seem to be able to reverse the adverse effects of malnutrition and sedentary lifestyle on patients and even their offspring [87].

The cardiac output and the exercise capacity of skeletal muscle will decrease significantly after exercise suspension, especially high-level athletes [88]. The exercise discontinuity could result in serious impact on adipose tissue, liver tissue, and other metabolic organs, which reveals the downregulation of PPAR, as a critical regulator of adipogenesis, thus leading to the reprisal accumulation of lipids [89]. Therefore, it is highly necessary for emphasizing the obvious benefits of exercise on metabolic diseases through regular lifelong exercise.

4 Exercise Prescriptions

Although the benefits of exercise on reducing fat content in liver fat are well documented, the effects vary among the studies using different exercise protocols. Thus, it is important to focus on optimizing exercise prescription for individuals with NAFLD. Many scholars advocate aerobic exercise at a moderate intensity (50–70% VO_{2max}) for at least 3 days a week with total exercise time of 150–300 min every week, which should be beneficial to liver health in patients with NAFLD [90]. Recently, HIIT has emerged as a time-effective substitute for traditional continuous exercise and has been validated to be beneficial to NAFLD [73, 91–96]. In animal models, HIIT has a certain corrective effect on metabolic disorders induced by high-fat diets. During the exercise intervention, HIIT as the optional choice of exercise style for NAFLD is the key to lipogenesis when compared with continuous endurance training [73], which exhibits the more obviously inhibitory effects on FAS, ACC, and SREBP-1c. In human exercise experiments, people focus on the evaluation of the overall metabolic level [93], insulin sensitivity [95] in peripheral tissues (mainly liver and adipose tissue), liver fat content [91], activity of liver enzymes [93], and cardiovascular function. The efforts to screen inflammatory markers during HIIT intervention are made to explore a breakthrough for the prevention and treatment of NAFLD [94].

A systematic review has reported that aerobic exercise, resistance exercise, and HIIT all effectively reduce fat content in liver [97]. The major difference between aerobic exercise and resistance exercise may be more obviously improved strength and cross-sectional area of skeletal muscle fibers from resistance exercise [98], and the enhanced function and content of mitochondria in skeletal muscle from aerobic endurance exercise [99].

Furthermore, exercise alone or combined with diet intervention can improve the levels of serum metabolic enzymes, the fat contents in liver, and the histological changes in liver. Even without body weight loss, exercise has a beneficial effect on intrahepatic triglycerides; however, it is difficult to evaluate the best exercise mode, and it is difficult to analyze the type, intensity, and amount of exercise at the same time [100].

In order to explore the optimal dose of aerobic exercise for the intervention of NAFLD, an 8-week randomized placebo-controlled experiment is conducted. Non-invasive examination techniques such as magnetic resonance imaging are used, and the effects of common exercise doses (low to medium intensity, high-volume aerobic exercise; high-intensity, low-volume aerobic exercise; low to medium intensity, low-volume aerobic exercise; and placebo) on reducing liver fat and visceral fat are tested. Aerobic exercise can also reduce liver fat despite no body weight loss [98]. In order to thoroughly understand the improvement of NAFLD by different types of exercise prescriptions, a summary is listed in Table 5.1.

For aerobic exercise, the median effective regimen includes exercise training for 40 min each time at the intensity of 4.8 METs, three times a week, and with the exercise duration of 12 weeks. Similarly, for resistance exercise, the median

Table 5.1 Exercise prescriptions for NAFLD

Exercise prescription	Description	NAFLD parameters	Evidence type
Aerobic exercise [98] (cycle ergometer)	50% VO_{2max}, 60 min, 4 days/week, 8 weeks 70% VO_{2max}, 45 min, 3 days/week, 8 weeks 50% VO_{2max}, 45 min, 3 days/week, 8 weeks	$2.38 \pm 0.73\%$ reduction in hepatic fat content $2.62 \pm 1.00\%$ reduction in hepatic fat content $0.84 \pm 0.47\%$ reduction in hepatic fat content	Adults, randomized controlled trial
Aerobic exercise [101] (Nordic walking)	60–75% VO_{2max}, 30–60 min, 2–3 days/week, 8.6 months	24.4% reduction in hepatic fat content	115 individuals, randomized controlled trial
Aerobic exercise [102] (cycle ergometer)	60–70% VO_{2max}, 40–50 min, 3 days/week, 8 weeks	IHTG $12.9 \pm 4.2\%$ vs. control $11.2 \pm 5.1\%$	Diabetic obese individuals
HIIT [102] (cycle ergometer)	4-min cycling sessions at 80–85% VO_{2max} with 2-min interval at 50% VO_{2max}, 3 sets, 8 weeks	IHTG $12.4 \pm 4.5\%$ vs. control $11.2 \pm 5.1\%$	
HIIT [103] (cycle ergometer)	4-min cycling sessions at 80–85% VO_{2max}, with 2-min interval at 50% VO_{2max}, 3 sets, 8 weeks	IHTG pre $12.4 \pm 4.5\%$ vs. post $10.1 \pm 1.3\%$	Diabetic obese patients with NAFLD
HIIT [91]	2–3 min and 50 s/week, total time 30–40 min, 12 weeks	Liver fat $11 \pm 5\%$ vs. $8 \pm 2\%$, whole body fat mass 35 ± 7 kg vs. 33 ± 8 kg	A single-center parallel design randomized trial
Resistance training [104] (RT)	(1) Sit-ups, (2) leg presses, (3) leg extensions, (4) leg curls, (5) chest presses, (6) seated rows, (7) pull-downs. 1 RM, 12 weeks	14.3% reduction in hepatic fat content	Sedentary obese men with NAFLD
High-intensity interval aerobic training [104] (HIAT)	3-min cycling sessions at 80–85% VO_{2max} with a 2-min active rest at 50% VO_{2max}, 3 sets, 12 weeks	13.7% reduction in hepatic fat content	
Moderate-intensity continuous aerobic training [104] (MICT)	40-min cycling at 60–65% VO_{2max}, 12 weeks	14.3% reduction in hepatic fat content	
Short-term aerobic exercise [105]	85% HR_{max}, 60 min/day, 7 days	Hepatic triglyceride content, $22.6 \pm 3.1\%$ vs. control $21.5 \pm 3.0\%$	Obese patients with NAFLD
Short-term aerobic exercise [106]	Treadmill walking at 85% HR_{max}, 60 min/day, 7 days	Increased lipid polyunsaturation index and resting fat oxidation rate	Obese individuals
Aerobic exercise [107]	12 weeks, 3–5 days/week	VO_{2max} increased by 17%, reduced fibrosis and hepatocyte ballooning by one stage in 58% and 67% of patients	MAFLD patients

(continued)

Table 5.1 (continued)

Exercise prescription	Description	NAFLD parameters	Evidence type
Aerobic exercise [108]	3-min exercise with HR_{max}, 3 days/week, 3 months	Decreased intrahepatic ALT, AST and TG	MAFLD patients
Aerobic exercise [109] (treadmill)	60% maximal speed, 5 days/week, 60 min/day, 8 weeks	No significant difference in serum triglyceride, decreased adiposity index	Male C57BL/6J mice
Aerobic exercise [110] (treadmill)	Treadmill running, 20 m/min, 60 min/day, 15% inclination, 5 days/week	Reduced fasting serum TG and FFA	Hyperphagic, obese, T2DM, Otsuka long-Evans fatty (OLETF) rats

effective regimen includes exercise for 45 min at the intensity of 3.5 METs, three times a week for 12 consecutive weeks. Both aerobic exercise and resistance exercise can improve hepatic steatosis [51].

During human exercise intervention, especially for high-level athletes, a clear phenomenon for the optimization of body compositions is observed, that is, when men and women accept similar intensity despite different exercise modes, there will be some significant differences in exercise intervention efficacy and body compositions. Of course, competitive athletes hardly show obvious signs of excessive lipid accumulation in liver, but when applied to exercise intervention for chronic diseases, there are certain gender differences in aerobic training and resistance training. Therefore, the specific role of exercise for different genders in the recommended exercise prescriptions should be considered; body weight management and appropriate physical activity for the prevention and treatment of NAFLD as the easiest and the most effective intervention methods should also be considered [97].

From the perspective of the development of exercise prescriptions, the most classic elements such as exercise intensity, exercise type, exercise frequency, and exercise volume need to be included. The basic exercise prescriptions should be designed based on the average level. People strictly abide by the gold standard to use. However, with the continuous development of disciplines such as exercise physiology, people have gradually realized that the human body has the same general apparent index, but the metabolic level, aerobic capacity, and other aspects may still be greatly different, which puts forward higher requirements for the formulation of exercise prescriptions.

Personalized exercise prescription, first of all, must be different from person to person, and a detailed exercise prescription must be established based on a large number of exercise rehabilitation evaluation. Through a comprehensive test of the parameters for participants' balance capability, aerobic endurance, and other exercise capacity, combined with the course of the metabolic diseases for people suffering from, a safe, reasonable, and effective exercise program can be developed. Developing a reasonably accurate exercise prescription requires extensive clinical practice among physicians. In addition, a further step for exploring underlying mechanisms to understand more detailed and microscopic effects of exercise interventions on chronic diseases by exercise physiologists and other scholars in other fields of exercise is also needed. Taken together, the combinatorial procedures for the development of exercise prescriptions will truly lead to the development of a set of complete and feasible interventional system for chronic diseases.

5 Conclusion and Future Perspectives

In summary, exercise has great potential in the prevention and treatment of NAFLD, while moderate-to-high exercise intensity is generally recommended for individuals to have benefits on alleviating NAFLD, more future studies are required to further optimize exercise prescription including exercise type, exercise intensity, exercise frequency, and exercise duration. Autophagy is emerged as a therapeutic target to develop novel and effective treatment strategies for metabolic diseases. However, the regulatory role of exercise-mediated autophagy in NAFLD and the underlying molecular mechanisms involving the regulation of NAFLD have not been fully elucidated and clarified, which should be further explored in order to optimize the design and development of exercise prescriptions.

References

1. Younossi ZM, Koenig AB, Abdelatif D et al (2016) Global epidemiology of nonalcoholic fatty liver disease-meta-analytic assessment of prevalence, incidence, and outcomes. Hepatology 64:73–84
2. Williams CD, Stengel J, Asike MI et al (2011) Prevalence of nonalcoholic fatty liver disease and nonalcoholic steatohepatitis among a largely middle-aged population utilizing ultrasound and liver biopsy: a prospective study. Gastroenterology 140:124–131
3. Younossi ZM, Golabi P, De Avila L et al (2019) The global epidemiology of NAFLD and NASH in patients with type 2 diabetes: a systematic review and meta-analysis. J Hepatol 71:793–801
4. Mann JP, Valenti L, Scorletti E et al (2018) Nonalcoholic fatty liver disease in children. Semin Liver Dis 38:1–13
5. Cairns SR, Peters TJ (1983) Biochemical analysis of hepatic lipid in alcoholic and diabetic and control subjects. Clin Sci (Lond) 65:645–652
6. Neuschwander-Tetri BA, Caldwell SH (2003) Nonalcoholic steatohepatitis: summary of an AASLD single topic conference. Hepatology 37:1202–1219
7. Wu WKK, Zhang L, Chan MTV (2018) Autophagy, NAFLD and NAFLD-related HCC. Adv Exp Med Biol 1061:127–138
8. Huang DQ, El-Serag HB, Loomba R (2021) Global epidemiology of NAFLD-related HCC: trends, predictions, risk factors and prevention. Nat Rev Gastroenterol Hepatol 18:223–238
9. Targher G, Bertolini L, Poli F et al (2005) Nonalcoholic fatty liver disease and risk of future cardiovascular events among type 2 diabetic patients. Diabetes 54:3541–3546

10. Targher G, Bertolini L, Padovani R et al (2006) Increased prevalence of cardiovascular disease in type 2 diabetic patients with non-alcoholic fatty liver disease. Diabet Med 23:403–409
11. Enomoto H, Bando Y, Nakamura H et al (2015) Liver fibrosis markers of nonalcoholic steatohepatitis. World J Gastroenterol 21:7427–7435
12. Fabbrini E, Sullivan S, Klein S (2010) Obesity and nonalcoholic fatty liver disease: biochemical, metabolic, and clinical implications. Hepatology 51:679–689
13. Magkos F (2012) Putative factors that may modulate the effect of exercise on liver fat: insights from animal studies. J Nutr Metab 2012:827417
14. Buzzetti E, Pinzani M, Tsochatzis EA (2016) The multiple-hit pathogenesis of non-alcoholic fatty liver disease (NAFLD). Metabolism 65:1038–1048
15. Day CP (2002) Pathogenesis of steatohepatitis. Best Pract Res Clin Gastroenterol 16:663–678
16. Mehta K, Van Thiel DH, Shah N et al (2002) Nonalcoholic fatty liver disease: pathogenesis and the role of antioxidants. Nutr Rev 60:289–293
17. Engin A (2017) Non-alcoholic fatty liver disease. Adv Exp Med Biol 960:443–467
18. Khan RS, Bril F, Cusi K et al (2019) Modulation of insulin resistance in nonalcoholic fatty liver disease. Hepatology 70:711–724
19. Machado MV, Diehl AM (2016) Pathogenesis of nonalcoholic steatohepatitis. Gastroenterology 150:1769–1777
20. Guarino M, Kumar P, Felser A et al (2020) Exercise attenuates the transition from fatty liver to steatohepatitis and reduces tumor formation in mice. Cancers (Basel) 12:1407
21. Trenell MI (2015) Sedentary behaviour, physical activity, and NAFLD: curse of the chair. J Hepatol 63:1064–1065
22. Trovato GM, Catalano D, Martines GF et al (2013) Western dietary pattern and sedentary life: independent effects of diet and physical exercise intensity on NAFLD. Am J Gastroenterol 108:1932–1933
23. Ganguli S, Deleeuw P, Satapathy SK (2019) A review of current and upcoming treatment modalities in non-alcoholic fatty liver disease and non-alcoholic steatohepatitis. Hepat Med 11:159–178
24. Delorme-Axford E, Guimaraes RS, Reggiori F et al (2015) The yeast Saccharomyces cerevisiae: an overview of methods to study autophagy progression. Methods 75:3–12
25. Chen Y, Yu L (2013) Autophagic lysosome reformation. Exp Cell Res 319:142–146
26. Martinez-Lopez N, Singh R (2015) Autophagy and lipid droplets in the liver. Annu Rev Nutr 35:215–237
27. Singh R, Kaushik S, Wang Y et al (2009) Autophagy regulates lipid metabolism. Nature 458:1131–1135
28. Khawar MB, Gao H, Li W (2019) Autophagy and lipid metabolism. Adv Exp Med Biol 1206:359–374
29. Wang X, Zhang X, Chu ESH et al (2018) Defective lysosomal clearance of autophagosomes and its clinical implications in nonalcoholic steatohepatitis. FASEB J 32:37–51
30. Yang L, Li P, Fu S et al (2010) Defective hepatic autophagy in obesity promotes ER stress and causes insulin resistance. Cell Metab 11:467–478
31. Choi AM, Ryter SW, Levine B (2013) Autophagy in human health and disease. N Engl J Med 368:651–662
32. Tao T, Xu H (2020) Autophagy and obesity and diabetes. Adv Exp Med Biol 1207:445–461
33. Ueno T, Komatsu M (2017) Autophagy in the liver: functions in health and disease. Nat Rev Gastroenterol Hepatol 14:170–184
34. Boon MR, Van Marken Lichtenbelt WD (2016) Brown adipose tissue: a human perspective. Handb Exp Pharmacol 233:301–319
35. Tao Z, Liu L, Zheng LD et al (2019) Autophagy in adipocyte differentiation. Methods Mol Biol 1854:45–53
36. Ko MS, Yun JY, Baek IJ et al (2021) Mitophagy deficiency increases NLRP3 to induce brown fat dysfunction in mice. Autophagy 17:1205–1221

37. Chouchani ET, Kazak L, Spiegelman BM (2019) New advances in adaptive thermogenesis: UCP1 and beyond. Cell Metab 29:27–37
38. Yang K, Cai J, Pan M et al (2020) Mark4 inhibited the browning of white adipose tissue by promoting adipocytes autophagy in mice. Int J Mol Sci 21:2752
39. Kim D, Kim JH, Kang YH et al (2019) Suppression of brown adipocyte autophagy improves energy metabolism by regulating mitochondrial turnover. Int J Mol Sci 20:3520
40. Goldman S, Zhang Y, Jin S (2010) Autophagy and adipogenesis: implications in obesity and type II diabetes. Autophagy 6:179–181
41. Zhang X, Wu D, Wang C et al (2020) Sustained activation of autophagy suppresses adipocyte maturation via a lipolysis-dependent mechanism. Autophagy 16:1668–1682
42. Igawa H, Kikuchi A, Misu H et al (2019) p62-mediated autophagy affects nutrition-dependent insulin receptor substrate 1 dynamics in 3T3-L1 preadipocytes. J Diabetes Invest 10:32–42
43. Allaire M, Rautou PE, Codogno P et al (2019) Autophagy in liver diseases: time for translation? J Hepatol 70:985–998
44. Chun SK, Lee S, Yang MJ et al (2017) Exercise-induced autophagy in fatty liver disease. Exerc Sport Sci Rev 45:181–186
45. Flores-Toro JA, Go KL, Leeuwenburgh C et al (2016) Autophagy in the liver: cell's cannibalism and beyond. Arch Pharm Res 39:1050–1061
46. Diniz TA, De Lima Junior EA, Teixeira AA et al (2021) Aerobic training improves NAFLD markers and insulin resistance through AMPK-PPAR-α signaling in obese mice. Life Sci 266:118868
47. Li R, Jia Z, Trush MA (2016) Defining ROS in biology and medicine. React Oxyg Species (Apex) 1:9–21
48. Yang Y, Karakhanova S, Hartwig W et al (2016) Mitochondria and mitochondrial ROS in cancer: novel targets for anticancer therapy. J Cell Physiol 231:2570–2581
49. Mittler R (2017) ROS are good. Trends Plant Sci 22:11–19
50. Braud L, Battault S, Meyer G et al (2017) Antioxidant properties of tea blunt ROS-dependent lipogenesis: beneficial effect on hepatic steatosis in a high fat-high sucrose diet NAFLD obese rat model. J Nutr Biochem 40:95–104
51. Gonçalves IO, Passos E, Diogo CV et al (2016) Exercise mitigates mitochondrial permeability transition pore and quality control mechanisms alterations in nonalcoholic steatohepatitis. Appl Physiol Nutr Metab 41:298–306
52. Rosa-Caldwell ME, Lee DE, Brown JL et al (2017) Moderate physical activity promotes basal hepatic autophagy in diet-induced obese mice. Appl Physiol Nutr Metab 42:148–156
53. Poblete-Aro C, Russell-Guzman J, Parra P et al (2018) Exercise and oxidative stress in type 2 diabetes mellitus. Rev Med Chil 146:362–372
54. Van Der Windt DJ, Sud V, Zhang H et al (2018) The effects of physical exercise on fatty liver disease. Gene Expr 18:89–101
55. Hamasaki H (2018) Interval exercise therapy for type 2 diabetes. Curr Diabetes Rev 14:129–137
56. Petersen AM, Pedersen BK (2005) The anti-inflammatory effect of exercise. J Appl Physiol (1985) 98:1154–1162
57. Oh S, Maruyama T, Eguchi K et al (2015) Therapeutic effect of hybrid training of voluntary and electrical muscle contractions in middle-aged obese women with nonalcoholic fatty liver disease: a pilot trial. Ther Clin Risk Manag 11:371–380
58. Henkel J, Buchheim-Dieckow K, Castro JP et al (2019) Reduced oxidative stress and enhanced FGF21 formation in livers of endurance-exercised rats with diet-induced NASH. Nutrients 11:2709
59. Yu Q, Xia Z, Liong EC et al (2019) Chronic aerobic exercise improves insulin sensitivity and modulates Nrf2 and NFkappaB/IkappaBalpha pathways in the skeletal muscle of rats fed with a high fat diet. Mol Med Rep 20:4963–4972
60. Di Mauro S, Scamporrino A, Petta S et al (2019) Serum coding and non-coding RNAs as biomarkers of NAFLD and fibrosis severity. Liver Int 39:1742–1754

61. Zhu M, Wang Q, Zhou W et al (2018) Integrated analysis of hepatic mRNA and miRNA profiles identified molecular networks and potential biomarkers of NAFLD. Sci Rep 8:7628
62. Jin X, Feng CY, Xiang Z et al (2016) CircRNA expression pattern and circRNA-miRNA-mRNA network in the pathogenesis of nonalcoholic steatohepatitis. Oncotarget 7:66455–66467
63. Gorden DL, Myers DS, Ivanova PT et al (2015) Biomarkers of NAFLD progression: a lipidomics approach to an epidemic. J Lipid Res 56:722–736
64. Hosoya T, Hashiyada M, Funayama M (2016) Acute physical stress increases serum levels of specific microRNAs. Microrna 5:50–56
65. Xiao J, Bei Y, Liu J et al (2016) miR-212 downregulation contributes to the protective effect of exercise against non-alcoholic fatty liver via targeting FGF-21. J Cell Mol Med 20:204–216
66. Liu Y, Zhou X, Xiao Y et al (2020) miR-188 promotes liver steatosis and insulin resistance via the autophagy pathway. J Endocrinol 245:411–423
67. Liu X, Chen S, Zhang L (2020) Downregulated microRNA-130b-5p prevents lipid accumulation and insulin resistance in a murine model of nonalcoholic fatty liver disease. Am J Physiol Endocrinol Metab 319:E34–E42
68. Lai YC, Kondapalli C, Lehneck R et al (2015) Phosphoproteomic screening identifies Rab GTPases as novel downstream targets of PINK1. EMBO J 34:2840–2861
69. Kim YS, Nam HJ, Han CY et al (2020) LXRα activation inhibits autophagy and lipophagy in hepatocytes by dysregulating ATG4B and Rab-8B, reducing mitochondrial fuel oxidation. Hepatology 73:1307–1326
70. Ali O, Darwish HA, Eldeib KM et al (2018) miR-26a potentially contributes to the regulation of fatty acid and sterol metabolism in vitro human HepG2 cell model of nonalcoholic fatty liver disease. Oxid Med Cell Longev 2018:8515343
71. Han W, Fu X, Xie J et al (2015) MiR-26a enhances autophagy to protect against ethanol-induced acute liver injury. J Mol Med (Berl) 93:1045–1055
72. Ghareghani P, Shanaki M, Ahmadi S et al (2018) Aerobic endurance training improves nonalcoholic fatty liver disease (NAFLD) features via miR-33 dependent autophagy induction in high fat diet fed mice. Obes Res Clin Pract 12:80–89
73. Kalaki-Jouybari F, Shanaki M, Delfan M et al (2020) High-intensity interval training (HIIT) alleviated NAFLD feature via miR-122 induction in liver of high-fat high-fructose diet induced diabetic rats. Arch Physiol Biochem 126:242–249
74. Sun HJ, Wu ZY, Nie XW et al (2021) Implications of hydrogen sulfide in liver pathophysiology: mechanistic insights and therapeutic potential. J Adv Res 27:127–135
75. Wang YD, Li JY, Qin Y et al (2020) Exogenous hydrogen sulfide alleviates-induced intracellular inflammation in HepG2 cells. Exp Clin Endocrinol Diabetes 128:137–143
76. Wu D, Zhong P, Wang Y et al (2020) Hydrogen sulfide attenuates high-fat diet-induced non-alcoholic fatty liver disease by inhibiting apoptosis and promoting autophagy via reactive oxygen species/phosphatidylinositol 3-kinase/AKT/mammalian target of rapamycin signaling pathway. Front Pharmacol 11:585860
77. Sun L, Zhang S, Yu C et al (2015) Hydrogen sulfide reduces serum triglyceride by activating liver autophagy via the AMPK-mTOR pathway. Am J Physiol Endocrinol Metab 309:E925–E935
78. Liu Z, Liu M, Fan M et al (2021) Metabolomic-proteomic combination analysis reveals the targets and molecular pathways associated with hydrogen sulfide alleviating NAFLD. Life Sci 264:118629
79. Ren R, Yang Z, Zhao A et al (2018) Sulfated polysaccharide from Enteromorpha prolifera increases hydrogen sulfide production and attenuates non-alcoholic fatty liver disease in high-fat diet rats. Food Funct 9:4376–4383
80. Kwanten WJ, Martinet W, Michielsen PP et al (2014) Role of autophagy in the pathophysiology of nonalcoholic fatty liver disease: a controversial issue. World J Gastroenterol 20:7325–7338

81. Baumeister SE, Leitzmann MF, Linseisen J et al (2019) Physical activity and the risk of liver cancer: a systematic review and meta-analysis of prospective studies and a bias analysis. J Natl Cancer Inst 111:1142–1151
82. Martinet W, Agostinis P, Vanhoecke B et al (2009) Autophagy in disease: a double-edged sword with therapeutic potential. Clin Sci (Lond) 116:697–712
83. Hallal PC, Andersen LB, Bull FC et al (2012) Global physical activity levels: surveillance progress, pitfalls, and prospects. Lancet 380:247–257
84. Hunter GR, Brock DW, Byrne NM et al (2010) Exercise training prevents regain of visceral fat for 1 year following weight loss. Obesity (Silver Spring) 18:690–695
85. Vilar-Gomez E, Martinez-Perez Y, Calzadilla-Bertot L et al (2015) Weight loss through lifestyle modification significantly reduces features of nonalcoholic steatohepatitis. Gastroenterology 149:367–378.e365; quiz e314–e365
86. Ajmera V, Belt P, Wilson LA et al (2018) Among patients with nonalcoholic fatty liver disease, modest alcohol use is associated with less improvement in histologic steatosis and steatohepatitis. Clin Gastroenterol Hepatol 16:1511–1520.e1515
87. Stevanović J, Beleza J, Coxito P et al (2020) Physical exercise and liver "fitness": role of mitochondrial function and epigenetics-related mechanisms in non-alcoholic fatty liver disease. Mol Metab 32:1–14
88. Convertino VA, Bloomfield SA, Greenleaf JE (1997) An overview of the issues: physiological effects of bed rest and restricted physical activity. Med Sci Sports Exerc 29:187–190
89. Sertie RA, Andreotti S, Proença AR et al (2013) Cessation of physical exercise changes metabolism and modifies the adipocyte cellularity of the periepididymal white adipose tissue in rats. J Appl Physiol (1985) 115:394–402
90. Keating SE, George J, Johnson NA (2015) The benefits of exercise for patients with non-alcoholic fatty liver disease. Expert Rev Gastroenterol Hepatol 9:1247–1250
91. Hallsworth K, Thoma C, Hollingsworth KG et al (2015) Modified high-intensity interval training reduces liver fat and improves cardiac function in non-alcoholic fatty liver disease: a randomized controlled trial. Clin Sci (Lond) 129:1097–1105
92. Hamasaki H (2019) Perspectives on interval exercise interventions for non-alcoholic fatty liver disease. Medicines (Basel) 6:83
93. Iraji H, Minasian V, Kelishadi R (2021) Changes in liver enzymes and metabolic profile in adolescents with fatty liver following exercise interventions. Pediatr Gastroenterol Hepatol Nutr 24:54–64
94. Khalafi M, Symonds ME (2020) The impact of high-intensity interval training on inflammatory markers in metabolic disorders: a meta-analysis. Scand J Med Sci Sports 30:2020–2036
95. Marcinko K, Sikkema SR, Samaan MC et al (2015) High intensity interval training improves liver and adipose tissue insulin sensitivity. Mol Metab 4:903–915
96. Winn NC, Liu Y, Rector RS et al (2018) Energy-matched moderate and high intensity exercise training improves nonalcoholic fatty liver disease risk independent of changes in body mass or abdominal adiposity—a randomized trial. Metabolism 78:128–140
97. Hashida R, Kawaguchi T, Bekki M et al (2017) Aerobic vs. resistance exercise in non-alcoholic fatty liver disease: a systematic review. J Hepatol 66:142–152
98. Keating SE, Hackett DA, Parker HM et al (2015) Effect of aerobic exercise training dose on liver fat and visceral adiposity. J Hepatol 63:174–182
99. Gollnick PD, Armstrong RB, Saubert CWT et al (1972) Enzyme activity and fiber composition in skeletal muscle of untrained and trained men. J Appl Physiol 33:312–319
100. Katsagoni CN, Georgoulis M, Papatheodoridis GV et al (2017) Effects of lifestyle interventions on clinical characteristics of patients with non-alcoholic fatty liver disease: a meta-analysis. Metabolism 68:119–132
101. Cheng S, Ge J, Zhao C et al (2017) Effect of aerobic exercise and diet on liver fat in pre-diabetic patients with non-alcoholic-fatty-liver-disease: a randomized controlled trial. Sci Rep 7:15952

102. Abdelbasset WK, Tantawy SA, Kamel DM et al (2020) Effects of high-intensity interval and moderate-intensity continuous aerobic exercise on diabetic obese patients with nonalcoholic fatty liver disease: a comparative randomized controlled trial. Medicine (Baltimore) 99:e19471
103. Abdelbasset WK, Tantawy SA, Kamel DM et al (2019) A randomized controlled trial on the effectiveness of 8-week high-intensity interval exercise on intrahepatic triglycerides, visceral lipids, and health-related quality of life in diabetic obese patients with nonalcoholic fatty liver disease. Medicine (Baltimore) 98:e14918
104. Oh S, So R, Shida T et al (2017) High-intensity aerobic exercise improves both hepatic fat content and stiffness in sedentary obese men with nonalcoholic fatty liver disease. Sci Rep 7:43029
105. Kullman EL, Kelly KR, Haus JM et al (2016) Short-term aerobic exercise training improves gut peptide regulation in nonalcoholic fatty liver disease. J Appl Physiol (1985) 120:1159–1164
106. Haus JM, Solomon TP, Kelly KR et al (2013) Improved hepatic lipid composition following short-term exercise in nonalcoholic fatty liver disease. J Clin Endocrinol Metab 98:E1181–E1188
107. O'Gorman P, Naimimohasses S, Monaghan A et al (2020) Improvement in histological endpoints of MAFLD following a 12-week aerobic exercise intervention. Aliment Pharmacol Ther 52:1387–1398
108. Khaoshbaten M, Gholami N, Sokhtehzari S et al (2013) The effect of an aerobic exercise on serum level of liver enzymes and liver echogenicity in patients with non alcoholic fatty liver disease. Gastroenterol Hepatol Bed Bench 6:S112–S116
109. Batatinha HA, Lima EA, Teixeira AA et al (2017) Association between aerobic exercise and rosiglitazone avoided the NAFLD and liver inflammation exacerbated in PPAR-α knockout mice. J Cell Physiol 232:1008–1019
110. Linden MA, Fletcher JA, Morris EM et al (2014) Combining metformin and aerobic exercise training in the treatment of type 2 diabetes and NAFLD in OLETF rats. Am J Physiol Endocrinol Metab 306:E300–E310

Chapter 6
Exercise-Mediated Autophagy and Brain Aging

Xianjuan Kou, Hu Zhang, Yuan Guo, Michael Kirberger, and Ning Chen

1 Autophagy in the Aging Process

Autophagy can remove abnormal proteins and damaged organelles in cells through a degradation process to maintain normal physiological and metabolic functions of the cells. At the same time, autophagy is also a self-protective mechanism in cells during stress states. Autophagy occurs during various physiological and pathological states of the body and is a critical function in cellular homeostasis for health promotion and regulating the occurrence and development of a series of chronic diseases. Autophagy is also closely correlated with senescence. During the aging process, the functional status of autophagy in cells progressively declines, thereby leading to reduced capacity to discharge harmful substances in cells, and the removal of oxidative stress-induced products in the body. The accumulation of these toxic substances in the cells will result in toxic modifications or cell death, which will eventually shorten the lifespan of the organism. Conversely, lifespan can be extended by inducing autophagy [1]. Aging cells lose function over time due to reduced basal level of autophagy. This deficiency, attributed to the aging process, is highly correlated with the truncation of telomere length, abnormal aggregation of proteins, and destruction of mitochondrial functions. Therefore, autophagy is involved in cell aging and anti-aging processes.

X. Kou · H. Zhang · Y. Guo · N. Chen (✉)
Tianjiu Research and Development Center for Exercise Nutrition and Foods, Hubei Key Laboratory of Exercise Training and Monitoring, College of Health Science, Wuhan Sports University, Wuhan, China

M. Kirberger
School of Science and Technology, Georgia Gwinnett College, Lawrenceville, GA, USA

N. Chen (ed.), *Exercise, Autophagy and Chronic Diseases*,
https://doi.org/10.1007/978-981-16-4525-9_6

1.1 Autophagy Protects Against Metabolic Stress

Autophagy is a defense mechanism of the body in response to its own adverse environmental stimuli. In order to maintain energy metabolism and cope with stress changes in the body, autophagy is activated under severe conditions that include nutritional deficiency, hypoxia, and growth factor depletion. Autophagy can maintain cell integrity by regenerating metabolic precursor cells and removing subcellular debris. A large number of abnormal substances in the cytoplasm will produce free amino acids and fatty acids, which can be recycled in the body or transported to other organelles. This process can not only maintain the homeostasis of the internal environment, but also regulate the development of higher organisms and prevent aging-related diseases.

1.2 Autophagy Acts as a Cell Housekeeper

Autophagy is generally considered the key to the completion of cell metabolism and plays a daily housekeeping function. It can remove damaged and aged organelles, clear foreign microorganisms, and degrade misfolded and denatured proteins. This process not only can delay the aging process and suppress aging-induced chronic diseases, but also has great significance for the survival of cells. The functional status of autophagy appears to be closely related to prevention and responses to aging, cardiovascular and cerebrovascular diseases, cancers, neurodegenerative diseases, and inflammation-related diseases. Although both autophagy and the ubiquitin–proteasome system (another proteolysis system) share similar functions, only autophagy can degrade organelles such as mitochondria, the Golgi complex, rough endoplasmic reticulum, viruses, and bacteria. Meanwhile, autophagy can also effectively prevent the accumulation of abnormal proteins in cells.

1.3 Autophagy as a Genome Guardian

Autophagy maintains cell homeostasis by removing harmful substances and plays an important role in protecting the standard transcription and translation of genes. DNA damage can activate autophagy, and the inhibition of autophagy aggravates DNA damage. Autophagy maintains DNA integrity by removing intracellular ROS, providing deoxyribonucleoside triphosphate (dNTP) and ATP. The loss of autophagy genes Beclin1 and Atg5 induces an increase in DNA damage and decreases chromosome stability, and even tumors and other diseases occur [2, 3]. For instance, the deficiency of autophagy in ATG-deficient immortalized epithelial cells can lead to DNA damage and limited chromosomal instability [1]. These cells have defects in DNA checkpoint and apoptotic signal pathways, indicating that autophagy can maintain the stability of the normal cell genome, and prevent genetic instability. This is consistent with the known functions of autophagy in protein degradation, energy balance, and organelle clearance.

2 Autophagy Regulators During Aging Process

2.1 mTOR

As one of the critical evolutionarily conserved kinases in cells, mTOR regulates intracellular physiological mechanisms in response to changes in the external environment, nutritional status, and cytokines. It is also an important regulator of autophagy [4]. mTOR is the catalytic subunit found in mTORC1 and mTORC2 complexes. The mTORC1 complex is comprised of mTOR catalytic subunit raptors, PRAS40, mLST8, and Deptor, while the composition of mTORC2 includes Rictor, Protor, mSIN1, mLST8, and Deptor [5, 6]. The mTORC1 mainly regulates cell growth, apoptosis, energy metabolism, and autophagy, while mTORC2 is related to cytoskeletal reorganization and cell survival. Autophagy starts from the induction of the Atg1 kinase complex, which is negatively regulated by mTORC1. Therefore, mTOR not only regulates cell autophagy, but also regulates the process of cell aging. It is currently believed that mTOR is an important confluence point for regulating both. The activity of mTOR is suppressed by upstream signal factors, thereby activating the downstream ULK/Atg1 complex and inducing autophagy [7]. Current studies on autophagy and neurodegenerative diseases mainly focus on mTORC1 [8]. In mammals, mTORC1 inhibits autophagy by phosphorylating the Ser758 sites of ULK1 and AMBRA1, respectively, thus blocking AMPK phosphorylation and disrupting the stability of ULK1 [9, 10]. In addition, mTORC1 can also suppress the activity of lipid kinase VPS34 (an important regulator of autophagy formation) to inhibit autophagy [11, 12]. Transcription factor EB (TFEB), as one of the regulators of lysosomal function, can be phosphorylated by mTORC1 and can be present in the nucleus to regulate autophagy. Under conditions involving stimulation from exercise or starvation, reduced mTORC1 activity is an indicator of suppressed phosphorylation of TFEB. TFEB promotes the transcription and translation of autophagy-related genes after entering the nucleus, thereby confirming its vital role in regulating the formation of autophagosomes and the binding of lysosomes [13, 14].

As the "central regulator" of aging, mTOR regulates protein synthesis and protein degradation, mitochondrial quality control, and functional regulation and maintenance of stem cells, by affecting the level of autophagy. The mutations or knockout of mTOR, and the suppression of mTORC1-dependent S6K protein, reportedly prolong the lifespan of *Caenorhabditis elegans* and mice, respectively [15–17]. mTOR also has a certain regulatory effect on chronic diseases. For instance, mTOR activation can induce the accumulation of β-amyloid in the nervous system and the abnormal hyperphosphorylation of Tau protein, thus aggravating abnormal changes in the nervous system [18, 19]. In cancer, mTOR exhibits excessive activation and mutation or loss of negative regulatory factors. Moreover, the moderate activation of mTOR is conducive to adipogenic differentiation that inhibits obesity [20]. Regulating functional status of autophagy through mTOR has shown a positive effect in diseases such as AD and cancers [21, 22]. Therefore, mTOR is an important factor for regulating autophagy in the aging process and chronic diseases, and targeted regulation of mTOR is of great significance in lifespan extension and disease prevention and treatment.

2.2 Sirtuin (SIRT1)

SIRT1 is located in the cell nucleus and is a nicotinamide adenine dinucleotide (NAD^+)-dependent enzyme. It uses NAD substrates to remove acetyl groups from histones. It is also known as the longevity gene and is a component of the cellular senescence signaling pathway. SIRT1 is a silencing factor in mammals that has antioxidative stress and anti-aging functions, and can prolong the lifespan of organisms. SIRT1 is widely expressed in a variety of organisms, and its structure is stable and highly conserved. SIRT1 is an auxiliary protein necessary to induce autophagy, but its expression in senescent cells is reduced, which reduces the activity of autophagy [23]. In the process of cell senescence, the functional status of autophagy presents a gradual decline, and SIRT1 can induce autophagy and directly participate in the autophagic process, indicating that the activation of SIRT1 can enhance the functional status of autophagy to reverse senescence and prolong the lifespan of organisms.

SIRT1, as a NAD^+-dependent deacetylase, participates in various signal pathways for regulating cell aging, and disease initiation and progression. The protein expression level of SIRT1 shows a decreasing trend with the extension of age [24, 25], while conversely, overexpression of SIRT1 may retard the aging process, based on a 30% increase in lifespan reportedly observed in *Saccharomyces cerevisiae* [26]. Further evidence for this was observed in polymorphisms found in sirtuin proteins from a study involving long-lived elderly patients in Europe [27]. The transient increase in SIRT1 can stimulate an increase in basal autophagy, while conversely, SIRT1 knockout mice exhibited the accumulation of damaged organelles, disruption of cellular homeostasis, and high early perinatal mortality, due to restricted activation of autophagy [28]. Recent studies have found that SIRT1 participates in the anti-aging process mainly by regulating cell metabolism, oxidative stress, DNA repairing, apoptosis, and autophagy [29, 30]. SIRT1-induced autophagy is also one of the crucial ways to delay aging [31]. SIRT1 can upregulate AMPK to phosphorylate Beclin1 at the Thr388 site and promote the dissociation of Beclin1 and Bcl-2 to trigger autophagy [32, 33]. Abnormally low SIRT1 will also inhibit autophagy due to induction of oxidative stress [34], which may be the reason for accelerated aging associated with a low SIRT1 state. However, other studies have reported that the overexpression of SIRT1 may also inhibit corticosterone-induced autophagy to mitigate neurotoxicity [35]. Therefore, SIRT1 may act as an important regulator in autophagy.

Regulating SIRT1 to affect autophagy also shows positive effects in alleviating chronic diseases. Exercise activates autophagy and promotes the combination of lysosomes and autophagic vesicles through the AMPK/SIRT1/TFEB signaling pathway, which is conducive to producing smooth autophagic flux [36]. The consumption of external dietary supplements also produces a significant regulatory effect. For example, polyunsaturated fatty acids, such as omega-3, upregulate SIRT1-mediated deacetylation of Beclin1, thereby inducing autophagy and relieving neuronal apoptosis [37]. Similarly, oxymatrine improves cognitive capacity through SIRT1-mediated autophagy [38], and berberine relieves steatosis by activating SIRT1 deacetylation activity, based on Atg5, to induce autophagy [39]. SIRT1 is also beneficial for the hypertrophy and repairing of skeletal muscle, and the activation of muscle satellite cells, thus preventing and treating aging-induced skeletal muscle atrophy and cachexia [40, 41]. Therefore, SIRT1 has potential value in preventing and treating diseases by regulating autophagy. It may also provide new opportunities for the discovery of natural products that activate autophagy in anti-aging strategies and disease treatments.

2.3 p53

As an important tumor suppressor, p53 exhibits a regulatory effect in anti-aging by participating in cell growth, apoptosis, metabolism, and gene transcription. The posttranscriptional activity of p53 in the cytoplasm regulates the transcription of related genes to relieve and eliminate stimuli, which is involved in the regulation of autophagy [42–44]. Previous studies have reported that p53 plays an important role in prolonging lifespan by regulating autophagy through gene transcription [45]. These events include increasing transcription of Dram and Sestrin2 genes to induce autophagy [46, 47]; promoting TSC2 and AMPK to inhibit mTOR [48, 49]; enhancing the release of apoptotic proteins, including the release of anti-apoptotic proteins to activate Beclin1 [50]; and upregulating death-associated protein kinase (DAPK) and Atg5 to increase autophagy level [51, 52]. Additionally, p53 appears to induce autophagy and delay cell senescence by inhibiting mTOR activity or the mTOR signaling pathway [49]. Transgenic mice carrying p53 fragments exhibit suppressed autophagy and premature aging, while in transgenic mice carrying the intact p53 gene and its *allele* read frequency (ARF), p53 is moderately expressed and senescence is delayed, indicating that the regulatory role of p53 in the aging process may be conditionally dependent with expression at a specific location and moderate levels. Since p53 and autophagy can improve the internal environment of cells, the crosstalk between p53 and autophagy may jointly regulate the process of aging [53]. Therefore, systematically sorting out the relationship between p53 and autophagy may be of great significance in prolonging the lifespan of elderly patients suffering from different diseases, especially cancers.

3 Brain Aging

Aging is a comprehensive manifestation of changes in the function, structure, and morphology of the organism, including the decline in cellular homeostasis in the body, degenerative changes in the center of the structure, and decline of functions (such as immune function). The aging of the body is highly correlated with the aging of cells, and the major manifestations are inflammation, decreased immunity, and senescence (cell and biological senescence) [54]. Simultaneously, the change in mitochondrial quality control with aging will decrease oxidative phosphorylation capacity. The increase in ROS is also essential for inducing senescent metabolic diseases [55, 56]. Cell senescence exhibits obvious cell cycle stagnation, which aggravates the remodeling and regeneration capacity of tissues and organs after injury is decreased [57]. Moderate aging of body cells can also activate the innate natural immune response to a certain extent, which is beneficial to maintaining cellular homeostasis. However, as extension of body age, there may be functional degradation of immune organs, thus leading to immune failure and accelerating disease progression. Intensification in the aging process can partially explain the increased rate of chronic diseases in the elderly. At the cellular and molecular levels, these changes include impaired autophagy, mitochondrial dysfunction, oxidative stress, hypoxia, and telomere shortening.

During aging, cellular homeostasis in brain tissue is imbalanced, thus causing the damage to biological macromolecules, the impairment of adaptive neuroplasticity and elasticity, the aggregation of toxic modifications, oxidative stress, inflammation, and the accumulation of damaged organelles and denatured proteins. Mitochondrial injury in particular can lead to oxidative phosphorylation disorders in cells, the accumulation of ROS, and the accumulation of defective proteins. These changes in cells can lead to increased incidence of cognitive deficits, depression, and neurodegenerative diseases. Moreover, aging brains are also prone to Alzheimer's disease (AD), Parkinson's disease (PD), and stroke. In addition, brain aging is closely related to lifestyle, as sedentary or excessive indulgence can accelerate aging of the brain.

3.1 Characteristics of Brain Aging

3.1.1 Cell Senescence

The morphological changes in senescent cells are mainly manifested in cell shrinkage, increased membrane permeability and fragility, nuclear membrane invagination, decreased mitochondrial number, increased mitochondrial volume, decreased intracellular glycogen, and accumulation of lipids and deposition of abnormal substances such as lipofuscin. Eventually, apoptosis or necrosis occurs. In short, the various structures of cells undergo degenerative changes.

3.1.2 Accumulation of Damaged Cellular Contents

Aging cells lose the capability to rapidly eliminate damaged and abnormally accumulated lipids, proteins, and DNA. This will affect cell metabolism in various ways, including inhibition of DNA replication and transcription (the removal of individual genes will be abnormally activated), telomere DNA loss, specific loss of mitochondrial DNA, reduced DNA oxidation, DNA fragmentation, DNA methylation, DNA cross-linking, decreased levels of mRNA and tRNA, reduced protein levels, loss of protein stability, antigenicity, and digestibility, increased protein denaturation, inactivated enzymes, and oxidation of unsaturated fatty acids. All these conditions can contribute to a decrease in membrane fluidity.

3.1.3 Structural Changes in Brain During Aging Process

In the natural aging process, the human brain will gradually shrink, gray matter and white matter will decrease, and the ventricles will enlarge. Longitudinal magnetic resonance imaging (MRI) studies have shown that aging-related gray matter reduction is the most prominent in the temporal and frontal lobes. During the aging process, it may be possible to predict whether someone will develop cognitive impairment and AD based on the rate of brain atrophy [58]. The analysis of cross-sectional slices of brain tissue has demonstrated that brain atrophy is the combinatorial effect of dendritic degeneration and neuronal death [54]. Some studies suggest that, although there are individual differences in the rate of brain atrophy during aging process, brain imaging data can be used to determine the approximate "biological age" of a person's brain [55]. The rate of change in the brain structure during aging process can be affected by environmental factors. For example, aerobic exercise can increase hippocampal volume [59, 60], while excessive energy intake and obesity can accelerate hippocampal atrophy [61].

3.2 Underlying Mechanisms of Brain Aging

Aging is a complex process involving multiple organs, systems, and factors. Advances in cell molecular biology, biochemistry, and genetics have improved our understanding of brain aging. Our current understanding of the underlying mechanism for brain aging focuses on the interaction between energy metabolism and cellular antistress, repairing, and growth signaling pathways.

Studies at the cellular and molecular levels have identified several mechanisms associated with brain aging. (1) Mitochondrial dysfunction: Mitochondria are mainly distributed in the axons and dendrites of neurons, and can maintain cell energy metabolism. Calcium homeostasis can be used as a signal source to regulate gene transcription to participate in the process of cell aging. (2) The accumulation of oxidative damaged molecules (proteins, nucleic acids, and lipids) in cells: During the aging process, oxidative stress imbalance (including increased ROS production or decreased antioxidant defense capacity) will cause the accumulation of abnormally aggregated proteins and dysfunctional mitochondria in neurons. Additionally, damaged antioxidant capacity and weakened capacity to remove oxidative damage molecules in cells will accelerate aging. (3) Disordered energy metabolism: During the process of aging, glucose metabolism and lipid metabolism in the cells are disturbed, and the capacity to transport glucose in the cells is also impaired, thus leading to increased glucose levels in the blood and insulin resistance, which can easily induce diabetes and cardiovascular diseases. Insulin resistance in the peripheral system may induce cognitive dysfunction during the aging process. (4) Impaired cellular "waste disposal" mechanism (autophagosome and proteasome function): Proteasome dysfunction and overloading and proteasome degradation will cause brain aging. Oxidative damage to the membrane ATPase can impair lysosome function, thereby resulting in the accumulation of waste products in the cells and the induction of cell death. (5) Impaired adaptive stress response signal transduction: This can lead to nerve dysfunction, aging-related cognitive impairment, and damaged neuron plasticity. (6) Damaged DNA repairing: The inability to repair damaged DNA can lead to the rapid development of a variety of aging phenotypes. (7) Abnormal neural network activity: Neural network activity is related to natural aging and aging-related cognitive impairment. (8) Disordered neuronal calcium ion processing: During the aging process, the normal capability of neurons to regulate Ca^{2+} is impaired. (9) Stem cell failure: During the normal aging process, hippocampal and olfactory neurogenesis activities are reduced, the DNA repairing function is damaged, and oxidative stress and inflammation can lead to the reduction in aging-related neurogenesis. (10) Inflammation: Local inflammation is a characteristic of the process of brain aging, with microglia in an activated state [60, 62].

3.3 Brain Aging and Neurodegenerative Diseases

The major characteristics of neurodegenerative diseases such as AD, PD, and amyotrophic lateral sclerosis (ALS) include abnormal accumulation of misfolded and denatured proteins in brain tissue. Abnormal protein aggregates come from intracellular or extracellular compartments (such as senile plaques in AD), neurons (Lewy bodies in PD and neurofibrillary tangles in AD), or oligodendrocytes. Therefore, autophagy plays an important role in clearing a variety of abnormally aggregated, misfolded, or denatured proteins, associated with neurodegenerative diseases.

Specific proteins such as p62 and microtubule-associated protein 1 light chain 3 (LC3) in the autophagy signal pathway frequently accumulate as abnormal aggregates in patients with neurodegenerative diseases, indicating that autophagy may be involved in the occurrence and development of neurodegenerative diseases. Similarly, in patients with AD, Beclin1 is reduced and the mTORC1 complex is hyperactivated, indicating that the functional status of autophagy is impaired in the pathological process of AD [63]. In patients with PD, lysosomal depletion is associated with the accumulation of nondegraded autophagosomes [64].

Excessive autophagosomes are observed in brain tissues of patients with neurodegenerative diseases, which may be due to several reasons. First, the increase in autophagosomes can induce autophagy to eliminate abnormally aggregated, misfolded, and denatured proteins in the body, thereby executing cellular homeostasis as a self-defense mechanism. Another possibility is the accumulation of nondegraded autophagosomes due to impaired lysosomal function, thereby resulting in the weakened degradation of vacuoles, the accumulation of autophagic vacuoles [64], and dysfunctional excess degradation and accumulation of autophagosomes. Lysosomal depletion can be a characteristic of both PD and AD, and the origin of autophagosome accumulation may be due to defective fusion between autophagosomes and lysosomes [65]. All living things go through the aging process, and aging-related diseases, such as AD or PD, have been observed in animals such as human, monkeys, and mice. Therefore, it is not clear whether aging promotes the onset of neurodegenerative diseases. While many studies have reported that aging neurons are more sensitive to degeneration, the aging process and its role in pathogenesis, and development of neurodegenerative diseases for impaired life expectancy of organisms, are still not fully elucidated [66, 67]. Therefore, elucidating the aging process and understanding its underlying mechanisms could lead to the development of anti-aging and lifespan-extending strategies, thereby delaying the onset and progression of neurodegenerative diseases, which will have both scientific significance and social significance.

4 Exercise-Induced Autophagy and Brain Aging

Brain tissue is mainly responsible for exercise, thinking, endocrine, and other functions. In the meantime, with the increase in age, the brain function also shows further decline. The major characteristics of brain aging include reduction in brain volume, degenerative changes in neurons, and slower metabolic waste removal. These changes manifest themselves in declined cognition, learning, and memory capacity, and exercise-executing capability. The molecular mechanism is mainly related to deficient autophagy, a decline in mitochondrial quality control, weakened DNA repairing, and increased inflammation [60]. Therefore, exercise as an important intervention method to slow brain aging has gained wide support within the scientific community.

Based on previous studies, the activity of autophagy and chaperone-mediated autophagy (CMA) reveals a gradual decrease during the natural aging process [16], especially reduced CMA efficiency, which can reduce the binding capability of damaged proteins to lysosomal membrane, and transport them into the lysosomes. CMA is mainly in charge of the transportation to the lysosomal membrane through specific recognition and binding of heat shock protein 70 (HSP70) and degraded proteins. The receptor on the lysosomal membrane is lysosomal-associated membrane protein 2A (Lamp2A), which facilitates the entrance into the lysosome to execute the degradation process. One possible reason for the decreased CMA efficiency is that the expression of the receptor Lamp2A on the lysosome membrane gradually decreases during the aging process, thereby reducing the transport of CMA into the lysosome, and reducing the capacity for degradation. In addition to the impact on CMA, cellular autophagy is also inhibited to varying degrees, especially for the degradation of functionally defective mitochondria.

4.1 Exercise and Brain Aging in Basic Research

4.1.1 Autophagy

During the aging process, autophagy deficiency or dysfunction might be a trigger for neurodegenerative diseases. Recent studies have shown that AD and PD, as high-incidence neurological diseases in the elderly, are mainly caused by Aβ deposition and the abnormal phosphorylation of Tau protein, thereby resulting in dysfunctional clearance, degeneration, and necrosis of dopaminergic neurons, decreased dopamine secretion, and the accumulation of α-synuclein [68, 69]. Exercise induces an increase in the level of AMPK in brain cells, and the activation of autophagy has a crucial effect on the integrity of neuronal structure and function [70–72]. Relevant studies have shown that Beclin1 decreases by 35% and 70% in mild and moderate AD models, respectively. The deposition of Aβ in Beclin1$^{+/-}$ mice is twice that of normal mice, but Beclin1 overexpression models significantly reduce the accumulation of Aβ protein, indicating that autophagy may mitigate the progression of AD by accelerating the clearance of Aβ and alleviating nerve cell toxicity [73–75]. Studies have also confirmed that some genes associated with autophagy in AD show abnormal functions [76]. Exercise-induced autophagy can promote the clearance of Aβ, inhibit excessive phosphorylation of Tau protein [77, 78], and improve hippocampal structure and the regeneration of nerve synapses [79]. In APP/PS1 transgenic AD model mice, aerobic exercise can significantly upregulate Beclin1, increase LC3-II/LC3-I ratio, and downregulate p62 and Lamp1 in hippocampal tissues, which is accompanied by a decrease in Aβ deposition [80]. Abnormal autophagy will also aggravate the dysfunction of PD gene-encoded proteins, and pathological changes in α-synuclein and Tau protein [81]. In early stage of PD patients, autophagy-related genes such as Beclin1, Atg5, LC3-II, and Lamp2 are reduced in the cerebrospinal fluid [82]. Neuronal synapses are essential for PD, and nearly half of disease-causing genes are closely related to synaptic changes [83]. The knockout of Atg7 or Atg5 can result in neurosynaptic dystrophy, reduced dopamine secretion, and decreased motor function, as well as pathological characteristics of PD. In contrast, the overexpression of Atg5 can alleviate PD symptoms [81, 84, 85]. Regulating the expression of Beclin1 and TFEB associated with autophagy exerts a neuroprotective effect on the PD model. Rapamycin, trehalose, and lysosomal modulators can activate autophagy to execute neuroprotective effects on PD models

[86]. Exercise, which can induce autophagy, has significant potential value for the prevention and treatment of neurodegenerative diseases. Although studies on exercise-induced autophagy for the regulation on PD are limited, numerous studies have confirmed that exercise can improve aging-induced neurological diseases.

4.1.2 Mitochondria

The weight of the human brain accounts for about 2% of total body mass, but consumes approximately 20% of total energy. Thus, the amount of ATP in brain tissue is extremely important, and this is extremely sensitive to the changes in the quantity and quality of mitochondria. With the extension of age, the process of brain aging can be accelerated by a series of factors, known collectively as the "mitochondrial cascade hypothesis," that include abnormal mitochondrial oxidative respiratory chain, increased ROS, mtDNA mutation, and abnormal mitochondrial dynamics for inducing mitochondrial dysfunction [87–90]. Mitochondria are the "controllers" of aging, and the functional degradation of mitochondria is also an important manifestation of brain failure [91, 92]. Relevant studies have also shown that the knockout of mitochondrial fission protein Drp1 can cause abnormal synaptic transmission in adult mice, thus leading to reduced memory capacity and hippocampal shrinkage [93]. Similarly, the knockdown of mitochondrial fusion protein Mfn2 exhibits imbalanced mitochondrial dynamics, which hinders synapse formation and generation of nerve cells. In contrast, the overexpression of Mfn2 can promote mitochondrial differentiation and maturation, and can significantly improve neurodegenerative changes and motor dysfunction induced by 1-methyl-4-phenyl-1,2,3,6-tetrahydropyridine (MPTP) [94, 95]. In addition to its role in health promotion or disease intervention, exercise may also alleviate brain aging by improving the quality of mitochondria in brain tissue. Exercise can effectively increase PGC-1α, SIRT1, NRF1, TFAM, and mtDNA in brain cells, and promote mitochondrial biogenesis [96–98]. Appropriate treadmill exercise can effectively balance mitochondrial fusion and division, suppress mitochondrial swelling, and increase mitochondrial cristae in the hippocampal tissue of AD model (APP/PS1) mice [99]. Exercise also increases mitochondrial autophagy and reduces apoptosis in the cerebral cortex and cerebellum, thereby increasing the functional activity of mitochondria, as evidenced by the increased levels of OXPHOS complex I, III, and V subunits [100]. Even in the mtDNA mutation model, exercise can improve exercise capacity and reduce mtDNA mutation, to a certain extent [101]. Therefore, exercise has been shown to have a positive effect on improving the quantity and quality of mitochondria and preventing brain aging. However, other studies have indicated that high-intensity vigorous exercise may be harmful to mitochondrial function in the brain [102], which needs to be further explored and validated.

4.1.3 Inflammation

During the aging process, chronic inflammation of the nervous system is an important factor that promotes brain aging. Neuroinflammation is also considered an important factor in improving chronic neurodegenerative diseases [103]. Relevant studies have reported that damage to the blood–brain barrier in the elderly and AD populations induces brain inflammation and is associated with a decline in cognitive capacity [104]. Aging itself may aggravate chronic inflammatory states. The combination of TNF-α, IL-1β, and IFN-γ produces neuronal toxicity, which affects βAPP metabolism and leads to increased Aβ production in the brain. However, excessive Aβ accumulation will increase inflammation and cause the loss of hippocampal neurons in the CA3 subregion, thereby triggering a decrease in brain volume and cognitive capacity [105–107]. The knock-in of Tau gene in brain tissue has also been found to produce a significant increase in the level of inflammation [108]. Inflammatory factors such as IL-1β, TNF-α, and IL-12p40, in the brain tissue of the elderly, are significantly increased. Exercise can improve cognitive capacity by reducing inflammation and effectively improving self-care capability [109–111]. Aerobic exercise can also improve the interaction between IL-12A, IL-12B, and IL-12RB2 genes, reduce TNF-α, IL-1β, and NF-κB, increase the level of anti-inflammatory factor IL-10, reduce neuronal apoptosis, and alleviate or rescue impaired cognitive damage and reduced spatial learning and memory capacity during the aging process [112–115]. During interval training at moderate intensity, PD patients present a significant decrease in TNF-α [116]. Moderate resistance training also can inhibit neuroinflammation in 3xTg AD mice and delay the progression of disease [117]. Not all exercises achieve the same effects on the nervous system. For instance, one bout of acute exercise can increase the expression levels of TNF-α and Caspase-3/7 in the hippocampal tissues of young and old mice [118], while high-intensity exercise may increase hippocampal neuronal inflammation and decrease spatial learning capacity [119]. Therefore, when considering the role of exercise to relieve inflammation and improve neurodegenerative diseases caused by aging, it may be necessary to pay more attention to optimizing and monitoring exercise intensity.

4.1.4 DNA Repairing

The view that DNA repairing capability declines with the increase in age has gradually been confirmed by experimental studies [120]. Aging may lead to cumulative damage to some nuclear-related DNA. A significant increase in single-stranded nuclear-related DNA breaks in hippocampal pyramidal cells and cerebellar granule cells of elderly mice indicates an imbalance between the rate of DNA breaks and the rate of DNA repairing [121, 122]. Oxidative damage of DNA in the brain tissue of old mice is aggravated, as evidenced by increased DNA methylation, which induces the changes in chromatin conformation and gene transcription [123]. DNA methylation is closely related to neuropathological changes in AD patients [124]. Comprehensive physical exercise training can significantly reduce DNA strand breaks [125], and aerobic exercise increases DNA repairing capability by increasing BDNF levels, and the activity of DNA repairing-related enzymes such as APE1 and ERCC1, in the brain [126, 127]. Although a few studies have reported that exercise may increase resistance to neurodegenerative changes by improving DNA repairing, the role of exercise in brain aging has not yet been firmly established.

4.1.5 Lifespan Extension

The degradation of proteins is essential for maintaining the protein homeostasis network, a series of functions for controlling protein fate. There are two major pathways in the degradation system: the ubiquitin–proteasome system (UPS) and the autophagy–lysosome system (ALS). UPS is mainly involved in the degradation of short-lived proteins, while ALS is mainly involved in the degradation of long-lived proteins, protein complexes, and organelles such as mitochondria, Golgi complex, and endoplasmic reticulum. The conserved mechanism of autophagy is involved in the degradation and recycling of cellular components. A dysfunctional autophagic process, involving more than 1000 proteins, can destroy the proteolytic network and cause a variety of diseases, including neurodegenerative diseases, metabolic diseases, and cardiovascular diseases.

A study on nematodes reported that the functionally defective mutations of dauer formation (DAF-2) can prolong the lifespan of nematodes [128]. Similarly, nematode *bec-1*, a gene homologous to the mammalian autophagy gene Beclin1, is essential for the development of mutation tolerance and the longevity of nematodes with DAF-2 functional defect [129], indicating a correlation between the lifespan of nematodes and autophagy. Recently, it was discovered that mTORC1 is involved in the aging process and is also closely correlated with lifespan. In animal and cell models, long-term inhibition of the mTORC1 signal pathway can prolong the lifespan of organisms [130]. During the aging process, the suppression of the mTORC1 signal pathway can reduce the accumulation of harmful cytoplasmic proteins by promoting degradation through autophagy [131]. Generally, mTORC1 is an inhibitor of autophagy, so mTORC1 can participate in the aging process mainly by suppressing autophagy [132]. Therefore, the activated mTORC1 signal pathway can inhibit autophagy during the aging process, accelerate the body's aging, shorten the lifespan of the organism, and even lead to premature aging of the body.

Since the suppression of autophagy can lead to premature aging [23], mutations in selective Atg proteins such as Atg1, Atg7, Atg18, and Beclin1, and impairment of their functions, reportedly shorten the lifespan of *Caenorhabditis elegans* (*C. elegans*) [133]. Similarly, in *Drosophila* [134], the downregulated expression of Atg1, an essential protein for the formation of autophagosomes, can significantly shorten lifespan. In mice, aging-induced damage can be stimulated by the knockout of Atg, including the accumulation of damaged organelles, abnormal folding and aggregation of proteins, mitochondrial dysfunction, and endoplasmic reticulum stress. Optimal functional status of autophagy is beneficial to longevity. Inducing autophagy in *C. elegans* can inhibit the insulin-like growth factor signal pathway, and inhibiting autophagy can prevent the extension of lifespan through mutations of critical genes such as Atg [135]. Calorie restriction (CR), or reduced calorie intake under normal physiological conditions, is a key anti-aging intervention. Current research indicates that CR can prolong the lifespan of organisms in most cases and reduce the incidence of brain atrophy, diabetes, and cardiovascular diseases [136]. Epidemiological studies have also shown that CR can promote human health by activating the autophagic signaling pathway, thereby enhancing metabolism and waste elimination in the body. Therefore, CR is the best physiological inducer of autophagy [137].

4.2 Exercise and Brain Aging in Clinical Research

4.2.1 Brain Volume

The human brain begins to exhibit signs of atrophy with increasing age, especially in the frontal, parietal, and temporal lobes. The hippocampal tissue also decreases in volume by 1–2% every year after the age of 50 [138]. However, exercise can have a positive effect on the brain at any age, thereby resulting in increased brain volume. Exercise enhances the structure and strength of nerve synapses and further enhances the plasticity of synapses by affecting neurogenesis and brain metabolism [139]. Previous studies have shown that aerobic exercise can effectively improve white matter microstructure and suppress the atrophy of both gray and white matter in the elderly [140, 141]. At the same time, exercise for at least 10 years can significantly slow down the atrophy of the medial temporal lobe caused by aging in the brain [142]. Hippocampal volume will also increase due to exercise training, along with an improvement of memory capacity [59]. This may be related to increased BDNF mRNA transcription level and IGF-1 [143]. Based on a comprehensive analysis, aerobic exercise training for 1–16 weeks can promote the acceleration of bilateral anterior cingulate cortex and cerebral blood flow, thereby benefiting nerve efficiency in the frontal and temporal lobes, and improving cognitive capacity. Similarly, 24–40 weeks of exercise training have been shown to improve brain volume (especially frontal and temporal lobes). After 52 weeks of exercise, white matter integrity and hippocampal volume were promoted, and results of this study indicated that the appropriate combination of resistance exercise could be more beneficial for the release of BDNF and IGF-1, which would result in an improvement in brain aging [144].

4.2.2 Cognitive Capacity

Moderate exercise has been shown to be beneficial to the improvement of adolescent brain structure and cognitive capacity [145]. In the elderly, the decline of cognitive capacity is an important manifestation of the brain aging process, and this cognitive decline can be reduced through exercise [146]. Older people engaged in physical exercise for extended periods exhibit more vital brain recruitment capacity, and cognitive functions also exhibit significant improvement [147]. A relevant retrospective analysis has documented that exercise can improve language fluency, plot memory, exercise-executing capability, and overall cognitive capacity of the elderly [148]. After resistance exercise, spatial awareness, vision, and body reaction time of the elderly could be increased by 40.0%, 14.6%, and 14.0%, respectively [149]. Even in people with cognitive impairment and AD, exercise intervention has been shown to reverse decline in cognitive capacity [150, 151]. Hippocampal tissue is the primary functional component involved in cognitive capacity. The benefits of exercise to brain tissue may be partly attributed to the increase in the number and reorganization of hippocampal neurons, and improved capacity to elucidate time and space-encoding information [152]. Exercise can also partially rescue the impairment of cognitive capacity of quiet ε4 mice, as evidenced by improved performance in the radial-arm water maze [153]. The underlying mechanism associated with this may be related to an increase in the expression of tropomyosin receptor kinase B (Trk-B), stimulated by exercise, and the change in synaptophysin expression [153]. However, current research suggests that there may be crosstalk between different exercise types. Resistance exercise may inhibit the increase in β-hydroxybutyrate and BDNF mRNA levels caused by aerobic exercise, manifested by a weakening effect on the improvement of cognitive capacity and hippocampal neurogenesis [154]. In addition, long-term high-intensity endurance exercise appears to produce an apparent white matter axial diffusivity, without apparent differences in episodic memory, information processing, and working memory [155].

However, the beneficial efficacy of exercise-induced autophagy on brain aging still needs to be further clarified, and its underlying mechanisms need to be further elucidated, to fully understand the extent to which exercise interventional strategies can delay brain aging and the onset and progression of neurodegenerative diseases.

5 Outlook and Prospects

Aging is an unavoidable risk factor affecting the health of the elderly and is closely correlated with the occurrence and development of many chronic diseases. The incidence of aging-related diseases has increased in longer lifespan. As the elderly population increases, there has also been an increase in the number of studies evaluating the correlation between functional status of autophagy and brain aging.

The autophagic signal pathway has become an interventional target for delaying brain aging and neurodegenerative diseases, to the extent that optimizing the functional status of autophagy is expected to inhibit the accumulation of damaged proteins or cellular organelles, promote brain health, and extend the lifespan of humans. Meanwhile, clarifying precise mechanisms for optimizing the functional status of autophagy can also be beneficial for the development of novel and effective strategies for delaying brain aging, preventing and treating aging-related diseases, and prolonging lifespan, which may be achieved by developing a more scientific approach to physical fitness, and novel and effective nutrition and drug interventions.

References

1. Nakamura S, Yoshimori T (2018) Autophagy and longevity. Mol Cells 41:65–72
2. Mathew R, Kongara S, Beaudoin B et al (2007) Autophagy suppresses tumor progression by limiting chromosomal instability. Genes Dev 21:1367–1381
3. Vessoni AT, Filippi-Chiela EC, Menck CF et al (2013) Autophagy and genomic integrity. Cell Death Differ 20:1444
4. Weichhart T (2018) mTOR as regulator of lifespan, aging, and cellular senescence: a mini-review. Gerontology 64:127–134
5. Laplante M, Sabatini DM (2012) mTOR signaling in growth control and disease. Cell 149:274–293
6. Zoncu R, Efeyan A, Sabatini DM (2011) mTOR: from growth signal integration to cancer, diabetes and ageing. Nat Rev Mol Cell Biol 12:21–35
7. Partridge L, Fuentealba M, Kennedy BK (2020) The quest to slow ageing through drug discovery. Nat Rev Drug Discov 19:513–532
8. Chang YF, Hu WM (2019) Roles of mammalian target of rapamycin signaling and autophagy pathway in Alzheimer's disease. Zhongguo Yi Xue Ke Xue Yuan Xue Bao (Acta Acad Med Sin) 41:248–255
9. Kim J, Kundu M, Viollet B et al (2011) AMPK and mTOR regulate autophagy through direct phosphorylation of Ulk1. Nat Cell Biol 13:132–141
10. Nazio F, Strappazzon F, Antonioli M et al (2013) mTOR inhibits autophagy by controlling ULK1 ubiquitylation, self-association and function through AMBRA1 and TRAF6. Nat Cell Biol 15:406–416
11. Yuan HX, Russell RC, Guan KL (2013) Regulation of PIK3C3/VPS34 complexes by MTOR in nutrient stress-induced autophagy. Autophagy 9:1983–1995
12. Kim YC, Guan KL (2015) mTOR: a pharmacologic target for autophagy regulation. J Clin Invest 125:25–32
13. Settembre C, Zoncu R, Medina DL et al (2012) A lysosome-to-nucleus signalling mechanism senses and regulates the lysosome via mTOR and TFEB. EMBO J 31:1095–1108
14. Settembre C, Di Malta C, Polito VA et al (2011) TFEB links autophagy to lysosomal biogenesis. Science 332:1429–1433
15. Anonymous (2009) Ribosomal protein S6 kinase 1 signaling regulates mammalian life span. Science 326:140–144
16. Jia K, Chen D, Riddle DL (2004) The TOR pathway interacts with the insulin signaling pathway to regulate C. elegans larval development, metabolism and life span. Development 131:3897–3906
17. Vellai T, Takacs-Vellai K, Zhang Y et al (2003) Genetics: influence of TOR kinase on lifespan in C. elegans. Nature 426:620

18. Cai Z, Chen G, He W et al (2015) Activation of mTOR: a culprit of Alzheimer's disease? Neuropsychiatr Dis Treat 11:1015–1030
19. Jiang TF, Zhang YJ, Zhou HY et al (2013) Curcumin ameliorates the neurodegenerative pathology in A53T α-synuclein cell model of Parkinson's disease through the downregulation of mTOR/p70S6K signaling and the recovery of macroautophagy. J Neuroimmune Pharmacol 8:356–369
20. Yeh WC, Bierer BE, Mcknight SL (1995) Rapamycin inhibits clonal expansion and adipogenic differentiation of 3T3-L1 cells. Proc Natl Acad Sci U S A 92:11086–11090
21. Gremke N, Polo P, Dort A et al (2020) mTOR-mediated cancer drug resistance suppresses autophagy and generates a druggable metabolic vulnerability. Nat Commun 11:4684
22. Shan SR, Jiang F, Xu SM (2019) Effects of H102 on the memory recognition ability and AMPK-mTOR autophagy-related pathway in AD mice. Zhongguo Ying Yong Sheng Li Xue Za Zhi 35:1–4
23. Rubinsztein DC, Mariño G, Kroemer G (2011) Autophagy and aging. Cell 146:682–695
24. Michan S, Sinclair D (2007) Sirtuins in mammals: insights into their biological function. Biochem J 404:1–13
25. Chen C, Zhou M, Ge Y et al (2020) SIRT1 and aging related signaling pathways. Mech Ageing Dev 187:111215
26. Kaeberlein M, Mcvey M, Guarente L (1999) The SIR2/3/4 complex and SIR2 alone promote longevity in Saccharomyces cerevisiae by two different mechanisms. Genes Dev 13:2570–2580
27. Rose G, Dato S, Altomare K et al (2003) Variability of the SIRT3 gene, human silent information regulator Sir2 homologue, and survivorship in the elderly. Exp Gerontol 38:1065–1070
28. Lee IH, Cao L, Mostoslavsky R et al (2008) A role for the NAD-dependent deacetylase Sirt1 in the regulation of autophagy. Proc Natl Acad Sci U S A 105:3374–3379
29. Sergi C, Shen F, Liu SM (2019) Insulin/IGF-1R, SIRT1, and FOXOs pathways—an intriguing interaction platform for bone and osteosarcoma. Front Endocrinol (Lausanne) 10:93
30. Almeida M, Porter RM (2019) Sirtuins and FoxOs in osteoporosis and osteoarthritis. Bone 121:284–292
31. Morselli E, Maiuri MC, Markaki M et al (2010) The life span-prolonging effect of sirtuin-1 is mediated by autophagy. Autophagy 6:186–188
32. Luo G, Jian Z, Zhu Y et al (2019) Sirt1 promotes autophagy and inhibits apoptosis to protect cardiomyocytes from hypoxic stress. Int J Mol Med 43:2033–2043
33. Zhang Q, Yang H, An J et al (2016) Therapeutic effects of traditional Chinese medicine on spinal cord injury: a promising supplementary treatment in future. Evid Based Complement Alternat Med 2016:8958721
34. Ou X, Lee MR, Huang X et al (2014) SIRT1 positively regulates autophagy and mitochondria function in embryonic stem cells under oxidative stress. Stem Cells 32:1183–1194
35. Jiang Y, Botchway BOA, Hu Z et al (2019) Overexpression of SIRT1 inhibits corticosterone-induced autophagy. Neuroscience 411:11–22
36. Huang J, Wang X, Zhu Y et al (2019) Exercise activates lysosomal function in the brain through AMPK-SIRT1-TFEB pathway. CNS Neurosci Ther 25:796–807
37. Chen X, Pan Z, Fang Z et al (2018) Omega-3 polyunsaturated fatty acid attenuates traumatic brain injury-induced neuronal apoptosis by inducing autophagy through the upregulation of SIRT1-mediated deacetylation of Beclin-1. J Neuroinflammation 15:310
38. Zhou S, Qiao B, Chu X et al (2018) Oxymatrine attenuates cognitive deficits through SIRT1-mediated autophagy in ischemic stroke. J Neuroimmunol 323:136–142
39. Sun Y, Xia M, Yan H et al (2018) Berberine attenuates hepatic steatosis and enhances energy expenditure in mice by inducing autophagy and fibroblast growth factor 21. Br J Pharmacol 175:374–387

40. Lee D, Goldberg AL (2013) SIRT1 protein, by blocking the activities of transcription factors FoxO1 and FoxO3, inhibits muscle atrophy and promotes muscle growth. J Biol Chem 288:30515–30526
41. Myers MJ, Shepherd DL, Durr AJ et al (2019) The role of SIRT1 in skeletal muscle function and repair of older mice. J Cachexia Sarcopenia Muscle 10:929–949
42. White E (2016) Autophagy and p53. Cold Spring Harb Perspect Med 6:a026120
43. Zhang XD, Qi L, Wu JC et al (2013) DRAM1 regulates autophagy flux through lysosomes. PLoS One 8:e63245
44. Carter S, Vousden KH (2008) p53-Ubl fusions as models of ubiquitination, sumoylation and neddylation of p53. Cell Cycle 7:2519–2528
45. Tavernarakis N, Pasparaki A, Tasdemir E et al (2008) The effects of p53 on whole organism longevity are mediated by autophagy. Autophagy 4:870–873
46. Crighton D, Wilkinson S, O'Prey J et al (2006) DRAM, a p53-induced modulator of autophagy, is critical for apoptosis. Cell 126:121–134
47. Maiuri MC, Malik SA, Morselli E et al (2009) Stimulation of autophagy by the p53 target gene Sestrin2. Cell Cycle 8:1571–1576
48. Arico S, Petiot A, Bauvy C et al (2001) The tumor suppressor PTEN positively regulates macroautophagy by inhibiting the phosphatidylinositol 3-kinase/protein kinase B pathway. J Biol Chem 276:35243–35246
49. Feng Z, Zhang H, Levine AJ et al (2005) The coordinate regulation of the p53 and mTOR pathways in cells. Proc Natl Acad Sci U S A 102:8204–8209
50. Pimkina J, Humbey O, Zilfou JT et al (2009) ARF induces autophagy by virtue of interaction with Bcl-xl. J Biol Chem 284:2803–2810
51. Rosenbluth JM, Pietenpol JA (2009) mTOR regulates autophagy-associated genes downstream of p73. Autophagy 5:114–116
52. Kenzelmann Broz D, Spano Mello S, Bieging KT et al (2013) Global genomic profiling reveals an extensive p53-regulated autophagy program contributing to key p53 responses. Genes Dev 27:1016–1031
53. Sica V, Kroemer G (2020) A bidirectional crosstalk between autophagy and TP53 determines the pace of aging. Mol Cell Oncol 7:1769434
54. Dodig S, Čepelak I, Pavić I (2019) Hallmarks of senescence and aging. Biochem Med (Zagreb) 29:030501
55. Kwon SM, Hong SM, Lee YK et al (2019) Metabolic features and regulation in cell senescence. BMB Rep 52:5–12
56. Korolchuk VI, Miwa S, Carroll B et al (2017) Mitochondria in cell senescence: is Mitophagy the weakest link? EBioMedicine 21:7–13
57. Hernandez-Segura A, Nehme J, Demaria M (2018) Hallmarks of cellular senescence. Trends Cell Biol 28:436–453
58. Mungas D, Gavett B, Fletcher E et al (2018) Education amplifies brain atrophy effect on cognitive decline: implications for cognitive reserve. Neurobiol Aging 68:142–150
59. Erickson KI, Voss MW, Prakash RS et al (2011) Exercise training increases size of hippocampus and improves memory. Proc Natl Acad Sci U S A 108:3017–3022
60. Mattson MP, Arumugam TV (2018) Hallmarks of brain aging: adaptive and pathological modification by metabolic states. Cell Metab 27:1176–1199
61. Cherbuin N, Sargent-Cox K, Fraser M et al (2015) Being overweight is associated with hippocampal atrophy: the PATH through life study. Int J Obes (Lond) 39:1509–1514
62. Li J, Cai D, Yao X et al (2016) Protective effect of ginsenoside Rg1 on hematopoietic stem/progenitor cells through attenuating oxidative stress and the Wnt/β-catenin signaling pathway in a mouse model of d-galactose-induced aging. Int J Mol Sci 17:849
63. Li X, Alafuzoff I, Soininen H et al (2005) Levels of mTOR and its downstream targets 4E-BP1, eEF2, and eEF2 kinase in relationships with tau in Alzheimer's disease brain. FEBS J 272:4211–4220

64. Dehay B, Bové J, Rodríguez-Muela N et al (2010) Pathogenic lysosomal depletion in Parkinson's disease. J Neurosci 30:12535–12544
65. Nixon RA (2007) Autophagy, amyloidogenesis and Alzheimer disease. J Cell Sci 120:4081–4091
66. Mattson MP, Magnus T (2006) Ageing and neuronal vulnerability. Nat Rev Neurosci 7:278
67. Verheijen BM, Vermulst M, Van Leeuwen FW (2018) Somatic mutations in neurons during aging and neurodegeneration. Acta Neuropathol 135:811–826
68. Burré J, Sharma M, Südhof TC (2018) Cell biology and pathophysiology of α-synuclein. Cold Spring Harb Perspect Med 8:a024091
69. Chandra A, Valkimadi P, Pagano G et al (2019) Applications of amyloid, tau, and neuroinflammation PET imaging to Alzheimer's disease and mild cognitive impairment. Hum Brain Mapp 40:5424–5442
70. Tavernarakis N (2020) Regulation and roles of autophagy in the brain. Adv Exp Med Biol 1195:33
71. Liu W, Wang Z, Xia Y et al (2019) The balance of apoptosis and autophagy via regulation of the AMPK signal pathway in aging rat striatum during regular aerobic exercise. Exp Gerontol 124:110647
72. Curry DW, Stutz B, Andrews ZB et al (2018) Targeting AMPK signaling as a neuroprotective strategy in Parkinson's disease. J Parkinsons Dis 8:161–181
73. Rocchi A, Yamamoto S, Ting T et al (2017) A Becn1 mutation mediates hyperactive autophagic sequestration of amyloid oligomers and improved cognition in Alzheimer's disease. PLoS Genet 13:e1006962
74. Nixon RA, Yang DS (2011) Autophagy failure in Alzheimer's disease—locating the primary defect. Neurobiol Dis 43:38–45
75. Pickford F, Masliah E, Britschgi M et al (2008) The autophagy-related protein beclin 1 shows reduced expression in early Alzheimer disease and regulates amyloid beta accumulation in mice. J Clin Invest 118:2190–2199
76. Yoon SY, Kim DH (2016) Alzheimer's disease genes and autophagy. Brain Res 1649:201–209
77. Ohia-Nwoko O, Montazari S, Lau YS et al (2014) Long-term treadmill exercise attenuates tau pathology in P301S tau transgenic mice. Mol Neurodegener 9:54
78. Liu Z, Tao L, Ping L et al (2015) The ambiguous relationship of oxidative stress, tau hyperphosphorylation, and autophagy dysfunction in Alzheimer's disease. Oxid Med Cell Long 2015:352723
79. Glatigny M, Moriceau S, Rivagorda M et al (2019) Autophagy is required for memory formation and reverses age-related memory decline. Curr Biol 29:435–448.e438
80. Zhao N, Zhang X, Song C et al (2018) The effects of treadmill exercise on autophagy in hippocampus of APP/PS1 transgenic mice. Neuroreport 29:819–825
81. Hou X, Watzlawik JO, Fiesel FC et al (2020) Autophagy in Parkinson's disease. J Mol Biol 432:2651–2672
82. Youn J, Lee SB, Lee HS et al (2018) Cerebrospinal fluid levels of autophagy-related proteins represent potentially novel biomarkers of early-stage Parkinson's disease. Sci Rep 8:16866
83. Soukup SF, Vanhauwaert R, Verstreken P (2018) Parkinson's disease: convergence on synaptic homeostasis. EMBO J 37:e98960
84. Ahmed I, Liang Y, Schools S et al (2012) Development and characterization of a new Parkinson's disease model resulting from impaired autophagy. J Neurosci 32:16503–16509
85. Hu ZY, Chen B, Zhang JP et al (2017) Up-regulation of autophagy-related gene 5 (ATG5) protects dopaminergic neurons in a zebrafish model of Parkinson's disease. J Biol Chem 292:18062–18074
86. Lu J, Wu M, Yue Z (2020) Autophagy and Parkinson's disease. Adv Exp Med Biol 1207:21–51
87. Grimm A, Eckert A (2017) Brain aging and neurodegeneration: from a mitochondrial point of view. J Neurochem 143:418–431

88. Swerdlow RH (2011) Brain aging, Alzheimer's disease, and mitochondria. Biochim Biophys Acta 1812:1630–1639
89. Stefanatos R, Sanz A (2018) The role of mitochondrial ROS in the aging brain. FEBS Lett 592:743–758
90. Swerdlow RH, Khan SM (2004) A "mitochondrial cascade hypothesis" for sporadic Alzheimer's disease. Med Hypotheses 63:8–20
91. Crouch PJ, Cimdins K, Duce JA et al (2007) Mitochondria in aging and Alzheimer's disease. Rejuvenation Res 10:349–357
92. Reddy PH, Reddy TP (2011) Mitochondria as a therapeutic target for aging and neurodegenerative diseases. Curr Alzheimer Res 8:393–409
93. Oettinghaus B, Schulz JM, Restelli LM et al (2016) Synaptic dysfunction, memory deficits and hippocampal atrophy due to ablation of mitochondrial fission in adult forebrain neurons. Cell Death Differ 23:18–28
94. Fang D, Yan S, Yu Q et al (2016) Mfn2 is required for mitochondrial development and synapse formation in human induced pluripotent stem cells/hiPSC derived cortical neurons. Sci Rep 6:31462
95. Zhao F, Austria Q, Wang W et al (2021) Mfn2 overexpression attenuates MPTP neurotoxicity in vivo. Int J Mol Sci 22:601
96. Lezi E, Burns JM, Swerdlow RH (2014) Effect of high-intensity exercise on aged mouse brain mitochondria, neurogenesis, and inflammation. Neurobiol Aging 35:2574–2583
97. Ferreira AFF, Binda KH, Singulani MP et al (2020) Physical exercise protects against mitochondria alterations in the 6-hidroxydopamine rat model of Parkinson's disease. Behav Brain Res 387:112607
98. Steiner JL, Murphy EA, Mcclellan JL et al (2011) Exercise training increases mitochondrial biogenesis in the brain. J Appl Physiol (1985) 111:1066–1071
99. Yan QW, Zhao N, Xia J et al (2019) Effects of treadmill exercise on mitochondrial fusion and fission in the hippocampus of APP/PS1 mice. Neurosci Lett 701:84–91
100. Marques-Aleixo I, Santos-Alves E, Balça MM et al (2015) Physical exercise improves brain cortex and cerebellum mitochondrial bioenergetics and alters apoptotic, dynamic and auto (mito)phagy markers. Neuroscience 301:480–495
101. Ross JM, Coppotelli G, Branca RM et al (2019) Voluntary exercise normalizes the proteomic landscape in muscle and brain and improves the phenotype of progeroid mice. Aging Cell 18: e13029
102. Aguiar AS Jr, Tuon T, Pinho CA et al (2008) Intense exercise induces mitochondrial dysfunction in mice brain. Neurochem Res 33:51–58
103. Mckenzie JA, Spielman LJ, Pointer CB et al (2017) Neuroinflammation as a common mechanism associated with the modifiable risk factors for Alzheimer's and Parkinson's diseases. Curr Aging Sci 10:158–176
104. Bowman GL, Dayon L, Kirkland R et al (2018) Blood-brain barrier breakdown, neuroinflammation, and cognitive decline in older adults. Alzheimers Dement 14:1640–1650
105. Sala-Llonch R, Idland AV, Borza T et al (2017) Inflammation, amyloid, and atrophy in the aging brain: relationships with longitudinal changes in cognition. J Alzheimers Dis 58:829–840
106. Blasko I, Stampfer-Kountchev M, Robatscher P et al (2004) How chronic inflammation can affect the brain and support the development of Alzheimer's disease in old age: the role of microglia and astrocytes. Aging Cell 3:169–176
107. Nell HJ, Whitehead SN, Cechetto DF (2015) Age-dependent effect of β-amyloid toxicity on basal forebrain cholinergic neurons and inflammation in the rat brain. Brain Pathol 25:531–542
108. Hull C, Dekeryte R, Koss DJ et al (2020) Knock-in of mutated hTAU causes insulin resistance, inflammation and proteostasis disturbance in a mouse model of frontotemporal dementia. Mol Neurobiol 57:539–550

109. Papenberg G, Ferencz B, Mangialasche F et al (2016) Physical activity and inflammation: effects on gray-matter volume and cognitive decline in aging. Hum Brain Mapp 37:3462–3473
110. Balter LJT, Higgs S, Aldred S et al (2019) Inflammation mediates body weight and ageing effects on psychomotor slowing. Sci Rep 9:15727
111. Gelinas DS, Mclaurin J (2005) PPAR-alpha expression inversely correlates with inflammatory cytokines IL-1beta and TNF-alpha in aging rats. Neurochem Res 30:1369–1375
112. Lin E, Kuo PH, Liu YL et al (2019) Association and interaction effects of interleukin-12 related genes and physical activity on cognitive aging in old adults in the Taiwanese population. Front Neurol 10:1065
113. Lovatel GA, Elsner VR, Bertoldi K et al (2013) Treadmill exercise induces age-related changes in aversive memory, neuroinflammatory and epigenetic processes in the rat hippocampus. Neurobiol Learn Mem 101:94–102
114. Gomes Da Silva S, Simões PS, Mortara RA et al (2013) Exercise-induced hippocampal anti-inflammatory response in aged rats. J Neuroinflammation 10:61
115. Ko YJ, Ko IG (2020) Voluntary wheel running improves spatial learning memory by suppressing inflammation and apoptosis via inactivation of nuclear factor kappa B in brain inflammation rats. Int Neurourol J 24:96–103
116. Zoladz JA, Majerczak J, Zeligowska E et al (2014) Moderate-intensity interval training increases serum brain-derived neurotrophic factor level and decreases inflammation in Parkinson's disease patients. J Physiol Pharmacol 65:441–448
117. Liu Y, Chu JMT, Yan T et al (2020) Short-term resistance exercise inhibits neuroinflammation and attenuates neuropathological changes in 3xTg Alzheimer's disease mice. J Neuroinflammation 17:4
118. Packer N, Hoffman-Goetz L (2015) Acute exercise increases hippocampal TNF-α, Caspase-3 and Caspase-7 expression in healthy young and older mice. J Sports Med Phys Fitness 55:368–376
119. Sun LN, Li XL, Wang F et al (2017) High-intensity treadmill running impairs cognitive behavior and hippocampal synaptic plasticity of rats via activation of inflammatory response. J Neurosci Res 95:1611–1620
120. Gensler HL, Bernstein H (1981) DNA damage as the primary cause of aging. Q Rev Biol 56:279–303
121. Rutten BP, Schmitz C, Gerlach OH et al (2007) The aging brain: accumulation of DNA damage or neuron loss? Neurobiol Aging 28:91–98
122. Chetsanga CJ, Tuttle M, Jacoboni A et al (1977) Age-associated structural alterations in senescent mouse brain DNA. Biochim Biophys Acta 474:180–187
123. Dorszewska J, Adamczewska-Goncerzewicz Z (2004) Oxidative damage to DNA, p53 gene expression and p53 protein level in the process of aging in rat brain. Respir Physiol Neurobiol 139:227–236
124. Huo Z, Zhu Y, Yu L et al (2019) DNA methylation variability in Alzheimer's disease. Neurobiol Aging 76:35–44
125. Soares JP, Silva AI, Silva AM et al (2015) Effects of physical exercise training in DNA damage and repair activity in humans with different genetic polymorphisms of hOGG1 (Ser326Cys). Cell Biochem Funct 33:519–524
126. Yang JL, Lin YT, Chuang PC et al (2014) BDNF and exercise enhance neuronal DNA repair by stimulating CREB-mediated production of apurinic/apyrimidinic endonuclease 1. NeuroMolecular Med 16:161–174
127. Ji N, Zhao W, Qian H et al (2019) Aerobic exercise promotes the expression of ERCC1 to prolong lifespan: a new possible mechanism. Med Hypotheses 122:22–25
128. Apfeld J, Kenyon C (1998) Cell nonautonomy of C. elegans daf-2 function in the regulation of diapause and life span. Cell 95:199–210
129. Braeckman BP, Vanfleteren JR (2007) Genetic control of longevity in C. elegans. Exp Gerontol 42:90–98

130. Lamming DW, Ye L, Sabatini DM et al (2013) Rapalogs and mTOR inhibitors as anti-aging therapeutics. J Clin Invest 123:980–989
131. Xu S, Cai Y, Wei Y (2014) mTOR signaling from cellular senescence to organismal aging. Aging Dis 5:263–273
132. Pani G (2011) From growing to secreting: new roles for mTOR in aging cells. Cell Cycle 10:2450–2453
133. Tóth ML, Sigmond T, Borsos E et al (2008) Longevity pathways converge on autophagy genes to regulate life span in Caenorhabditis elegans. Autophagy 4:330–338
134. Lee JH, Budanov AV, Park EJ et al (2010) Sestrin as a feedback inhibitor of TOR that prevents age-related pathologies. Science 327:1223–1228
135. Meléndez A, Tallóczy Z, Seaman M et al (2003) Autophagy genes are essential for dauer development and life-span extension in C. elegans. Science 301:1387–1391
136. Colman RJ, Anderson RM, Johnson SC et al (2009) Caloric restriction delays disease onset and mortality in rhesus monkeys. Science 325:201–204
137. Levine B, Kroemer G (2008) Autophagy in the pathogenesis of disease. Cell 132:27–42
138. Raz N, Lindenberger U, Rodrigue KM et al (2005) Regional brain changes in aging healthy adults: general trends, individual differences and modifiers. Cereb Cortex 15:1676–1689
139. Cotman CW, Berchtold NC, Christie LA (2007) Exercise builds brain health: key roles of growth factor cascades and inflammation. Trends Neurosci 30:464–472
140. Colcombe SJ, Erickson KI, Scalf PE et al (2006) Aerobic exercise training increases brain volume in aging humans. J Gerontol A Biol Sci Med Sci 61:1166–1170
141. Clark CM, Guadagni V, Mazerolle EL et al (2019) Effect of aerobic exercise on white matter microstructure in the aging brain. Behav Brain Res 373:112042
142. Bugg JM, Head D (2011) Exercise moderates age-related atrophy of the medial temporal lobe. Neurobiol Aging 32:506–514
143. Kramer AF, Erickson KI, Colcombe SJ (2006) Exercise, cognition, and the aging brain. J Appl Physiol (1985) 101:1237–1242
144. Cabral DF, Rice J, Morris TP et al (2019) Exercise for brain health: an investigation into the underlying mechanisms guided by dose. Neurotherapeutics 16:580–599
145. Herting MM, Chu X (2017) Exercise, cognition, and the adolescent brain. Birth Defects Res 109:1672–1679
146. Blondell SJ, Hammersley-Mather R, Veerman JL (2014) Does physical activity prevent cognitive decline and dementia?: a systematic review and meta-analysis of longitudinal studies. BMC Public Health 14:510
147. Hatta A, Nishihira Y, Kim SR et al (2005) Effects of habitual moderate exercise on response processing and cognitive processing in older adults. Jpn J Physiol 55:29–36
148. Barha CK, Davis JC, Falck RS et al (2017) Sex differences in exercise efficacy to improve cognition: a systematic review and meta-analysis of randomized controlled trials in older humans. Front Neuroendocrinol 46:71–85
149. Fragala MS, Beyer KS, Jajtner AR et al (2014) Resistance exercise may improve spatial awareness and visual reaction in older adults. J Strength Cond Res 28:2079–2087
150. Karssemeijer EGA, Aaronson JA, Bossers WJ et al (2017) Positive effects of combined cognitive and physical exercise training on cognitive function in older adults with mild cognitive impairment or dementia: a meta-analysis. Ageing Res Rev 40:75–83
151. Kwak YS, Um SY, Son TG et al (2008) Effect of regular exercise on senile dementia patients. Int J Sports Med 29:471–474

152. Vivar C, Peterson BD, Van Praag H (2016) Running rewires the neuronal network of adult-born dentate granule cells. NeuroImage 131:29–41
153. Nichol K, Deeny SP, Seif J et al (2009) Exercise improves cognition and hippocampal plasticity in APOE epsilon4 mice. Alzheimers Dement 5:287–294
154. Lan Y, Huang Z, Jiang Y et al (2018) Strength exercise weakens aerobic exercise-induced cognitive improvements in rats. PLoS One 13:e0205562
155. Young JC, Dowell NG, Watt PW et al (2016) Long-term high-effort endurance exercise in older adults: diminishing returns for cognitive and brain aging. J Aging Phys Act 24:659–675

Chapter 7
Exercise-Mediated Autophagy and Alzheimer's Disease

Xianjuan Kou, Meng Zhang, Hu Zhang, Michael Kirberger, and Ning Chen

Alzheimer's disease (AD), a nervous system disease typically diagnosed in the elderly (aged 65 and above), is characterized by multiple symptoms including progressive memory loss, amnesia, aphasia, agnosia, and visual–spatial dysfunction, eventually leading to death [1]. AD has two typical pathological features: neuroinflammatory plaques formed by the deposition of extracellular amyloid-beta (Aβ) protein in brain tissue; and neurofibrillary tangles (NFTs) formed by the hyperphosphorylation of Tau protein in neurons, eventually manifested as continuous shrinkage of the cortex, and massive loss of neurons in hippocampal and other brain tissues [2]. The etiology and pathogenesis of AD are still unclear. Multiple factors are involved in the occurrence of AD, including Aβ toxicity, abnormal phosphorylation of the Tau protein, neuroinflammation, gene mutation, neurotransmitter loss, oxidative stress, and mitochondrial dysfunction [3]. An increase in the elderly population has also resulted in an increase in the incidents of AD. This trend has prompted new research efforts, including the development of donepezil, galantamine, and memantine, and these drugs can improve the symptoms of patients with moderate and mild AD, but there are currently no therapeutics capable of preventing, stopping, or reversing the development of AD.

X. Kou · M. Zhang · H. Zhang · N. Chen (✉)
Tianjiu Research and Development Center for Exercise Nutrition and Foods, Hubei Key Laboratory of Exercise Training and Monitoring, College of Health Science, Wuhan Sports University, Wuhan, China

M. Kirberger
School of Science and Technology, Georgia Gwinnett College, Lawrenceville, GA, USA

N. Chen (ed.), *Exercise, Autophagy and Chronic Diseases*,
https://doi.org/10.1007/978-981-16-4525-9_7

1 The Epidemiology of AD

AD is the most common type of aging-related disease, accounting for approximately 50–75% of the total number of dementia patients [4]. According to data from China's National Bureau of Statistics, elderly people aged 60 years and above account for 17.3% of the total population of 1.4 billion, while the population aged 65 years or older accounts for 11.4%, or 158.3 million people. This increase in the elderly population has become a very serious problem in China [5], resulting in a steady increase in the incidents of neurodegenerative diseases, dominated by AD [6]. Epidemiological studies have shown that AD is the most common cause of dementia, which accounts for approximately 50% of dementia patients worldwide. Age is the most significant risk factor for AD. For elderly people aged 65 or older, the prevalence of AD doubles for each successive 5-year period of age [7]. An epidemiological survey in China in 2005 reported that the prevalence of dementia in those aged 65 or older is 7.8%, with AD representing 4.8%. And it increases with age. In China, the prevalence rate is higher in the north than in the south, and the age of onset in the west is higher than in the east [8]. At the same time, AD is also closely related to environmental, economic conditions, genetics, and other factors [9]. AD causes a heavy economic burden on both the families of the patients, and society, making it a major public health problem. According to the World Alzheimer Report 2015, there are more than 46 million people worldwide suffering from dementia, including 9.4, 10.5, 22.9, and 4.0 million from the Americas, Europe, Asia, and Africa, respectively; and this number could increase to 74.7 and 131.5 million by 2030 and 2050, respectively, based on the estimation, which could reveal a double within every 20 years. It is well known that dementia also has a huge economic impact on patents' families and society. Currently, the total cost of dementia worldwide is estimated to be 818 billion U.S. dollars, and it may climb to a trillion U.S. dollars by 2018 and increase to 2 trillion U.S. dollars in 2030, which presents an expenditure trend as an order from high-income, upper-middle-income, lower-middle-income, and low-income countries, with the distribution of 715.1, 86.3, 15.3, and 1.2 billion U.S. dollars, respectively [10]. Prior to 2016, the number of AD patients in the USA was reported at 5.4 million, with a prevalence rate of 11% for people over 65 years old [11]. According to statistics, the number of AD patients in China had reached 9.19 million by 2010 [12]. Current estimates project as many as 36 million dementia patients globally by 2050 [13]. At present, the pathogenesis of AD is still unclear, and its etiology may be related to genetic factors, education level, dietary factors, smoking, reduced estrogen levels, thyroid hormones, vascular diseases, or other factors [14, 15].

2 The Pathogenesis of AD

Current studies have shown that the pathogenesis of AD is still unclear, but Tau protein hyperphosphorylation and the imbalance of Aβ production and clearance are the main pathological mechanisms for leading to neuronal degeneration and dementia [16, 17]. Senile plaques (SPs), formed by the aggregation of extracellular Aβ and NFTs resulting from hyperphosphorylation of Tau proteins, are also major pathological features of AD [18]. However, hypotheses regarding the pathogenesis of AD remain inconclusive, indicating the need for further research into the prevention and treatment of AD.

2.1 Amyloid Cascade Hypothesis

One of the current promising hypotheses is the Aβ cascade hypothesis, which suggests that the deposition of Aβ is the initial pathological event of AD, followed by formation of senile plaques and neurofibrillary tangles, nerve cell death, and eventually, dementia. The Aβ protein is the core component of neuroinflammatory plaques and constitutes a characteristic pathological change in AD. The plaques are surrounded by various glial cells, such as astrocytes and microglia. Aβ is composed of 39–43 amino acids, and is the decomposition product of Aβ precursor protein (APP), after its digestion by proteolytic enzymes. APP belongs to type I transmembrane proteins, which are widely distributed in various tissues, especially neuronal axons and dendrites [19]. APP located on the cell membrane can be degraded in two different ways: nonamyloid and amyloid. APP can be decomposed by α-, β-, and γ-proteases, and the continuous action of β-protease and γ-protease can result in the degradation of APP to produce Aβ. Under normal circumstances, the production and degradation of Aβ protein will maintain a certain balance. APP is cleaved between residues 16 and 17 by α-secretase, without producing Aβ fragments, but instead forming large soluble fragments that do not induce neurotoxicity. The smaller fragments are then cleaved by β-secretase and γ-secretase, resulting in the formation of soluble Aβ fragments. However, excessive production and deposition of Aβ protein can occur for other reasons, including the action of certain pathogenic factors, such as APP, PSEN, and other genes with multiple site mutations [20]; aging; dietary factors [21]; or brain injury. The most common residue subtypes are Aβ40 and Aβ42. Aβ40 is more common than Aβ42, but Aβ42 is more likely to accumulate in the brain, and its toxicity is stronger than Aβ40 [22]. When Aβ is produced in large quantities and cannot be eliminated by the body, it will accumulate and eventually lead to AD. This is the premise of the "amyloid cascade hypothesis" [23]. Additionally, the molecular configurations and the states of the Aβ proteins are closely related to neurotoxicity. Aβ can spontaneously aggregate into a variety of forms, such as an intermediate formed by the aggregation of 2–6 peptides, which are soluble oligomers. Fibrils can also be formed, which then arrange themselves into β-sheets and become insoluble amyloid plaques. Both Aβ oligomers are neurotoxic. Soluble Aβ oligomers are more toxic than fibrous insoluble amyloid plaques and can lead to neuronal death. Therefore, inhibiting the accumulation of Aβ and increasing its clearance are the major strategies for the prevention and treatment of AD [24]. The development of β-secretase inhibitors may also provide a new direction for the treatment of AD. Alternatively, a molecule capable of binding the Aβ monomer could be used as targeted therapy for Aβ-induced AD [25].

2.2 Tau Protein Hyperphosphorylation

NFTs, another characteristic pathological change associated with AD, are composed of intracellular double-helix filaments (PHFs) formed by hyperphosphorylation of microtubule-associated Tau protein [26]. During the early stages of AD, NFTs accumulate in the neurons of the entorhinal cortex (EC), especially in layers II and III of the EC, and then spread to the limbic lobes and corresponding cortex of the brain through presynapses. Structural imaging studies have confirmed that EC is the main site of early AD dysfunction. There is a direct, positive correlation between the number of NFTs and the severity of AD. Under normal circumstances, the abundant soluble proteins in axons can promote the assembly and stability of microtubules, while the affinity of hyperphosphorylated Tau protein to microtubules exhibits a decreasing trend, which makes microtubules more unstable. During infection, metabolic disease, or chronic inflammation, microtubule-associated protein can be activated by glycogen synthase kinase 3β (GSK-3β), cyclin-dependent kinase and its activated subunit p25, phosphorylated mitogen-activated protein kinase, overenhanced kinase activity, reduced phosphatase activity, or some combination of these activities, thereby causing the overphosphorylation of Tau protein and the depolymerization of intracellular microtubules. Protein kinases and protein phosphatases (PPs) can directly regulate the degree of hyperphosphorylation of Tau protein in AD. Of all various kinases and phosphatases, protein phosphatase 2A (PP2A) and GSK-3β are the most important regulatory enzymes. In the brains of AD patients, PP2A directly regulates Tau phosphorylation, and it can also indirectly regulate Tau phosphorylation by activating GSK-3β. PP2A knockout can inhibit the phosphorylation levels of GSK-3β at the site of Ser9 [27]. Tyr307 phosphorylation and Leu309 methylation can regulate PP2A activity, and both can induce the hyperphosphorylation of Tau protein by inhibiting PP2A activity. For example, hyperhomocysteinemia inhibits PP2A activity by increasing Leu309 demethylation, so hyperhomocysteinemia is an important factor in the development of AD [28]. Tyr307 phosphorylation can catalyze the PP2A subunit PP2Ac and ultimately can significantly increase the hyperphosphorylation of Tau protein [29]. GSK-3β can inhibit the hyperphosphorylation of Tau protein at multiple AD-related sites by activating the phosphatidylinositol-3-kinase (PI3K)/Akt signal pathway. When GSK-3β is activated, PP2A activity decreases. Increased oxidative stress, impaired endoplasmic reticulum protein folding function, and decreased protein clearance mediated by proteasomes and autophagy can all accelerate the deposition of Tau protein in AD. In fact, abnormal intermediate molecular aggregates of Tau protein are toxic and can impair cognitive function. Similarly, the pathogenic Tau protein can bind to synaptic vesicles through its N-terminal domain, interfere with the migration and release of synaptic vesicles, reduce the transmission efficiency of neurotransmitters, and impair neuronal synaptic transmission [30].

2.3 Neuroinflammation

Due to the participation and influence of microglia, acute or chronic systemic infections will greatly increase the risk of AD. Under the action of the central nervous immune system, microglia can continuously activate pro-inflammatory cells through the conduction of pro-inflammatory signals. Microglia cannot regulate the loss of anti-inflammatory cytokines and lipid mediators, thus leading to neurodegeneration, neuron damage, an increase in inflammatory responses, and increasing Tau protein hyperphosphorylation. Therefore, neuroinflammation has been established as a feature of AD. Inhibiting the pro-inflammatory signal interleukin-1β (IL-1β) can successfully improve cognitive deficits and reduce neuroinflammation, and restore pathological changes in AD mouse models. Similar to the effect of IL-1β, inhibiting interleukin-10 (IL-10) and interleukin-6 (IL-6) can also have the same effect. Persistent inflammation caused by periodontitis can increase the risk of AD [31]. Brain infections caused by gram-negative bacteria are related to AD, especially late sporadic AD. The application of nonsteroidal anti-inflammatory drugs (NSAIDs) can prevent the incidence of AD [32] by suppressing the synthesis of prostaglandins, thereby inhibiting the toxic effects of prostaglandins and neuronal damage. NSAIDs such as ibuprofen can reduce the production of Aβ42 [33]. Aβ can induce neuroinflammation, and inflammation can in turn promote the production of Aβ and NFTs, thus interacting to promote the development of AD [34]. Similarly, the accumulation of IL-1β can damage nerve cells and cause IL-1-β-dependent damage to hippocampal tissues responsible for learning and memory capacity. Thus, the inhibition of the pro-inflammatory signal IL-1β can reduce neuroinflammation and improve cognitive deficits in AD mice [34, 35].

2.4 Mitophagy Dysfunction

Autophagy is one of the important ways for animal cells to eliminate dysfunctional or denatured proteins and maintain cell stability. In neurons, the accumulation of Aβ can inhibit critical enzymes in the mitochondrial metabolic chain, such as cytochrome C oxidase, which in turn leads to obstacles or damage to electron transmission, ATP production, and mitochondrial membrane potential. In the damaged brain structure of AD patients and transgenic animal models, Aβ can be isolated from mitochondria. These Aβ proteins can directly lead to mitochondrial dysfunction and have a significant impact on genes such as nuclear-induced OXPHOS [36]. Similarly, mitochondrial stimulant metformin hydrochloride can improve the cognitive function of patients with mild-to-moderate AD. Other studies have confirmed that mitochondrial autophagy can improve AD by removing damaged mitochondria and cytotoxic Aβ [37]. When the autophagy/lysosomal pathway in neurons is damaged, the clearance of damaged mitochondria in cells will be hindered, thus resulting in a gradual increase in abnormal mitochondria, enhanced oxidative stress, and increases in β-secretase, γ-secretase, and presenilin 1. The increase in Aβ and phosphorylated Tau protein, and the deposition of this pathological protein, causes further disordered mitochondrial autophagy [38]. The overexpression of PTEN-induced putative kinase (PINK1) can activate autophagy receptors (OPTN and NDP52) to enhance autophagy and reduce Aβ levels, thereby reducing synaptic loss and cognitive decline [39]. Conversely, the excessive activation of silent information regulator 2 (SIRT2) can cause dysfunctional mitochondrial autophagy and promote the expression of Aβ [40]. Mitochondrial dysfunction in damaged neurons in the brain of AD patients will lead to insufficient energy production, which will further aggravate the deposition of Aβ and Tau proteins, thereby forming a vicious circular mechanism.

2.5 Cholinergic Neurotransmitter Pathway Abnormalities

The cholinergic system is involved in the processes of learning and memory, and its abnormality is also considered to be one of the important factors involved in the occurrence and development of AD. The lack of acetylcholine can affect the processing of information in cortex and hippocampal tissues, which in turn affects cognition and behavior. The onset of AD is usually related to a reduction in neurotransmitters, including serotonin, norepinephrine, dopamine, and acetylcholine. Reducing acetylcholine levels can directly cause cell damage in the basal nucleus, temporal lobe, and parietal lobe, thereby reducing serotonin levels and aggravating the development of NFTs. Additionally, the reduction in acetylcholine, which is necessary to maintain normal brain function, will eventually lead to the loss of neurons. Acetylcholinesterase can hydrolyze acetylcholine; therefore, inhibiting acetylcholinesterase can increase the concentration of acetylcholine in the central synapse, and enhancing cholinergic function is currently the most effective way to treat AD. The loss of cholinergic neurons may be the major cause of AD-related psychiatric symptoms. The presynaptic α7 nicotinic acetylcholine receptor is essential for cognitive capacity, and its expression level increases during early stage of AD and then decreases at later stages [41, 42]. The powerful type II-positive allosteric modulator PNU-120596 can enhance and prolong the activation of α7 [43]. One study has reported that nicotinic acetylcholine receptors (nAChRs) on glial cells can affect the survival, synaptic plasticity, and memory behavior of AD neurons [44]. The nAChRs on astrocytes and microglia also can contribute to the metabolism of Aβ in brain tissues, and reduce Aβ-related oxidative stress by regulating the phagocytosis and degradation of Aβ, thereby reducing its neurotoxic hazards [45]. It has also been shown that Meynert basal cholinergic neurons are severely reduced during the onset of AD. The uptake of cholinergic transmitters is reduced in the hippocampal tissue of AD animals, as determined through quantitative autoradiography, indicating that the expression of muscarinic and nicotinic cholinergic receptors is downregulated. Acetylcholinesterase can bind to growing Aβ fibers, and their interaction can promote the formation of amyloid fibrils [46]. The above studies have

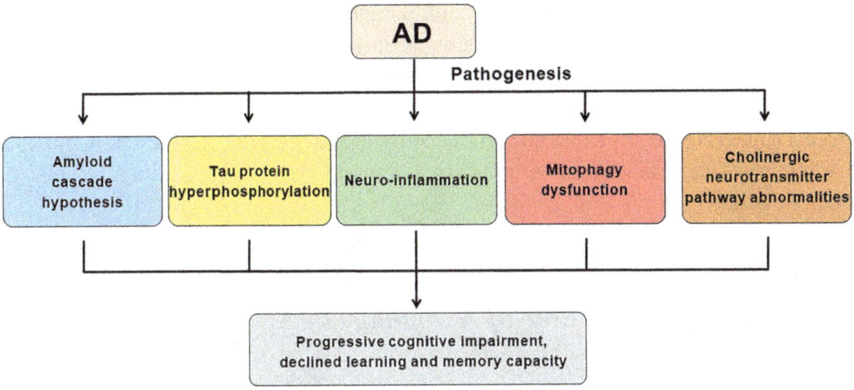

Fig. 7.1 Pathogenesis of AD

documented that the cholinergic neurotransmitter pathway also plays a critical role in the pathogenesis of AD (Fig. 7.1).

3 Autophagy and AD

3.1 Autophagy

Autophagy is a cellular physiological process that has been highly conserved during evolution. It is the major mechanism for clearing senescent or damaged organelles and long-lived proteins in eukaryotic cells. It can be used to remove abnormally aggregated or misfolded proteins and receptors, as well as damaged organelles. The recovery of some essential molecules (such as amino acids and fatty acids), or the clearance of organelles, plays important roles in maintaining cellular homeostasis and recycling cell components [47]. Autophagy is the process of transporting damaged, senescent, necrotic, or incorrectly aggregated proteins and organelles in cells, to lysosomes to form autophagic lysosomes, through lysosome degradation and recycling, to realize the process of cell self-metabolic demands and material recycling. In order to maintain cellular homeostasis, the process is involved in more than 40 autophagy-related genes or proteins [48]. Under normal circumstances, there are three types of autophagy: macroautophagy, microautophagy, and molecular chaperone-mediated autophagy. Of these, macroautophagy exhibits the more in-depth molecular mechanism and has the most obvious characteristics. References to autophagy usually imply microautophagy, as it is the most common process of the three types [49]. Autophagy is a nonselective degradation pathway comprising the following sequence of steps: (1) generation of the autophagy dual-layer membrane; (2) formation of autophagosomes; (3) transport of autophagosomes and formation of lysosomes; and (4) degradation of autophagosomes [50]. Recent studies have documented a close link between autophagy and neurodegenerative diseases [51]. As an important degradation pathway in cells, autophagy plays a significant role in maintaining the homeostasis of neurons. Even under physiological conditions, autophagy is still highly active [52]. Since neurons are permanent cells and cannot dilute the toxic substances produced by themselves through cell division, it is particularly important for neurons to maintain normal functional status of autophagy [53].

3.2 *The Regulatory Role of Autophagy in AD*

Autophagy is one of the important ways to eliminate and degrade erroneous proteins in animal cells, thus maintaining the stability of proteins and cells. Autophagy dysfunction can be manifested during early stages of AD [54]. Autophagy disorders can lead to the abnormal accumulation of proteins, including Aβ and abnormally phosphorylated Tau protein, thereby forming senile plaques and neurofibrillary tangles. Studies by Hara et al. [55] and Komatsu et al. [56] have reported that mice with knockouts of Atg5 or Atg7 exhibit abnormal accumulation of ubiquitinated proteins, the loss of a large number of neurons, and behavioral defects, within a few weeks. Thus, these studies clearly illustrate that abnormal autophagy plays an important role in the occurrence and development of AD.

3.2.1 Autophagy and Aβ

Autophagy is involved in the production of Aβ. Autophagic vacuoles (AVs) in mouse hepatocytes overexpressing APP contain large amounts of APP, β-C-terminal fragment (β-CTF), and β-site APP cleaving enzyme (BACE), suggesting that AVs may be potential sites for the production of Aβ [57]. Studies have further clarified that AVs in the brains of AD patients and model mice contain a large amount of APP, β-CTF, and γ-secretase complexes, suggesting that autophagy is activated at the early stages of AD, thereby leading to the amplification of AVs and the production of large amounts of Aβ [58]. After the SHSY5Y cells overexpressing APP are induced in vitro to undergo oxidative stress, their autophagy levels are significantly increased, and the AVs encapsulated with APP, and Aβ monomers and oligomers, are also significantly increased. In contrast, 3-methyladenine (3-MA) can significantly reduce the content of Aβ in SH-SY5Y cells [59]. These studies indicated that autophagy is an important regulatory way to produce Aβ. Not only AVs can degrade the APP wrapped by itself to produce Aβ, but also β-CTF in endosomes can be transported to autophagosomes and be hydrolyzed by γ-secretase to produce more Aβ [60].

Autophagy is involved in the clearance of Aβ. There is a complex correlation between Aβ and autophagy. In the brains of AD patients, swelling and degeneration of neurites result from a lack of nutritional neurofactors, and the abnormal aggregation of Aβ is also observed in autophagic vesicles, suggesting that autophagy may be related to the pathogenesis of AD [61]. The protein kinase AMP-activated catalytic subunit alpha1 (PRKAA1) signal pathway is involved in the degradation of Aβ, thereby further promoting the occurrence of autophagy [62]. Autophagy-related proteins such as Atg5, Beclin1, and Ulk1 are closely related to the degradation of Aβ and APP-CTF [63]. Small interfering RNA (siRNA) is used to knock out Beclin1 in N2a-APP cells, and Aβ40 and APP in cells, which significantly increases the level of CTF. Similarly, APP and its degradation products can be used as the substrates for autophagy, and the accumulation of APP and CTF is often accompanied by clearing obstacles to autophagosomes [64]. Conversely, rapamycin or other autophagy activators such as SMER28 can significantly increase the clearance of CTF and Aβ in cells. Pharmacological inhibition of autophagy can significantly weaken the effect of SMER28, which further confirms that SMER28 is based on the regulation of autophagy. Aβ clearance disorder plays an important role in the pathological process of AD. Under physiological conditions, Aβ-rich AVs need to be transported to lysosomes for degradation, because lysosomes contain cathepsins necessary for Aβ degradation, such as cathepsin B and cathepsin D [65, 66]. However, since lysosomes are mainly distributed in the perinuclear body, AVs formed on neurites must undergo long-distance reverse transport to reach the cell body and lysosomes [67]. Previous studies have also found that a large number of autophagosomes and other types of autophagy vesicles containing APP and related secretases are accumulated in the swollen axons of the cerebral cortex and hippocampal tissues of AD patients and model mice, further suggesting that the clearance of AVs in the AD brain is hindered [68, 69]. The retained AVs cannot be degraded by lysosomes, thus leading to a large accumulation of Aβ in the cells, and accelerating the pathological progression of AD. According to above studies, although autophagy does not play a complete role in Aβ metabolism, autophagy is dysfunctional in the regulation of pathological progression of AD. Therefore, autophagy can also affect the secretion of biomarkers for the regulation of AD [70], and the regulatory role of autophagy in Aβ metabolism is bifunctional.

Autophagy is involved in the secretion of Aβ. Autophagy can not only regulate the production and clearance of Aβ, but it also participates in the secretion process. Nilsson et al. [71] reported significant reductions in autophagy-related gene Atg7, Aβ secretion, and the number of extracellular senile plaques in the neurons of APP transgenic mice. Subsequently, the abovementioned neurons were infected with the lentivirus carrying Atg7 to restore Atg7 to normal levels, thereby reactivating autophagy and restoring the secretion of Aβ to previous levels. After low-dose rapamycin treatment of primary neurons in wild-type mice, intracellular autophagosomes exhibited significant increases, accompanied by increases in Aβ secretion. Conversely, the application of spautin-1 to inhibit autophagy significantly reduced the secretion of Aβ to the outside of the cell. In the brains of AD patients, AVs are retained in the dystrophic neurites, which increases the likelihood of fusion with the plasma membrane, thus resulting in more Aβ released outside the cell to form amyloid plaques. Therefore, extracellular Aβ is considered to be pathogenic. A subsequent study also confirmed that, in the absence of amyloid plaques, elevated intracellular Aβ levels can still cause cognitive dysfunction in AD model mice [72]. Even without the formation of senile plaques, the increase in Aβ oligomers in cells can not only cause synaptic change, but also lead to the abnormal phosphorylation of Tau protein [73]. This suggests that Aβ in cells is the real pathogenic factor of AD. The accumulation of Aβ in the cells not only affects the transport of autophagosomes and their fusion with lysosomes, but also destroys the stability of the lysosome membrane, thereby further affecting the degradation of autophagy substrates and aggravating the pathological changes in AD.

3.2.2 Autophagy and Tau Protein

Tau protein is the most abundant microtubule-related protein. Its gene is located on the long arm of chromosome 17, and can express 6 isoforms, mainly distributed in the axons of neurons in the central nervous system [74]. Tau protein is the protein containing phosphate groups, and Tau protein molecules in normal brains contain 2–3 phosphate groups. In the brains of AD patients, each molecule of Tau protein can contain 5–9 phosphate groups, and this abnormal, hyperphosphorylated Tau protein causes the cells to lose normal functions. At the same time, excessive amounts of Tau protein are found in the brains of AD patients, with an increase in the content of hyperphosphorylated Tau protein and a decrease in the content of normal Tau protein. In addition, the binding force of normal Tau protein and tubulin is 10 times higher than that of abnormal Tau protein, which causes Tau protein to lose its biological function of promoting the assembly and formation of microtubules, and its role in maintaining microtubule stability. An increasing body of evidence indicates that autophagic signal pathways are related to the phosphorylation of Tau protein under normal physiological and pathological conditions of neurons [75, 76]. Chronic cerebral hypoperfusion can enhance the hyperphosphorylation of Tau protein and reduce autophagy in model mice with AD [77]. Various forms of Tau protein can be degraded by both the ubiquitin–proteasome system (UPS) and the autophagy–lysosomal system (ALS) [78]. Similarly, Tau protein hyperphosphorylation can lead to autophagy dysfunction. Autophagosome transport mainly depends on its movement along microtubules during autophagy. Hyperphosphorylation of Tau protein leads to instability of neuronal microtubules, which affects the placement and function of lysosomes and mitochondria [79]. On the other hand, the activation of autophagy can inhibit Tau aggregation and eliminate its cytotoxicity. Autophagy dysfunction also can induce Tau protein aggregation and neurodegeneration. In experimental animal models with autophagy deficiency, although there is no NFT, the silencing of Atg7 expression can increase concentrations of phosphorylated Tau protein and GSK-3β [80]. The induction of autophagy-related protein NDP52 can reduce Tau protein phosphorylation in neurons [81], and increased expression of the multifunctional protein p62 can activate autophagy to remove Tau [82]. Therefore, there is a close connection between autophagy and pathological Tau clearance.

Protein phosphatase 2A (PP2A) is an important phosphatase of Tau protein, and its abnormal function will directly lead to the hyperphosphorylation of Tau protein, promote its self-aggregation and deposition, and ultimately aggravate the pathological progression of AD [83]. Phospholipase D1 (PLD1), as a negative regulator of Vps34, can hinder the maturation of autophagosomes and accelerate the accumulation of Tau protein in the brain [84]. After treating the cells with the proteasome inhibitor MG-132, the number of LC3-II and autophagosomes increases significantly, while the level of Tau protein in the cells exhibits an obvious decrease [85], which is due to the blocking of the proteasome pathway, thereby resulting in the complete upregulation of the autophagy–lysosomal pathway, and acceleration of hydrolysis of Tau protein. Conversely, after blocking the autophagic signal pathway of neurons, the degradation of Tau protein through UPS is minimal [86]. These data all demonstrate that autophagy is the most important way to degrade Tau protein in neurons, and the abnormal accumulation of Tau protein will adversely affect the autophagy process. The PP2A inhibitor okadaic acid can cause high phosphorylation of Tau protein, thus leading to the loss of Tau protein function of assembling and binding microtubules, hindering axon reverse transport and autophagosome–lysosome fusion, as well as the accumulation of autophagosomes in neurons in a large quantity, and finally hindering the clearance of abnormal proteins, and further aggravating the pathological progression of AD [87, 88].

3.2.3 Autophagy and Neuroinflammation

Neuronal inflammation is a prominent feature in the progression of AD [89]. Based on the self-protection mechanism, the inflammatory response is usually beneficial, but this process needs to be strictly controlled to avoid continuous inflammation, which may lead to self-damage. Many studies have concluded that neuroinflammation is related to neuronal damage in the brain during the AD process [90]. In recent years, more and more evidence has come to support the idea that chronic inflammation is closely related to the onset of AD. The ratio of anti-inflammatory factor IL-10 to pro-inflammatory factor IL-1 in the blood of AD patients is greatly reduced, and the incidence of AD is lower in people taking nonsteroidal anti-inflammatory drugs [91]. In addition, the accumulation of pro-inflammatory factors such as IL-1β and IL-18 may damage neuronal cells, thus leading to hippocampal-dependent damage of learning and memory capacity [35]. All these indicate that the pathogenesis of AD is closely related to neuroinflammation. At the same time, Aβ itself can promote the occurrence of inflammation, and neuroinflammation can promote the production of neurofibril tangles and Aβ, thereby further aggravating the cognitive dysfunction of AD patients [34]. Relevant studies have also shown that the autophagy activator rapamycin has a neuroprotective effect. It can reduce Aβ precipitation and promote Aβ clearance by activating autophagy, and improve the pathological process caused by Tau protein hyperphosphorylation and NFTs, thereby inhibiting AD. Another study has reported that the protective mechanism of AD is related to neuroinflammation [92]. These

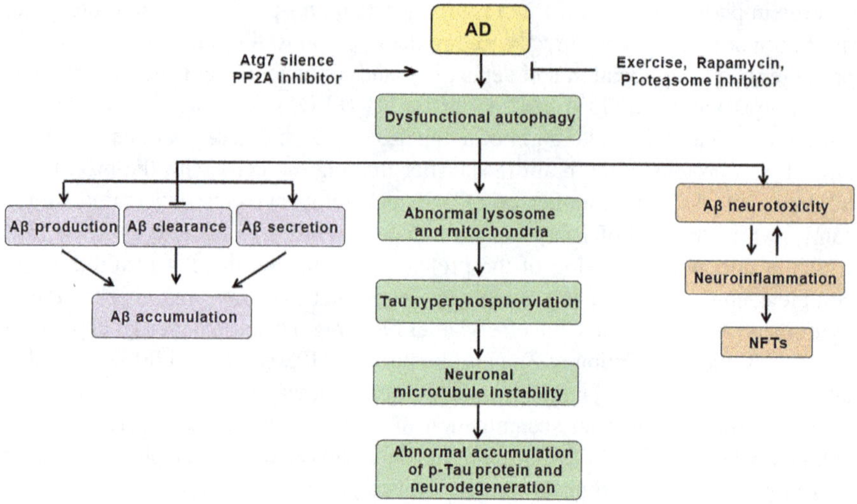

Fig. 7.2 Regulatory roles of autophagy in AD

studies suggest that rapamycin exerts neuroprotective effects through autophagy induction and anti-inflammatory mechanisms [93, 94]. Other studies have shown that autophagy is an important mechanism of Aβ-induced neuroinflammatory response in astrocytes. For example, neurosteroid progesterone (PG) inhibits the potential neuroprotective mechanism of neuroinflammatory response by enhancing autophagy [95]. Therefore, it can be speculated that there is a close connection between autophagy and neuroinflammatory response (Fig. 7.2).

4 Exercise-Induced Autophagy in AD

Autophagy can remove damaged organelles and misfolded proteins in the brain, prevent neuronal apoptosis, and maintain normal brain functions. Therefore, autophagy is known as the "garbage disposal worker" in the body [96, 97]. Nixon et al. [98] reportedly observed very few autophagosomes in nutrient-rich healthy neurons, but a large number of autophagosomes and autophagolysosomes wrapped with Aβ in the swollen brain neurons of AD patients. This indicated that autophagy degradation in AD brain was blocked, and autophagy activity was reduced. Animal experiments have also found that APP mutant mice, SAMP8, 3xTg-AD, and other AD model mice exhibit increased expression of p62 and aggregated autophagosomes in the brain, suggesting that autophagy activity in cells of AD brain tissue is reduced [99]. The injection of Aβ42 into SD rats can downregulate LC3-II expression, increase p62 expression, and upregulate the expression of pro-apoptotic proteins such as Bax and caspase-3, suggesting that the decreased autophagy in cells of AD brain tissue is closely related to the neurotoxicity of Aβ42

[100]. Exercise, as a special form of environmental stimulus, can affect the synthesis, release, and degradation of other important neurotransmitters such as dopamine in the brain, regulate the plasticity of cranial nerves, and effectively prevent neurological diseases such as AD. A large number of studies in the field of exercise science have demonstrated that appropriate exercise has a positive effect on the structure and function of the brain, and also has a significant promotion effect on learning and memory capacity, which is beneficial for the prevention or rehabilitation of AD [101]. Although previous studies on the effects of exercise on autophagy are focused on peripheral tissues, such as skeletal muscle, myocardium, and liver, He et al. [102] reported that treadmill exercise can not only increase the level of autophagy in peripheral tissues of normal mice, but also increase the level of autophagy in brain cells.

4.1 Exercise Can Improve the Level of Autophagy

4.1.1 Exercise Can Increase the Autophagy Level of Normal Brain Cells

Exercise as a nonpharmacological intervention can effectively improve the self-protection of cranial nerves [103], promote the activation of the hippocampal nervous system [104], and maintain brain metabolism [105], but its underlying mechanism has not been fully elucidated. He et al. [102] reported that 8 weeks of treadmill exercise can enhance the expression of LC3-II in the cerebral cortex of mice with knock-in nonphosphorylation mutation of BCL2 (Thr69Ala, Ser70Ala, and Ser84Ala) (BCL2AAA), reduce the expression of p62, and, at the same time, improve the learning and memory capacity of mice. This study was significant as it demonstrated for the first time that exercise can promote brain health. The mechanism of brain health may be related to the increased autophagy in brain tissues. In a related study, Zhang et al. [106] used 5-week-old SD male rats to study the effects of treadmill exercise for 36 weeks. Results of this study indicated that treadmill exercise inhibited the expression of mTOR in the cerebral cortex of rats, increased the LC3-II/LC3-I ratio (increase in autophagy flux), reduced p62 expression, and increased Lamp1 expression. The above studies indicated that exercise can induce autophagy of normal mouse brain cells, accelerate the formation of autophagosomes, and degrade misfolded proteins such as Aβ and pathological Tau protein in the brain by increasing autophagy levels to maintain brain health. Rocchi and He [107] also discovered that 2 weeks of treadmill running and voluntary wheel running training could significantly increase the expression of LC3-II and reduce the expression of p62 in the cerebral cortex of 2–3-month-old C57BL/6 mice. In another study that produced similar results, it was determined that 12 weeks of voluntary wheel running and treadmill running could also increase the expression of Beclin1 and LC3-II in the brain of 3-week-old male SD rats and reduce the expression of p62 [108]. The above studies have shown that both active exercise and passive exercise can induce the initiation of autophagy in brain cells and promote the formation of autophagosomes, as shown in the increased expression of LC3-II and the decreased expression of p62, thereby accelerating the degradation rate of damaged organelles and misfolded proteins in the brain in autophagolysosomes, and maintaining normal brain functions.

4.1.2 Exercise Can Improve the Functional Status of Autophagy in Brain with Nerve Damage

Exercise can protect brain health and resist aging or aging-induced cognitive damage and AD pathology. The aging of the brain is an important manifestation of body aging and an important cause of neurodegenerative diseases. Recent studies have shown that the functional status of autophagy in brain cells during the aging process tends to decrease with the extension of age, and the capacity to clear misfolded proteins or damaged organelles in the brain tissue also decreases, which leads to a decline in learning and memory capacity, followed by onset of various neurodegenerative diseases, especially AD [109]. In another study [110], 13-month-old NSE/APPsw transgenic AD mice exhibited improved learning and memory capacity and alleviated AD symptoms after 16 weeks of treadmill exercise. The loss of brain neurons is one of the major symptoms during early stage of AD. It was also reported that injection of Aβ25–35 into the lateral ventricles of 2-month-old C57BL/6 mice could cause loss of brain neurons and reduction in learning and memory capacity, while a 2-week voluntary wheel running exercise could significantly rescue the reduced learning and memory capacity and promote neurogenesis [111]. Adlard et al. [112] also documented that 1 month-old TgCRNDB transgenic AD mice exhibited increased learning and memory capacity, increased number of neuron axons, and reduced Aβ deposition, following 5 months of treadmill running. The above study hinted that both voluntary wheel running and treadmill training could promote the regeneration of brain neurons in mice, improve learning and memory capacity, and delay the onset of AD.

Recently, Marques-Aleixo et al. [113] injected doxorubicin (DOX) into 3-week-old SD rats to create an AD-like model. This resulted in reduced learning and memory capacity, and neuronal apoptosis in the cerebral cortex, coupled with overexpression of p62, suggesting that abnormal functional status of autophagy in rat brain cells can lead to neuronal damage. In contrast, treadmill running or voluntary wheel running for 5 weeks significantly increased learning and memory capacity of the DOX-stimulated rats, coupled with reduced accumulation of autophagosomes and reduced p62 expression in the brain tissues. Consistent with these results, Zhang et al. [114] also confirmed that long-term exercise can reduce the accumulation of autophagosomes in the brain of cerebral ischemic rats, promote neurogenesis, and improve learning and memory capacity of cerebral ischemic rats. These studies all suggest that exercise can improve learning and memory capacity by enhancing the autophagy level of brain tissues, thereby maintaining the normal function of neurons.

4.2 Exercise Can Improve AD Through Inducing Autophagy

Exercise has clearly proven to be an effective and low-cost intervention for neurodegenerative diseases and cognitive dysfunction. Exercise activates a series of neuroprotective mechanisms and promotes nerve regeneration to slow down memory loss and cognitive dysfunction. Exercise can increase the autophagy activity of hippocampal tissues in model rats with nerve injury, inhibit neuronal death, and protect brain health. Previous studies have documented that the expression of autophagosome marker protein LC3-II and expression of autophagy substrate protein p62 in AD brain tissues are both upregulated, and a large number of autophagosomes and autophagosomes are wrapped with Aβ. The aggregation of autophagolysosomes indicates that reduced or deficient autophagy in AD brain tissue can impair Aβ clearance, which is an important cause of Aβ aggregation [115]. Subsequent studies have confirmed that exercise can induce autophagy of AD mice, inhibit neuronal apoptosis, and mitigate cognitive dysfunction [116]. Herring et al. [117] confirmed the activation of autophagy in 7-month-old TgCRND8 transgenic AD female mice after 5-month voluntary wheel running, as evidenced by the upregulation of Beclin1, Atg5, and STX17, in cerebral cortex and hippocampal tissues, as well as the significantly reduced the deposition of Aβ, when compared with the controls without exercise training. Similarly, 1-month-old TgCRND8 transgenic AD female mice, after 5 months of treadmill running exercise, also exhibited significant increases in Beclin1 and Atg5 in the cerebral cortex, which further suggested that exercise could not only increase the activation of autophagy in the brain tissues of AD transgenic mice at the age of 7 months, but also induce autophagy in the brain tissues of transgenic AD mice at the age of 1 month. Again, these results indicated that exercise could improve brain function at different stages of AD. The induction of autophagy may also accelerate the clearance of Aβ and play an important role in the prevention and alleviation of AD. In addition, the phosphorylation levels of PI3K and Akt in the hippocampal tissues of mice in the exercise training group were significantly increased, suggesting that exercise may increase the autophagy level of brain tissues of transgenic AD mice by activating the PI3K/Akt/mTOR signaling pathway. In addition, Fang [118] found that treadmill exercise can mediate synaptic plasticity in the hippocampus of stressed rats through PI3K/Akt signaling pathway. Herring [117] and Chae [119] also found that exercise can activate PI3K/Akt/mTOR signal pathway that improves the learning and memory capacity of mice.

4.2.1 Exercise Can Improve AD by Increasing the Activation of Autophagy

Exercise can accelerate the initiation of autophagy in hippocampal tissue. According to the mechanism of autophagy, autophagy activity is mainly determined by autophagy initiation and autophagy–lysosome fusion degradation [115]. Exercise can target and regulate mTOR and AMPK signal pathways, and then induce autophagy activation. Experimental evidence has documented that 9 months of treadmill running can inhibit the overexpression of mTOR in the hippocampal tissue and increase the autophagy level in neurons of aging rats [116]. After 12 weeks of treadmill exercise, the PI3K/Akt/mTOR signal pathway was activated to induce autophagy in the hippocampal tissue of NSE/htau23 transgenic AD mice and relieve the symptoms of AD [120]. Similarly, as the "metabolic receptor" of eukaryotic cells, AMPK is very sensitive to energy changes due to stress from exercise stimulus. Eight weeks of voluntary wheel running training was found to upregulate AMPK/silent information regulator 2-related enzyme 1 (sirtuin1, SIRT1)/transcription factor EB (transcription factor EB, TFEB) signals, to activate the autophagy–lysosomal system in brain tissue of normal mice [121]. Additionally, 36 weeks of moderate-intensity treadmill running was observed to activate autophagy by enhancing AMPK regulation and increasing hippocampal synaptophysin levels [122]. These results suggested that under exercise stress, metabolic changes related to high energy demand may be correlated with the induction of neuronal autophagy, which in turn triggers neuronal adaptation and protective enhancement signals. Marques-Aleixo et al. [108] reported that 12 weeks of treadmill running and voluntary wheel running increased the expression of Beclin1, a key protein for autophagosome formation in the hippocampal tissue of SD rats, indicating that exercise could promote the formation of autophagosomes in the cortex of SD rats. Herring et al. [117] reported that exercise significantly upregulated the expression of Beclin1 and Atg5 in the cerebral cortex and hippocampal tissue, and increased the formation of autophagosomes, in 7-month-old TgCRND8 female AD mice, following 5 months of voluntary wheel running training. When cells have a sufficient energy supply, ULK1 activity is inhibited, thereby ensuring that autophagy functions at a basic level. When cells are in situations of starvation and high energy requirements, coupled with insufficient ATP production, AMPK can be activated to promote the phosphorylation of ULK1, and the phosphorylated ULK1 can promote the formation of autophagosomes to initiate autophagy [123]. Studies have demonstrated that exercise can increase AMPK expression in the brain tissue of AD model mice [124]. In addition, 12 weeks of treadmill exercise can also downregulate the expression of mTOR in hippocampal tissue of APP/PS1 transgenic mice and upregulate the expression of ULK1, thereby improving learning and memory capacity by initiating autophagy [125]. The above studies suggest that exercise, as an important intervention to regulate energy, can activate ULK1 by upregulating AMPK to induce autophagy initiation in the hippocampal tissue of mice.

4.2.2 Exercise Can Improve AD by Enhancing the Degradation Function of Lysosomes

Exercise can enhance the degradation function of hippocampal lysosomes in AD mice. Autophagy is a degradation process that depends on lysosomes. The normal fusion of autophagosomes and lysosomes and the normal degradation function of lysosomes are also important factors for autophagy activity. If the function of autophagosomes for degradation is impaired, even if autophagy is induced, autophagy activity is still limited, which may aggravate the aggregation of autophagosomes or cause neuron death [126]. Under physiological conditions, lysosomes with normal function can rapidly degrade Aβ wrapped in autophagosomes and mitochondria waiting to be degraded. However, under pathological conditions of AD, excessive Aβ in lysosomes can increase neurotoxicity and destroy lysozymes. The stability of the membrane can affect the leakage of lysosomal enzymes and reduce the content of degrading enzymes in lysosomes. In another aspect, researchers reported that the mRNA and protein expression levels of Rab7 in the hippocampal tissue of APP/PS1 mice were reduced, suggesting that the hippocampal lysosome maturation of APP/PS1 mice was impaired, and the fusion of autophagosomes and lysosomes was blocked, and in contrast, exercise significantly increased the expression levels of Rab7 mRNA and protein in the hippocampal tissue of APP/PS1 mice [127]. Similarly, the expression levels of Rab7 mRNA and protein in the hippocampal tissue of C57BL/6 mice were also upregulated following exercise training, indicating that exercise can promote the maturation of lysosomes and accelerate the fusion of autophagosomes and lysosomes. Blocking the fusion of autophagosomes and lysosomes with the autophagy inhibitor chloroquine was found to weaken the effect of exercise on the induction of autophagy activity in the hippocampal tissue of SD rats [128]. Similarly, the expression of Lamp1 in the hippocampal tissue of 6-month-old APP/PS1 mice was upregulated, suggesting that lysosomes in the hippocampal tissue of APP/PS1 mice were aggregated and that 3 months of treadmill running exercise could significantly reduce Lamp1 expression, an indication that exercise could reduce the accumulation of lysosomes in the hippocampal tissue of AD mice and accelerate the rate of lysosome degradation [129]. Results of the above studies suggest that the impaired function of hippocampal lysosomes in APP/PS1 mice is an important reason for decreased autophagy activity. However, exercise can promote the maturation of lysosomes in the hippocampal tissue, increase the content of lysosomal degrading enzymes, and accelerate the fusion of autophagosomes and lysosomes to enhance the degradation function of lysosomes.

4.2.3 Exercise Can Reduce the Deposition of AD-Like Aβ Through Autophagy

Aβ deposition and the accumulation of hyperphosphorylated Tau protein are two major pathological features of AD, so reducing Aβ deposition is believed to be one of the important targets of AD treatment. An exercise program of treadmill running and voluntary wheel running for 16 weeks was found to reduce the deposition of Aβ in the hippocampal tissue, improve learning and memory capacity, and relieve AD [130]. Similarly, 10 weeks of treadmill training was observed to significantly reduce Aβ deposition in the hippocampal tissue of TgAPP/PS1 AD mice and relieve AD symptoms [131]. Moderate- and high-intensity treadmill exercise reportedly reduced the levels of Aβ in the brain tissue of AD model mice as a dose-dependent effect of exercise [132]. Autophagy plays an important role in the clearance of Aβ, and abnormal autophagy can lead to abnormal deposition of Aβ [133]. In the brains of APP/PS1/Tau triple transgenic AD model mice, an increased LC3-II/LC3-I ratio, abnormal accumulation of autophagosomes in neurons, and disordered lysosomal clearance pathways were observed [134]. The induction of autophagy with rapamycin was found to improve the cognitive function of AD mice and reduce intracellular Aβ levels [135]. Similarly, increased Aβ levels, increased LC3-II/LC3-I ratios, and increased p62 and LAMP1 protein expression levels were also observed in 6-month-old APP/PS1 mouse hippocampal tissue [129], suggesting impaired autophagic flux. On the other hand, 12 weeks of aerobic exercise intervention resulted in activation of hippocampal Beclin1, increased LC3-II levels, and reduced p62 and LAMP1 protein expression levels, suggesting that aerobic exercise could rescue the impaired autophagic flux to promote the clearance of accumulated Aβ. This indicates that the anti-AD efficiency of exercise is related to the autophagy-mediated clearance of Aβ in neurons. Therefore, exercise may regulate the generation and deposition of Aβ in hippocampal tissue by activating autophagy signal pathways, thereby playing a critical role in prevention and treatment of AD.

4.2.4 Exercise Can Reduce the Abnormal Phosphorylation of Tau by Improving Autophagy

Neurofibrillary tangles formed by abnormal phosphorylation of Tau protein are another important pathological feature of AD. Hyperphosphorylated Tau protein is prone to form neurofibrillary tangles, thus disrupting basic functions such as information transmission and material transportation between synapses of neurons [136]. Exercise can alleviate AD by regulating the abnormal phosphorylation of Tau protein. It has been reported that exercise reduced the phosphorylation levels of Tau protein at residues Ser404, Ser202, Thr231, and other sites, in AD mice [137]. Long-term exercise promotes the content and activity of kinase for regulating Tau protein, thereby inhibiting the hyperphosphorylated Tau protein levels observed in Tau transgenic mice [138]. These studies indicate that the pathological transformation of Tau protein following exercise intervention may be another target for the prevention and treatment of AD. Autophagy is also involved in the movement regulation mechanism of Tau protein pathology. Dysfunction of the autophagy–lysosomal system can lead to the formation of Tau protein oligomers and insoluble aggregates, while the induction of autophagy can reduce this aggregation [139]. In addition, autophagy may also affect the phosphorylation status of Tau protein. Inoue et al. [140] reported that a large amount of hyperphosphorylated Tau protein accumulated in the brain tissue of autophagy-deficient mice, and after inducing the restoration of autophagy function, the accumulated hyperphosphorylated Tau protein was eliminated. The degradation of Tau protein through autophagy is regulated by mTOR signaling. In animal models with AD, abnormally activated mTOR signaling is closely related to abnormally aggregated hyperphosphorylated Tau protein. Long-term voluntary wheel running in Tau22 transgenic mice with AD appeared to promote the growth of mouse hippocampal neurons, enhance the expression of Tau protein clearing factors, and reduce pathological Tau protein [141]. In contrast, after treatment with mTOR inhibitor rapamycin in AD transgenic mice, neuronal autophagy was activated and the abnormal phosphorylation of Tau protein was reduced, indicating that inhibiting mTOR signaling could activate autophagy and reduce the aggregation of abnormally phosphorylated proteins, thereby alleviating Tau pathology [142]. Long-term exercise also appears to regulate mTOR signaling, which provides a potential explanation of the underlying mechanism of exercise-induced autophagy for regulating Tau protein. Similarly, 12 weeks of treadmill exercise appeared to improve the cognitive function of NSE/htau23 transgenic mice [131]. The molecular mechanism associated with these observed results may involve exercise-mediated inhibition of mTOR signaling, which initiates the expression of autophagy-related genes, thereby reducing the abnormal phosphorylation of Tau protein. Moreover, 12 weeks of treadmill exercise reduced the hyperphosphorylation of Tau protein in the spinal cord and hippocampal tissue of P301S Tau transgenic mice, but the autophagy-related protein LC3-II in the spinal cord and hippocampal tissue did not change, while p62 protein in spinal cord neurons was significantly reduced, indicating that exercise-induced autophagy has a certain regulatory effect on Tau pathology, but has brain tissue specificity [143]. It should be pointed out that the hyperphosphorylation of Tau protein can also induce autophagy dysfunction. Tau protein is the most abundant microtubule-associated protein in nerve cells and is involved in maintaining the formation and stability of microtubules. This is reversed in autophagosomes. Transportation and maturation of lysosomes play an important role in the fusion process [144].

In another study [120], after 12 weeks of treadmill exercise intervention in 18-month-old NSE/htau23 transgenic AD mice, researchers observed reduced hyperphosphorylated Tau protein in the hippocampal tissue, accompanied by inhibited mTOR, upregulated Beclin1, increased LC3-II expression and decreased p62 expression, and improved learning and memory capacity. It has been suggested that the induced autophagy in brain tissue, upon exercise intervention, could accelerate the clearance of hyperphosphorylated Tau protein, thereby alleviating the symptoms of AD. However, whether exercise regulates Tau hyperphosphorylation to modulate the anti-AD efficiency through autophagy, or corrects autophagy dysfunction by reducing the abnormal phosphorylation of Tau protein, is still unclear.

4.2.5 Exercise Can Regulate Synaptic Plasticity Through Autophagy

The most obvious clinical symptoms of AD patients are memory loss and cognitive impairment. The hippocampal tissue is the most important area in the brain for controlling learning and memory capacity, and an important transferring station for short-term memory to long-term memory. The long-term inhibition of synaptic transmission in the hippocampal tissue can cause memory damage. Synaptic plasticity refers to the change in specific number, structure, and function, of synapses under the influence of continuous neuron activity, which is the basis for the formation of learning and memory. Impaired synaptic integrity, abnormal plasticity, and decreased density in the joint area of the neocortex and hippocampal tissues are considered to be the bases for the pathogenesis of AD and cognitive impairment [145]. Therefore, elucidating synaptic plasticity to explore the pathogenesis and corresponding treatments of AD may provide novel strategies for the prevention and treatment of AD. The underlying mechanism for preventing and treating AD may be the possible involvement of the improved neuronal synaptic structure and enhanced functional plasticity. A recent study has documented that regular exercise can inhibit neuronal apoptosis and repair the damage of synapses in the hippocampal tissue, thereby restoring the efficiency of hippocampal nerve conduction, improving animal memory, and delaying the process of cognitive impairment [146]. In addition, 4 weeks of voluntary wheel running has been found to increase the expression of synaptic plasticity-related proteins in AD model rats [147]. Long-term regular treadmill exercise reportedly improved the learning and memory function of 8-month-old AD transgenic mice and was accompanied by enhanced synaptic plasticity [148]. Similarly, 4 weeks of moderate-intensity exercise may alleviate the pathological symptoms of AD models by correcting the disordered expression of synaptic plasticity-related proteins [149]. Synaptic plasticity is regulated by the autophagy signal pathway, and exercise, as an effective strategy of autophagy activation, can not only regulate autophagy in peripheral tissue cells, but also regulate the level of autophagy in nerve cells, and regulate synaptic plasticity. This suggests that explaining the anti-AD effect of exercise can start from the regulation of synaptic plasticity by autophagy. In fact, autophagy can regulate synaptic plasticity, and exercise can improve synaptic structure–function plasticity by activating the mTOR signal pathway, thereby improving cognitive function [150]. However, it has not yet been elucidated whether exercise-regulated mTOR signaling and the improvement of synaptic function are related to autophagy. In addition, autophagy also plays an important role in brain-derived neurotrophic factor (BDNF)-mediated synaptic plasticity. BDNF signaling can inhibit autophagy in the forebrain of adult mice through tyrosine kinase receptor B (TrkB) and PI3K/Akt signal pathways to protect synaptic plasticity. The knockout of BDNF can overactivate autophagy to cause synapse defects and impair LTP function. In autophagy-deficient mice, the accumulation of postsynaptic density-95 (PSD-95), protein kinase C-interacting protein 1 (PICK1), and postsynaptic skeleton protein Shank3 was also detected, indicating that autophagy, as an important regulatory element of BDNF signal, plays a key role in BDNF-induced synaptic plasticity [150]. Since BDNF is an important regulator of exercise-induced synaptic plasticity, exercise or pharmacological simulated exercise can alleviate the pathological features of AD by inducing the expression of BDNF [151]. Therefore, the increase in synaptic plasticity induced by exercise may represent a physiological paradigm; that is, increasing the level of BDNF and inhibiting excessive activation of autophagy in adult hippocampal tissue may enhance a healthy adaptation to exercise stress (Fig. 7.3).

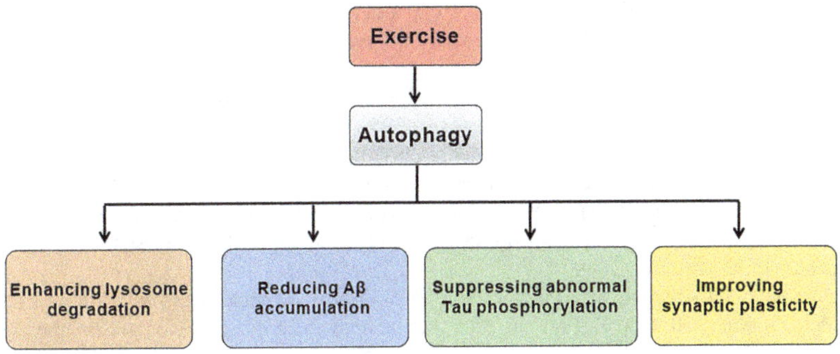

Fig. 7.3 Exercise improves AD through autophagy

References

1. Querfurth HW, LaFerla FM (2010) Alzheimer's disease. N Engl J Med 62(4):329–344
2. Pini L, Pievani M, Bocchetta M et al (2016) Brain atrophy in Alzheimer's disease and aging. Ageing Res Rev 30:25–48
3. Cohen TJ, Guo JL, Hurtado DE et al (2011) The acetylation of tau inhibits its function and promotes pathological tau aggregation. Nat Commun 2:252
4. Lane CA, Hardy J, Schott JM (2018) Alzheimer's disease. Eur J Neurol 25(1):59–70
5. Hu YT, Chen XL, Huang SH et al (2019) Early growth response-1 regulates acetylcholinesterase and its relation with the course of Alzheimer's disease. Brain Pathol 29(4):502–512
6. Espay AJ, Brundin P, Lang AE (2017) Precision medicine for disease modification in Parkinson disease. Nat Rev Neurol 13(2):119–126
7. Mount C, Downton C (2006) Alzheimer disease: progress or profit? Nat Med 12(7):780–784
8. Zhang ZX, Zahner GE, Román GC et al (2005) Dementia subtypes in China: prevalence in Beijing, Xian, Shanghai, and Chengdu. Arch Neurol 62(3):447–453
9. Wimo A, Winblad B, Jönsson L (2007) An estimate of the total worldwide societal costs of dementia in 2005. Alzheimers Dement 3(2):81–91
10. Prince M, Wimo A, Guerchet M et al (2015) The global impact of dementia: an analysis of prevalence, incidence, cost and trends. World Alzheimer Report 2015
11. Alzheimer's Association (2016) 2016 Alzheimer's disease facts and figures. Alzheimers Dement 12(4):459–509
12. Chan KY, Wang W, Wu JJ et al (2013) Epidemiology of Alzheimer's disease and other forms of dementia in China, 1990-2010: a systematic review and analysis. Lancet 381 (9882):2016–2023
13. Pistollato F, Ohayon EL, Lam A et al (2016) Alzheimer disease research in the 21st century: past and current failures, new perspectives and funding priorities. Oncotarget 7 (26):38999–39016
14. Gibbs RB (1997) Effects of estrogen on basal forebrain cholinergic neurons vary as a function of dose and duration of treatment. Brain Res 757(1):10–16
15. Fan N, He H, Xiao J (2002) Cognitive impairment in Alzheimer disease. Chin Ment Health J 16(9):590–591. 594
16. Kozlov S, Afonin A, Evsyukov I et al (2017) Alzheimer's disease: as it was in the beginning. Rev Neurosci 28(8):825–843
17. Bischof GN, Endepols H, van Eimeren T et al (2017) Tau-imaging in neurodegeneration. Methods 130:114–123

18. Jouanne M, Rault S, Voisin-Chiret AS (2017) Tau protein aggregation in Alzheimer's disease: an attractive target for the development of novel therapeutic agents. Eur J Med Chem 139:153–167
19. Wang X, Zhou X, Li G et al (2017) Modifications and trafficking of APP in the pathogenesis of Alzheimer's disease. Front Mol Neurosci 10:294
20. Cruchaga C, Haller G, Chakraverty S et al (2012) Rare variants in APP, PSEN1 and PSEN2 increase risk for AD in late-onset Alzheimer's disease families. PLoS One 7(2):e31039
21. Oh H, Madison C, Baker S et al (2016) Dynamic relationships between age, amyloid-β deposition, and glucose metabolism link to the regional vulnerability to Alzheimer's disease. Brain 139(Pt 8):2275–2289
22. Baranello RJ, Bharani KL, Padmaraju V et al (2015) Amyloid-beta protein clearance and degradation (ABCD) pathways and their role in Alzheimer's disease. Curr Alzheimer Res 12 (1):32–46
23. Selkoe DJ, Hardy J (2016) The amyloid hypothesis of Alzheimer's disease at 25 years. EMBO Mol Med 8(6):595–608
24. Benilova I, Karran E, De Strooper B (2012) The toxic Aβ oligomer and Alzheimer's disease: an emperor in need of clothes. Nat Neurosci 15(3):349–357
25. Liu YH, Giunta B, Zhou HD et al (2012) Immunotherapy for Alzheimer disease: the challenge of adverse effects. Nat Rev Neurol 8(8):465–469
26. Kelleher RJ, Soiza RL (2013) Evidence of endothelial dysfunction in the development of Alzheimer's disease: is Alzheimer's a vascular disorder? Am J Cardiovasc Dis 3(4):197–226
27. Qian W, Shi J, Yin X et al (2010) PP2A regulates tau phosphorylation directly and also indirectly via activating GSK-3beta. J Alzheimers Dis 19(4):1221–1229
28. Zhang Y, Ma RH, Li XC et al (2014) Silencing formula: see text. Rescues tau pathologies and memory deficits through rescuing PP2A and inhibiting GSK-3β signaling in human tau transgenic mice. Front Aging Neurosci 6:123
29. Zhou P, Chen Z, Zhao N et al (2011) Acetyl-L-carnitine attenuates homocysteine-induced Alzheimer-like histopathological and behavioral abnormalities. Rejuvenation Res 14 (6):669–679
30. Zhou L, McInnes J, Wierda K et al (2017) Tau association with synaptic vesicles causes presynaptic dysfunction. Nat Commun 8:15295
31. Gatz M, Mortimer JA, Fratiglioni L et al (2006) Potentially modifiable risk factors for dementia in identical twins. Alzheimers Dement 2(2):110–117
32. Hayden KM, Zandi PP, Khachaturian AS et al (2007) Does NSAID use modify cognitive trajectories in the elderly? The Cache County study. Neurology 69(3):275–282
33. Meister S, Zlatev I, Stab J et al (2013) Nanoparticulate flurbiprofen reduces amyloid-β42 generation in an in vitro blood-brain barrier model. Alzheimers Res Ther 5(6):51
34. Cai Z, Yan LJ, Ratka A (2013) Telomere shortening and Alzheimer's disease. Neuromolecular Med 15(1):25–48
35. Wang D, Zhang J, Jiang W et al (2017) The role of NLRP3-CASP1 in inflammasome-mediated neuroinflammation and autophagy dysfunction in manganese-induced, hippocampal-dependent impairment of learning and memory ability. Autophagy 13(5):914–927
36. Mastroeni D, Khdour OM, Delvaux E et al (2017) Nuclear but not mitochondrial-encoded oxidative phosphorylation genes are altered in aging, mild cognitive impairment, and Alzheimer's disease. Alzheimers Dement 13(5):510–519
37. Palikaras K, Tavernarakis N (2012) Mitophagy in neurodegeneration and aging. Front Genet 3:297
38. Lin MT, Beal MF (2006) Mitochondrial dysfunction and oxidative stress in neurodegenerative diseases. Nature 443(7113):787–795
39. Du F, Yu Q, Yan S et al (2017) PINK1 signalling rescues amyloid pathology and mitochondrial dysfunction in Alzheimer's disease. Brain 140(12):3233–3251

40. Silva DF, Esteves AR, Oliveira CR et al (2017) Mitochondrial metabolism power SIRT2-dependent deficient traffic causing Alzheimer's-disease related pathology. Mol Neurobiol 54 (6):4021–4040
41. Rosenberg PB, Nowrangi MA, Lyketsos CG (2015) Neuropsychiatric symptoms in Alzheimer's disease: what might be associated brain circuits? Mol Asp Med 43–44:25–37
42. Martorana A, Koch G (2014) Is dopamine involved in Alzheimer's disease? Front Aging Neurosci 6:252
43. Hurst RS, Hajós M, Raggenbass M et al (2005) A novel positive allosteric modulator of the alpha7 neuronal nicotinic acetylcholine receptor: in vitro and in vivo characterization. J Neurosci 25(17):4396–4405
44. Sadigh-Eteghad S, Majdi A, Talebi M et al (2015) Regulation of nicotinic acetylcholine receptors in Alzheimer's disease: a possible role of chaperones. Eur J Pharmacol 755:34–41
45. Yu W, Mechawar N, Krantic S et al (2012) Up-regulation of astrocytic α7 nicotinic receptors in Alzheimer's disease brain- possible relevant to amyloid pathology. Mol Neurodegener 7 (Suppl):7
46. De Ferrari GV, Canales MA, Shin I et al (2001) A structural motif of acetylcholinesterase that promotes amyloid beta-peptide fibril formation. Biochemistry 40(35):10447–10457
47. Ravanan P, Srikumar IF, Talwar P (2017) Autophagy: the spotlight for cellular stress responses. Life Sci 188:53–67
48. Mochida K, Oikawa Y, Kimura Y et al (2015) Receptor-mediated selective autophagy degrades the endoplasmic reticulum and the nucleus. Nature 522(7556):359–362
49. Zare-Shahabadi A, Masliah E, Johnson GV et al (2015) Autophagy in Alzheimer's disease. Rev Neurosci 26(4):385–395
50. Randhawa R, Sehgal M, Singh TR et al (2015) Unc-51 like kinase 1 (ULK1) in silico analysis for biomarker identification: a vital component of autophagy. Gene 562(1):40–49
51. Nah J, Yuan J, Jung YK (2015) Autophagy in neurodegenerative diseases: from mechanism to therapeutic approach. Mol Cells 38(5):381–389
52. Boland B, Kumar A, Lee S et al (2008) Autophagy induction and autophagosome clearance in neurons: relationship to autophagic pathology in Alzheimer's disease. J Neurosci 28 (27):6926–6937
53. Lee W, Kim SH (2019) Autophagy at synapses in neurodegenerative diseases. Arch Pharm Res 42(5):407–415
54. Li Q, Liu Y, Sun M (2017) Autophagy and Alzheimer's disease. Cell Mol Neurobiol 37 (3):377–388
55. Hara T, Nakamura K, Matsui M et al (2006) Suppression of basal autophagy in neural cells causes neurodegenerative disease in mice. Nature 441(7095):885–889
56. Komatsu M, Waguri S, Chiba T et al (2006) Loss of autophagy in the central nervous system causes neurodegeneration in mice. Nature 441(7095):880–884
57. Yu WH, Kumar A, Peterhoff C et al (2004) Autophagic vacuoles are enriched in amyloid precursor protein-secretase activities: implications for beta-amyloid peptide over-production and localization in Alzheimer's disease. Int J Biochem Cell Biol 36(12):2531–2540
58. Yu WH, Cuervo AM, Kumar A et al (2005) Macroautophagy—a novel Beta-amyloid peptide-generating pathway activated in Alzheimer's disease. J Cell Biol 171(1):87–98
59. Zheng L, Terman A, Hallbeck M et al (2011) Macroautophagy-generated increase of lysosomal amyloid β-protein mediates oxidant-induced apoptosis of cultured neuroblastoma cells. Autophagy 7(12):1528–1545
60. Luzio JP, Rous BA, Bright NA et al (2000) Lysosome-endosome fusion and lysosome biogenesis. J Cell Sci 113(Pt 9):1515–1524
61. Di Scala C, Chahinian H, Yahi N et al (2014) Interaction of Alzheimer's β-amyloid peptides with cholesterol: mechanistic insights into amyloid pore formation. Biochemistry 53 (28):4489–4502
62. Peters OM, Ghasemi M, Brown RH Jr (2015) Emerging mechanisms of molecular pathology in ALS. J Clin Invest 125(5):1767–1779

63. Tian Y, Bustos V, Flajolet M et al (2011) A small-molecule enhancer of autophagy decreases levels of Abeta and APP-CTF via Atg5-dependent autophagy pathway. FASEB J 25 (6):1934–1942
64. Jaeger PA, Pickford F, Sun CH et al (2010) Regulation of amyloid precursor protein processing by the Beclin 1 complex. PLoS One 5(6):e11102
65. Embury CM, Dyavarshetty B, Lu Y et al (2017) Cathepsin B improves ß-amyloidosis and learning and memory in models of Alzheimer's disease. J Neuroimmune Pharmacol 12 (2):340–352
66. Di Domenico F, Tramutola A, Perluigi M (2016) Cathepsin D as a therapeutic target in Alzheimer's disease. Expert Opin Ther Targets 20(12):1393–1395
67. Tammineni P, Ye X, Feng T et al (2017) Impaired retrograde transport of axonal autophagosomes contributes to autophagic stress in Alzheimer's disease neurons. eLife 6: e21776
68. Nixon RA, Wegiel J, Kumar A et al (2005) Extensive involvement of autophagy in Alzheimer disease: an immuno-electron microscopy study. J Neuropathol Exp Neurol 64(2):113–122
69. Nixon RA (2007) Autophagy, amyloidogenesis and Alzheimer disease. J Cell Sci 120 (Pt 23):4081–4091
70. Nilsson P, Saido TC (2014) Dual roles for autophagy: degradation and secretion of Alzheimer's disease Aβ peptide. BioEssays 36(6):570–578
71. Nilsson P, Loganathan K, Sekiguchi M et al (2013) Aβ secretion and plaque formation depend on autophagy. Cell Rep 5(1):61–69
72. Koistinaho M, Ort M, Cimadevilla JM et al (2001) Specific spatial learning deficits become severe with age in beta -amyloid precursor protein transgenic mice that harbor diffuse beta -amyloid deposits but do not form plaques. Proc Natl Acad Sci U S A 98(25):14675–14680
73. Tomiyama T, Matsuyama S, Iso H et al (2010) A mouse model of amyloid beta oligomers: their contribution to synaptic alteration, abnormal tau phosphorylation, glial activation, and neuronal loss in vivo. J Neurosci 30(14):4845–4856
74. Li K, Wei Q, Liu FF et al (2018) Synaptic dysfunction in Alzheimer's disease: Aβ, tau, and epigenetic alterations. Mol Neurobiol 55(4):3021–3032
75. Wang Y, Mandelkow E (2012) Degradation of tau protein by autophagy and proteasomal pathways. Biochem Soc Trans 40(4):644–652
76. Caccamo A, Magrì A, Medina DX et al (2013) mTOR regulates tau phosphorylation and degradation: implications for Alzheimer's disease and other tauopathies. Aging Cell 12 (3):370–380
77. Qiu L, Ng G, Tan EK et al (2016) Chronic cerebral hypoperfusion enhances Tau hyperphosphorylation and reduces autophagy in Alzheimer's disease mice. Sci Rep 6:23964
78. Hamano T, Gendron TF, Causevic E et al (2008) Autophagic-lysosomal perturbation enhances tau aggregation in transfectants with induced wild-type tau expression. Eur J Neurosci 27 (5):1119–1130
79. Liu Z, Li T, Li P et al (2015) The ambiguous relationship of oxidative stress, tau hyperphosphorylation, and autophagy dysfunction in Alzheimer's disease. Oxidative Med Cell Longev 2015:352723
80. Kim SI, Lee WK, Kang SS et al (2011) Suppression of autophagy and activation of glycogen synthase kinase 3beta facilitate the aggregate formation of tau. Korean J Physiol Pharmacol 15 (2):107–114
81. Jo C, Gundemir S, Pritchard S et al (2014) Nrf2 reduces levels of phosphorylated tau protein by inducing autophagy adaptor protein NDP52. Nat Commun 5:3496
82. Caccamo A, Ferreira E, Branca C et al (2017) p62 improves AD-like pathology by increasing autophagy. Mol Psychiatry 22(6):865–873
83. Gao Y, Tan L, Yu JT et al (2018) Tau in Alzheimer's disease: mechanisms and therapeutic strategies. Curr Alzheimer Res 15(3):283–300
84. Dall'Armi C, Hurtado-Lorenzo A, Tian H et al (2010) The phospholipase D1 pathway modulates macroautophagy. Nat Commun 1:142

85. Krüger U, Wang Y, Kumar S et al (2012) Autophagic degradation of tau in primary neurons and its enhancement by trehalose. Neurobiol Aging 33(10):2291–2305
86. Rodríguez-Martín T, Cuchillo-Ibáñez I, Noble W et al (2013) Tau phosphorylation affects its axonal transport and degradation. Neurobiol Aging 34(9):2146–2157
87. Zhao L, Xiao Y, Wang XL et al (2016) Original research: influence of okadaic acid on hyperphosphorylation of tau and nicotinic acetylcholine receptors in primary neurons. Exp Biol Med (Maywood) 241(16):1825–1833
88. Tian Q, Lin ZQ, Wang XC et al (2004) Injection of okadaic acid into the meynert nucleus basalis of rat brain induces decreased acetylcholine level and spatial memory deficit. Neuroscience 126(2):277–284
89. McGeer EG, McGeer PL (2010) Neuroinflammation in Alzheimer's disease and mild cognitive impairment: a field in its infancy. J Alzheimers Dis 19(1):355–361
90. Braidy N, Essa MM, Poljak A et al (2016) Consumption of pomegranates improves synaptic function in a transgenic mice model of Alzheimer's disease. Oncotarget 7(40):64589–64604
91. Ghavami S, Shojaei S, Yeganeh B et al (2014) Autophagy and apoptosis dysfunction in neurodegenerative disorders. Prog Neurobiol 112:24–49
92. Chen HC, Fong TH, Hsu PW et al (2013) Multifaceted effects of rapamycin on functional recovery after spinal cord injury in rats through autophagy promotion, anti-inflammation, and neuroprotection. J Surg Res 179(1):e203–e210
93. Marobbio CM, Pisano I, Porcelli V et al (2012) Rapamycin reduces oxidative stress in frataxin-deficient yeast cells. Mitochondrion 12(1):156–161
94. Espinosa-García C, Aguilar-Hernández A, Cervantes M et al (2014) Effects of progesterone on neurite growth inhibitors in the hippocampus following global cerebral ischemia. Brain Res 1545:23–34
95. Hong Y, Liu Y, Zhang G et al (2018) Progesterone suppresses Aβ(42)-induced neuroinflammation by enhancing autophagy in astrocytes. Int Immunopharmacol 54:336–343
96. He J, Liao T, Zhong GX et al (2017) Alzheimer's disease-like early-phase brain pathogenesis: self-curing amelioration of neurodegeneration from pro-inflammatory 'Wounding' to anti-inflammatory 'Healing'. Curr Alzheimer Res 14(10):1123–1135
97. Rahman MA, Rhim H (2017) Therapeutic implication of autophagy in neurodegenerative diseases. BMB Rep 50(7):345–354
98. Nixon RA, Yang DS, Lee JH (2008) Neurodegenerative lysosomal disorders: a continuum from development to late age. Autophagy 4(5):590–599
99. Reddy PH, Yin X, Manczak M et al (2018) Mutant APP and amyloid beta-induced defective autophagy, mitophagy, mitochondrial structural and functional changes and synaptic damage in hippocampal neurons from Alzheimer's disease. Hum Mol Genet 27(14):2502–2516
100. Yuan H, Jiang C, Zhao J et al (2018) Euxanthone attenuates Aβ(1-42)-induced oxidative stress and apoptosis by triggering autophagy. J Mol Neurosci 66(4):512–523
101. Friedland RP, Fritsch T, Smyth KA et al (2001) Patients with Alzheimer's disease have reduced activities in midlife compared with healthy control-group members. Proc Natl Acad Sci U S A 98(6):3440–3445
102. He C, Sumpter R Jr, Levine B (2012) Exercise induces autophagy in peripheral tissues and in the brain. Autophagy 8(10):1548–1551
103. García-Mesa Y, López-Ramos JC, Giménez-Llort L et al (2011) Physical exercise protects against Alzheimer's disease in 3xTg-AD mice. J Alzheimers Dis 24(3):421–454
104. Ke HC, Huang HJ, Liang KC et al (2011) Selective improvement of cognitive function in adult and aged APP/PS1 transgenic mice by continuous non-shock treadmill exercise. Brain Res 1403:1–11
105. Parachikova A, Nichol KE, Cotman CW (2008) Short-term exercise in aged Tg2576 mice alters neuroinflammation and improves cognition. Neurobiol Dis 30(1):121–129
106. Zhang L, Niu W, He Z et al (2014) Autophagy suppression by exercise pretreatment and p38 inhibition is neuroprotective in cerebral ischemia. Brain Res 1587:127–132

107. Rocchi A, He C (2017) Activating autophagy by aerobic exercise in mice. J Vis Exp 120:55099
108. Marques-Aleixo I, Santos-Alves E, Balça MM et al (2015) Physical exercise improves brain cortex and cerebellum mitochondrial bioenergetics and alters apoptotic, dynamic and auto (mito)phagy markers. Neuroscience 301:480–495
109. Orr ME, Oddo S (2013) Autophagic/lysosomal dysfunction in Alzheimer's disease. Alzheimers Res Ther 5(5):53
110. Um HS, Kang EB, Leem YH et al (2008) Exercise training acts as a therapeutic strategy for reduction of the pathogenic phenotypes for Alzheimer's disease in an NSE/APPsw-transgenic model. Int J Mol Med 22(4):529–539
111. Wang Q, Xu Z, Tang J et al (2013) Voluntary exercise counteracts Aβ25-35-induced memory impairment in mice. Behav Brain Res 256:618–625
112. Adlard PA, Perreau VM, Pop V et al (2005) Voluntary exercise decreases amyloid load in a transgenic model of Alzheimer's disease. J Neurosci 25(17):4217–4221
113. Marques-Aleixo I, Santos-Alves E, Balça MM et al (2016) Physical exercise mitigates doxorubicin-induced brain cortex and cerebellum mitochondrial alterations and cellular quality control signaling. Mitochondrion 26:43–57
114. Zhang L, Hu X, Luo J et al (2013) Physical exercise improves functional recovery through mitigation of autophagy, attenuation of apoptosis and enhancement of neurogenesis after MCAO in rats. BMC Neurosci 14:46
115. Ntsapi C, Lumkwana D, Swart C et al (2018) New insights into autophagy dysfunction related to amyloid beta toxicity and neuropathology in Alzheimer's disease. Int Rev Cell Mol Biol 336:321–361
116. Bayod S, Del Valle J, Pelegri C et al (2014) Macroautophagic process was differentially modulated by long-term moderate exercise in rat brain and peripheral tissues. J Physiol Pharmacol 65(2):229–239
117. Herring A, Münster Y, Metzdorf J et al (2016) Late running is not too late against Alzheimer's pathology. Neurobiol Dis 94:44–54
118. Fang ZH, Lee CH, Seo MK et al (2013) Effect of treadmill exercise on the BDNF-mediated pathway in the hippocampus of stressed rats. Neurosci Res 76(4):187–194
119. Chae CH, Kim HT (2009) Forced, moderate-intensity treadmill exercise suppresses apoptosis by increasing the level of NGF and stimulating phosphatidylinositol 3-kinase signaling in the hippocampus of induced aging rats. Neurochem Int 55(4):208–213
120. Kang EB, Cho JY (2015) Effect of treadmill exercise on PI3K/AKT/mTOR, autophagy, and Tau hyperphosphorylation in the cerebral cortex of NSE/htau23 transgenic mice. J Exerc Nutr Biochem 19(3):199–209
121. Huang J, Wang X, Zhu Y et al (2019) Exercise activates lysosomal function in the brain through AMPK-SIRT1-TFEB pathway. CNS Neurosci Ther 25(6):796–807
122. Bayod S, Del Valle J, Canudas AM et al (2011) Long-term treadmill exercise induces neuroprotective molecular changes in rat brain. J Appl Physiol (1985) 111(5):1380–1390
123. Tian W, Li W, Chen Y et al (2015) Phosphorylation of ULK1 by AMPK regulates translocation of ULK1 to mitochondria and mitophagy. FEBS Lett 589(15):1847–1854
124. Azimi M, Gharakhanlou R, Naghdi N et al (2018) Moderate treadmill exercise ameliorates amyloid-β-induced learning and memory impairment, possibly via increasing AMPK activity and up-regulation of the PGC-1α/FNDC5/BDNF pathway. Peptides 102:78–88
125. Zhao N, Zhang X, Xia J et al (2019) Effects of 12 weeks aerobic running on autophagy activity of hippocampal cells in APP/PS1 mice. China Sport Sci 39(12):43–53
126. Mindell JA (2012) Lysosomal acidification mechanisms. Annu Rev Physiol 74:69–86
127. Ling D, Magallanes M, Salvaterra PM (2014) Accumulation of amyloid-like Aβ1-42 in AEL (autophagy-endosomal-lysosomal) vesicles: potential implications for plaque biogenesis. ASN Neuro 6(2):e00139

128. Luo L, Dai JR, Guo SS et al (2017) Lysosomal proteolysis is associated with exercise-induced improvement of mitochondrial quality control in aged hippocampus. J Gerontol A Biol Sci Med Sci 72(10):1342–1351
129. Zhao N, Zhang X, Song C et al (2018) The effects of treadmill exercise on autophagy in hippocampus of APP/PS1 transgenic mice. Neuroreport 29(10):819–825
130. Yuede CM, Zimmerman SD, Dong H et al (2009) Effects of voluntary and forced exercise on plaque deposition, hippocampal volume, and behavior in the Tg2576 mouse model of Alzheimer's disease. Neurobiol Dis 35(3):426–432
131. Lin TW, Shih YH, Chen SJ et al (2015) Running exercise delays neurodegeneration in amygdala and hippocampus of Alzheimer's disease (APP/PS1) transgenic mice. Neurobiol Learn Mem 118:189–197
132. Moore KM, Girens RE, Larson SK et al (2016) A spectrum of exercise training reduces soluble Aβ in a dose-dependent manner in a mouse model of Alzheimer's disease. Neurobiol Dis 85:218–224
133. Sanchez-Varo R, Trujillo-Estrada L, Sanchez-Mejias E et al (2012) Abnormal accumulation of autophagic vesicles correlates with axonal and synaptic pathology in young Alzheimer's mice hippocampus. Acta Neuropathol 123(1):53–70
134. Lee JH, Yu WH, Kumar A et al (2010) Lysosomal proteolysis and autophagy require presenilin 1 and are disrupted by Alzheimer-related PS1 mutations. Cell 141(7):1146–1158
135. Caccamo A, Majumder S, Richardson A et al (2010) Molecular interplay between mammalian target of rapamycin (mTOR), amyloid-beta, and Tau: effects on cognitive impairments. J Biol Chem 285(17):13107–13120
136. Reddy PH (2011) Abnormal tau, mitochondrial dysfunction, impaired axonal transport of mitochondria, and synaptic deprivation in Alzheimer's disease. Brain Res 1415:136–148
137. Um HS, Kang EB, Koo JH et al (2011) Treadmill exercise represses neuronal cell death in an aged transgenic mouse model of Alzheimer's disease. Neurosci Res 69(2):161–173
138. Leem YH, Lim HJ, Shim SB et al (2009) Repression of tau hyperphosphorylation by chronic endurance exercise in aged transgenic mouse model of tauopathies. J Neurosci Res 87 (11):2561–2570
139. Congdon EE, Wu JW, Myeku N et al (2012) Methylthioninium chloride (methylene blue) induces autophagy and attenuates tauopathy in vitro and in vivo. Autophagy 8(4):609–622
140. Inoue K, Rispoli J, Kaphzan H et al (2012) Macroautophagy deficiency mediates age-dependent neurodegeneration through a phospho-tau pathway. Mol Neurodegener 7:48
141. Belarbi K, Burnouf S, Fernandez-Gomez FJ et al (2011) Beneficial effects of exercise in a transgenic mouse model of Alzheimer's disease-like Tau pathology. Neurobiol Dis 43 (2):486–494
142. Caccamo A, Maldonado MA, Majumder S et al (2011) Naturally secreted amyloid-beta increases mammalian target of rapamycin (mTOR) activity via a PRAS40-mediated mechanism. J Biol Chem 286(11):8924–8932
143. Ohia-Nwoko O, Montazari S, Lau YS et al (2014) Long-term treadmill exercise attenuates tau pathology in P301S tau transgenic mice. Mol Neurodegener 9:54
144. Lin WL, Lewis J, Yen SH et al (2003) Ultrastructural neuronal pathology in transgenic mice expressing mutant (P301L) human tau. J Neurocytol 32(9):1091–1105
145. Frere S, Slutsky I (2018) Alzheimer's disease: from firing instability to homeostasis network collapse. Neuron 97(1):32–58
146. Kim TW, Sung YH (2017) Regular exercise promotes memory function and enhances hippocampal neuroplasticity in experimental autoimmune encephalomyelitis mice. Neuroscience 346:173–181
147. Dao AT, Zagaar MA, Alkadhi KA (2015) Moderate treadmill exercise protects synaptic plasticity of the dentate gyrus and related signaling cascade in a rat model of Alzheimer's disease. Mol Neurobiol 52(3):1067–1076

148. Liu HL, Zhao G, Cai K et al (2011) Treadmill exercise prevents decline in spatial learning and memory in APP/PS1 transgenic mice through improvement of hippocampal long-term potentiation. Behav Brain Res 218(2):308–314
149. Dao AT, Zagaar MA, Levine AT et al (2016) Comparison of the effect of exercise on late-phase LTP of the dentate gyrus and CA1 of Alzheimer's disease model. Mol Neurobiol 53 (10):6859–6868
150. Chen K, Zheng Y, Wei JA et al (2019) Exercise training improves motor skill learning via selective activation of mTOR. Sci Adv 5(7):eaaw1888
151. Choi SH, Bylykbashi E, Chatila ZK et al (2018) Combined adult neurogenesis and BDNF mimic exercise effects on cognition in an Alzheimer's mouse model. Science 361(6406): eaan8821

Chapter 8
Exercise-Induced Autophagy and Parkinson's Disease

Xianjuan Kou, Shuangshuang Wu, Michael Kirberger, and Ning Chen

1 Overview of Parkinson's Disease (PD)

PD is the second most common neurodegenerative disease after Alzheimer's disease (AD). The occurrence of PD is more common in the elderly, and the average age of onset is 60 years. The major pathological change in PD is the degeneration and death of dopamine (DA) neurons in the substantia nigra of the midbrain, which causes a significant decrease in DA content in the striatum. The exact cause of this pathological change is still unclear. Genetic factors, environmental factors, aging, oxidative stress, and other relevant factors may be involved in the degeneration and death of dopaminergic neurons in PD. In 1817, British doctor James Parkinson first provided a detailed description of this disease. Its clinical manifestations mainly include resting tremor, bradykinesia, muscle rigidity, and postural and gait disorders. Patients may also experience symptoms of depression, constipation, and sleep disorders. Diagnosis of PD primarily depends on the history, clinical symptoms, and signs of this disease. There are usually no abnormal changes identified during general auxiliary examinations. Levodopa, a natural chemical converted to dopamine in the brain, combined with carbidopa, a compound that prevents premature conversion of levodopa, is the current primary drug treatment strategy for PD, while surgical treatment is an effective supplemental treatment. Rehabilitation, psychotherapy, and nursing care can also alleviate the symptoms of PD to a certain extent. These currently available treatments can mitigate symptoms and significantly improve the quality of life of PD patients, but cannot cure or halt the progression

X. Kou · S. Wu · N. Chen (✉)
Tianjiu Research and Development Center for Exercise Nutrition and Foods, Hubei Key Laboratory of Exercise Training and Monitoring, College of Health Science, Wuhan Sports University, Wuhan, China

M. Kirberger
School of Science and Technology, Georgia Gwinnett College, Lawrenceville, GA, USA

N. Chen (ed.), *Exercise, Autophagy and Chronic Diseases*,
https://doi.org/10.1007/978-981-16-4525-9_8

of this disease. The life expectancy of PD patients is shorter than general population, especially those who have been diagnosed before the age of 70 [1].

1.1 The Epidemiology of PD

PD is a typical chronic disease of the elderly. The increasing proportion of elderly people in our global population has led to an obvious increase in the prevalence of PD, with a currently estimated prevalence of approximately 0.1–0.2% in the general population [9]. However, this prevalence increases by 1–2% for those aged 65 or older and by 3–5% for those aged 85 or older [10]. Specifically, the prevalence rates of PD are 0.25% at age 60; 0.5% at age 65; 1% at age 70; 1.5% at age 75; 2.5% at age 80; and 3.5–4.0% at age 85 [11]. There are also differences in PD risk between genders. Related analysis shows that the risk of PD in men is much higher than that for women [12].

1.2 The Pathogenesis of PD

The prominent pathological changes in PD include the degeneration and death of DA neurons in the substantia nigra; a significant decrease in DA content in the striatum; and the appearance of eosinophilic inclusion bodies (i.e., Lewy bodies) in the cytoplasm of the substantia nigra remnant neurons. When clinical symptoms appear, the death of dopaminergic neurons in the substantia nigra is at least 50%, and the DA content in the striatum is reduced by more than 80%.

In addition to the dopaminergic system, the nondopaminergic system of PD patients is also significantly impaired, such as the cholinergic neurons in the basal nucleus of Meynert, noradrenergic neurons in the locus coeruleus, serotonergic neurons in the raphe nucleus of the brainstem, and nerves in the cerebral cortex, brainstem, spinal cord, and peripheral autonomic nervous system. The significant decrease in DA content in the striatum is closely related to the appearance of PD motor symptoms. Significant reduction in DA contents in both midbrain–limbic and midbrain–cortex systems is also closely correlated with the loss of intelligence and affective disorders in patients with PD [4].

1.3 Pathological Mechanisms of PD

Although numerous studies on PD have been conducted, the precise pathological mechanisms of PD are not yet fully understood. Factors that may be involved in the degeneration and death processes of PD dopaminergic neurons include neuropathology, genetic factors, environmental factors, immunity, inflammation, apoptosis, and oxidative stress.

1.3.1 Neuropathological Pathogenesis

The major pathological characteristics of PD are the destruction of DA neurons in the substantia nigra striatum and a reduction in protein-like substances in the cytoplasm of neurons, as well as the appearance of characteristic Lewy bodies. Neuropathological studies of PD have revealed part of the pathogenesis of PD. First, aging is an important and unavoidable risk factor for PD, and the degeneration of DA neurons in PD is its characteristic anatomical feature, which is significantly different from normal aging brain tissue. The loss of DA neurons in PD is concentrated in the ventrolateral and tail of the midbrain substantia nigra pars compacta (SNpc); however, with the extension of age, the affected regions also involve the middle and back parts of SNpc. Second, the loss of neurons in the terminal striatum of the projection fibers of SNpc neurons is more significant than that of SNpc, suggesting that the death of PD neurons is a process similar to axonal degeneration (i.e., dying back) beginning at the most distal regions. The degeneration of somatic dopamine neurons can gradually progress into the degeneration process of SNpc dopamine neurons. The degeneration of striatal nerve endings has been detected in monkey brains treated with 1-methyl-4-phenyl-1,2,3,6-tetrahydropyridine (MPTP) before the degeneration of SNpc neuron cell bodies. At the same time, experimental studies have found that in the MPTP-treated mouse model, minimizing the damage of the striatal nerve endings can reduce the loss of SNpc dopamine neurons [13].

A Lewy body is an intracellular inclusion with eosinophilic protein deposition. It has a dense central core surrounded by a filamentous halo, with a diameter of approximately 20μm. During early stage of PD, not all patients have Lewy bodies, but it is now known that Lewy bodies can be observed in the colored neurons in the substantia nigra of almost every PD patient [14]. The compositions of Lewy bodies in cytoplasm include α-synuclein, parkin, ubiquitin, and neurofilaments.

1.3.2 Genetic Factors

The role of genetic factors in PD cases remains unclear. Approximately 5–10% of patients have a family history, and PD cases in these families are genetically correlated. At least 10 loci are related to PD, including autosomal dominant genetic correlation. The mutations of some genes, including α-synaptic nucleus gene (PARK1), PARK3, PARK4, UchL1 (PARK5), and PARK8, can cause PD clinical symptoms similar to typical late and sporadic PD. The mutations of autosomal recessive genetic-related genes such as parkin (PARK2), PINK1 (PARK6), and DJ1 (PARK7) can also cause early-onset PD.

The α-synaptic gene was a breakthrough in the pathogenesis of PD in 1997. It was first identified in a single family in Italy, and three unrelated families in Greece, all exhibiting autosomal dominant familial PD associated with 4q21-q23 mutations in the α-synuclein gene [15]. Sequence analyses have also determined that the guanine at position 209 of α-synuclein was changed to adenine (G209A), thus resulting in substitution in the amino acid sequence of alanine by threonine at position 53 (A53T). A second mutation, G88C, thus resulting in substitution of alanine by proline at position 30 (A30P), was also observed. These PD-related variants (A53T and A30P) are located at the amino terminus of α-synuclein and can affect lipid binding. The A30P mutation alters the helical structure of the amino terminal, thereby preventing binding between α-synuclein and lipids. Major functions of the protein α-synuclein have not yet been fully elucidated, but studies have suggested that they may be related to the plasticity of synapses, the regulation of DA vesicle release, and the transport of fatty acids in the cytoplasm of neurons [16]. The protein is mainly located in the presynaptic nerve endings and nuclei, and is expressed in many parts of the brain tissue. It is also one of the major components of Lewy bodies. In the Lewy body, α-synuclein adopts a β-sheet structure and can bind to other proteins such as synphilin-1, parkin, and anti-apoptotic chaperone 14-3-3 [17].

The parkin gene has been cloned and identified as the disease-causing gene of autosomal recessive juvenile Parkinsonism (AR-JP) [18]. In 2000, parkin was first reported as an ubiquitin protein ligase. As an important member of the ubiquitin–proteasome system (UPS), parkin plays a critical role in the quality control of intracellular proteins. This important finding not only has clarified the pathogenesis of familial PD, but also renewed efforts to clarify the neurodegenerative mechanisms of sporadic PD. Some scholars have speculated that the mutation of parkin gene in PD patients could cause parkin protein to lose its activity as an ubiquitin protein ligase, thereby resulting in the abnormal accumulation of harmful proteins in cells, and ultimately leading to the death of dopaminergic neurons. Thus, the dysfunction of the UPS may be involved in the pathogenesis of familial PD and sporadic PD [19].

Most ubiquitin carboxyl-terminal hydrolase L1 (UCH-L1) genes with PD-associated mutations play important regulatory roles in UPS. Therefore, some relevant genes encoding UPS have been studied in pairs of identical twins and a mutation in the UCH-L1 gene (exon 4I1e93Met) has been detected. Ubiquitin is also an important component of the Lewy body. The mutation of UCH-Ll can reduce the catalysis of ubiquitin-hydrolyzing protease via an abnormal proteolytic pathway, thus leading to the obstacles in the ubiquitin cycle pathway, protein aggregation, and neuronal degeneration [20].

1.3.3 Environmental Factors

At present, the most toxic environmental substances capable of triggering PD are a class of substances derived from MPTP. The active metabolite of MPTP is 1-methyl-4-phenylpyridine ion (MPP^+), which can be taken up selectively by dopamine transporter (DAT) into neurons. In addition to inhibiting the mitochondrial respiratory chain complex, MPP^+ also has a high affinity for vesicular monoamine transporter 2, which can transport MPP^+ to the vesicles of dopaminergic neurons. Therefore, in addition to depleting intracellular ATP reserves, MPP^+ can also cause the redistribution of dopamine in the cytoplasm and induce oxidative stress in a dopamine-dependent manner. Pesticides, especially rotenone and paraquat, can selectively trigger the degeneration of dopaminergic neurons in the substantia nigra of rodents [21]. Rotenone and paraquat have structures similar to MPP^+, including the bipyridine structure, so they can inhibit the function of electron transport chain complex I and increase dopamine levels in the cytoplasm by changing the storage of monoamine vesicles. In addition, some substances similar to MPTP in nature can also cause PD, such as tetrahydroisoquinoline and β-carboline derivatives. These substances can usually be detected in plants and foods, and can easily enter the central nervous system through the blood–brain barrier, thereby inhibiting the activity of complex enzyme I on the mitochondrial respiratory chain to varying degrees, thus leading to mitochondrial dysfunction and neuronal death. Currently, many people may experience PD symptoms after being exposed to such substances. However, the existence of these substances in the substantia nigra of PD patients has not been confirmed. A study in Germany on the brain tissues of PD patients has reported the detection of an organochlorine insecticide, dieldrin, which is believed to cause PD. Similarly, previous studies have shown that dithiocarbamate and diethyldithiocarbamate are strongly correlated with the pathogenesis of PD, and both of these substances can increase the toxicity of MPTP. The results of these studies indicate that environmental factors play an important role in the pathogenesis of PD.

1.3.4 Other Factors

Immune Factors

Recent studies have also suggested that immunological mechanisms, especially humoral immunity, may play a critical role in the progression of PD during substantia nigra cell injury. In one study, the production of anti-α-synuclein autoantibody is found to be involved in the pathogenesis of familial PD [22]. Additionally, many symptoms of PD can be induced by the injection of proteolytic enzyme inhibitor epoxomicin or synthetic proteasome inhibitor into the brain tissues of rats, including progressive tremors, muscle stiffness, and motor retardation. Autopsies of rats exhibiting PD symptoms have revealed a reduction in DA levels in the striatum, an obvious inflammatory response, the selective loss of dopaminergic neurons in the substantia nigra compact area, and the presence of Lewy bodies in the remaining neurons. These results have also demonstrated that dopaminergic neurons in the substantia nigra are damaged in PD model. In addition, changes in interferon-gamma, tumor necrosis factor-alpha, interleukin 6, epithelial growth factor, metastatic growth factor-alpha, and beta 2-microglobulin may also correlate with the pathogenesis of PD.

The immune/inflammatory response is an important pathological characteristic of PD, which is closely correlated with the degeneration of dopaminergic neurons. It is unclear whether these immune or inflammatory responses serve only to eliminate degraded products and necrotic cells, or whether they can aggravate damage to the damaged nerve cells, and induce new damage to nerve cells. Since previous studies have shown that certain anti-inflammatory drugs can reduce the degeneration and necrosis of dopaminergic neurons caused by neurotoxicity [23], the identification of drugs targeting inflammation or immune response is expected to provide a novel effective strategy for the prevention and treatments of PD.

Oxidative Stress Factors

Immunohistochemical staining has confirmed that 8-hydroxyguanosine, as a biomarker of DNA and RNA oxidation, exhibits a significant increase in the substantia nigra neurons of PD models. In addition, the concentration of protein carbonyls in the substantia nigra, as biomarkers of protein oxidation, is also found to be much higher than in other brain tissues [24]. In contrast, glutathione (GSH), the substrate of GSH peroxidase in PD, is significantly reduced in the substantia nigra, and the reduction in activity in both GSH peroxidase and catalase indicates a functional decline in the antioxidant system. Oxidative stress can also affect mitochondrial DNA mutations, which can damage mitochondrial complexes and reduce ATP production. Similarly, oxidative stress can also increase cell excitatory amino acids and accelerate cell apoptosis [25].

1.4 Current Status of Treatments

Although the effective treatments of PD are being explored, the vast majority of PD patients still choose drug therapy as the major treatment strategy, and levodopa preparations are currently the most effective drugs. Surgical treatment is an effective supplement to drug therapy, and surgical operations may utilize gamma knife, X knife, or transplantation of brain tissue and genetically modified cells. Rehabilitation, psychotherapy, and supportive care can also alleviate PD symptoms to a certain extent [26]. However, current medications and surgical treatments can only relieve symptoms, but cannot prevent, terminate, or reverse the development of PD.

In recent years, multiple studies have found that autophagy has a significant effect on body stability, tissue remodeling, and disease prevention. Autophagy deficiency is closely correlated with neurodegeneration, aging, and other diseases. Research suggests that autophagy can be regulated by exercise, so exercise is receiving increasing attention as a potential nondrug intervention or treatment for aging-related neurodegenerative diseases, including PD.

2 Autophagy Is Involved in PD

Autophagy is a homeostatic mechanism where lysosomes degrade and recycle components of the cytoplasm. These components, including proteins and organelles, are enveloped in double membrane-bound vesicles to form autophagosomes, which in turn fuse with lysosomes to form autolysosomes that degrade the corresponding proteins and organelles (Fig. 8.1), thereby achieving intracellular homeostasis and organelle renewal. Autophagy, as a common, conserved cellular process observed in mammals, relates to energy renewal and substance metabolism. The autophagy–lysosomal system is an important degradation pathway that eliminates misfolded or aggregated proteins and damaged organelles in neurons, and this plays a very important role in the degeneration of neurons in PD. Nerve cells must have basic autophagy activity to ensure their normal cell function and maintain the dynamic stability of the central nervous system. The defect or deficiency of autophagy will lead to the accumulation of denatured proteins and damaged organelles, which is also an important reason for the occurrence of neurodegenerative diseases such as PD [27].

2.1 Mitophagy and PD

2.1.1 Parkinson's Gene-Encoded Protein Is Involved in Mitochondrial Autophagy

PTEN-Induced Kinase 1 (PINK1) and PARKIN Proteins Regulate Mitochondrial Autophagy

PINK1 and PARKIN genes are responsible for the expression of PINK1 and PARKIN proteins, respectively, and their cooperation can regulate intracellular mitochondrial quality control and energy production. PINK1 is abundantly expressed in the substantia nigra, exists in mitochondria (mainly located in the inner membrane of mitochondria, with a small amount located in the inner and outer membrane spaces or on the surface) [28], and participates in mitochondrial metabolism, oxidative stress, oxidative phosphorylation, and calcium stability, thereby controlling mitochondrial motility and proteasome degradation of abnormal proteins [29]. PARKIN mainly executes selective aggregation. In the outer membrane of old mitochondria, ubiquitin is linked to proteins to be degraded and participates in the regulation of autophagosome phagocytosis and the degradation of mitochondria. PARKIN can catalyze lysine 48-linked polyubiquitination, monoubiquitination, and lysine 63-linked polyubiquitination, and other regulatory pathways to control cellular metabolism, such as signal transduction, transcriptional regulation, and protein and membrane transport. PINK1 protein is the upstream

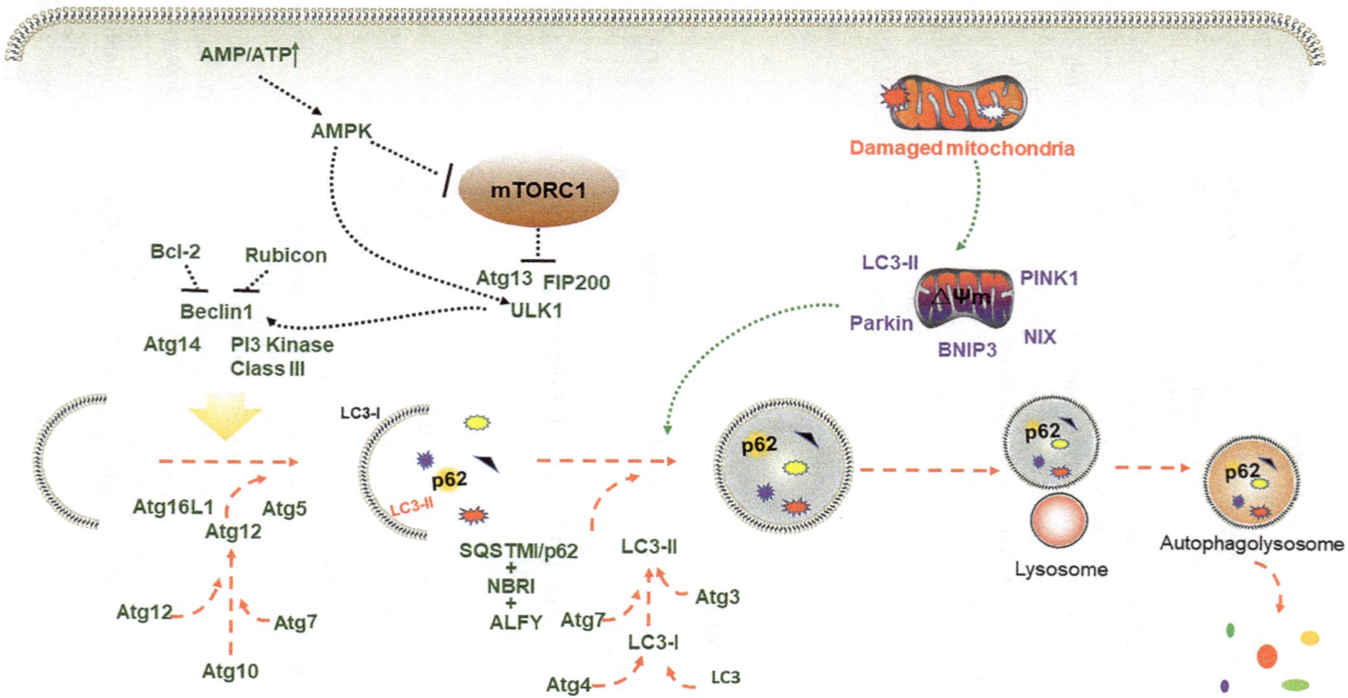

Fig. 8.1 Autophagy formation process. Autophagy mainly includes three stages: autophagy induction stage, autophagy formation stage, and mature degradation stage, among which mTOR and Beclin1 are the convergence points of various regulatory pathways

Table 8.1 Similarities and differences between PD and AD

	PD	AD
Major clinical manifestations	Stationary tremor, bradykinesia, myotonia, and postural gait disorder [2]	Memory disorders, aphasia, apraxia, agnosia, impaired visuospatial skills, executive dysfunction, and personality and behavioral changes [3]
Major pathological changes	Dopaminergic neuron degenerative death in mesencephalic substantia nigra, significant reduction in striatal DA content, and Lewy body formation [4]	Extracellular Aβ deposition, Tau protein hyperphosphorylation, neurofibrillary tangles, neuronal decline [5]
Pathological mechanisms	Neuropathology, genetic factors, environmental factors, immunity, inflammation, apoptosis, oxidative stress [6]	
Diagnosis	Diagnoses typically follow clinical manifestations [7, 8]	
Therapeutic methods	No effective and precise treatments have been successfully developed. Current treatments are mainly focused on alleviating symptoms	

molecule of PARKIN protein and together regulates mitochondrial function through a coordinated mechanism.

DJ-1 Protein Regulates Mitochondrial Autophagy

DJ-1 is mainly located throughout the cytoplasm of neurons, but is also present in the mitochondrial matrix and intermembrane cavity to regulate the activity and morphology of mitochondria. In an oxidative stress environment, Cys106 promotes the redistribution of DJ-1 from the cytoplasm to the outer mitochondrial membrane [30]. As an oxidative stress response protein, DJ-1 can compensate for the function of PINK1, but cannot compensate for the function of PARKIN, suggesting that DJ-1 can execute its function in parallel to PINK1, or act downstream of PINK1 [31]. On the other hand, mice with simultaneous knockout of PINK1/PARKIN/DJ-1 genes could not produce Parkinson-related phenotypes, suggesting that these three genes may play important protective roles in neurons, but not necessarily for neuronal development and function maintenance. The phosphatidylinositol 3 phosphate (PI3K)/Akt signaling pathway is involved in the specific regulation of DJ-1-related cell death, and downregulation of DJ-1α can lead to the damage of the PI3K/Akt signaling pathway, thereby blocking apoptosis [32]. In addition, as an upstream molecule of α-synuclein, DJ-1 is not involved in the regulation of α-synuclein-induced neurotoxicity. DJ-1 mutant mice have enlarged mitochondria and increased susceptibility to hydrogen peroxide and paraquat. In short, the DJ-1 mutation can inhibit mitochondrial autophagy, which in turn can cause the accumulation of damaged mitochondria in cells, and the degeneration of nerve cells, ultimately leading to the occurrence of PD.

SNCA (Gene Encoding α-Synuclein) and α-Synuclein Regulate Mitochondrial Autophagy

During conditions of cellular stress and cytoplasmic acidification, α-synuclein interferes with the outer mitochondrial membrane to prevent mitochondrial fusion, thus interfering with mitochondrial lysis [33]. Moreover, α-synuclein inclusion at high concentrations can lead to the reduction in mitochondrial complex I–IV activity and mitochondrial DNA damage [34]. In addition, α-synuclein can also induce oxidative stress and intracellular calcium and nitric oxide abnormalities, making it a driving factor for mitochondrial dysfunction [35]. Similarly, the occurrence of SNCA gene mutations can also increase the susceptibility of dopaminergic neurons to mitochondrial toxins, including MPP^+ and 6-hydroxydopamine [35]. The SNCA mutation inhibits the formation of autophagosomes, thus resulting in the accumulation of degenerating mitochondria in cells, which cannot be cleared in time, thereby affecting the normal function of nerve cells. The formation of α-synuclein aggregates may increase membrane permeability, thus directly leading to cell death or extracellular activation of microglia, and subsequently induce pro-inflammatory reactions and the production of reactive oxygen species. The overexpression of α-synuclein also interferes with Rabla (an important early regulator of the secretory pathway, associated with influencing factors and proteins such as SarlA, SarlB, and VDP) at the early stages of autophagosome formation [36, 37]. Megalin can inhibit the secretion of LC3-II to inhibit autophagy and reduce cellular homeostasis, while α-synuclein can induce autophagy [38]. These observations suggest that α-synuclein plays a regulatory role in the formation of autophagosomes [39]. Previous studies have also found that the ubiquitin–proteasome system and mitochondrial autophagy are involved in the regulation of α-synuclein [36], thus forming a vicious circle of α-synuclein overexpression and abnormal autophagy. In addition, heat shock proteins (HSPs) not only block the early stages of protein aggregate formation, but also reduce the formation of fibrosis, and play an active role in the pathological process of PD [37].

Leucine-Rich Repeat Kinase 2 (LRRK2) Regulates Mitochondrial Autophagy

The G2019S mutation of LRRK2 promotes the division of damaged mitochondria and triggers mitochondrial autophagy through ULK2 (via DLP1) and JNK signal pathways [40]. LRRK2 is involved in the regulation of α-synuclein fibrosis and Golgi fragmentation, mitochondrial morphology, and mutant α-synuclein-induced microtubule production, suggesting that LRRK2 expression can damage the transport pathway based on mitochondrial microtubules, thereby further aggravating the damage of Golgi transport induced by α-synuclein, after mutation [41]. The G2019S mutation of LRRK2 can also lead to calcium dyshomeostasis, which induces autophagy and causes degradation of mitochondria by increasing sensitivity to mitochondrial toxins, including rotenone and an oxidative stress environment [38]. LRRK2 protein contains a variety of protein domains including the WD40 repeat domain located in the outer membrane of the mitochondria, and participates in the regulation of mitochondrial autophagic degradation by acting as a scaffold on the mitochondrial membrane [42].

Glucocerebrosidase (GBA) Regulates Mitochondrial Autophagy

Deficiency of the GBA gene can lead to the deletion of lysosomal glucocerebrosidase, thus causing the accumulation of glucocerebrosides in lysosomes, which affects material degradation [43]. In addition, mutations in GBA can accelerate the accumulation of α-synuclein [44]. Similarly, abnormal mitophagy is also observed in GBA-deficient neurons, which may inhibit recruitment of PARKIN to locate damaged mitochondria [45].

ATPase Cation-Transporting 13A2 (ATP13A2) Regulates Mitochondrial Autophagy

ATP13A2 is involved in the process of transferring endocytic proteins or autophagy substrates to lysosomes [46]. The loss or downregulation of ATP13A2 caused by mutations or other factors can cause lysosomal membrane instability. The reduced activity of lysosomal enzymes attenuates their capability to degrade substances through autophagosomes [47], and further leads to the accumulation of α-synuclein and damaged mitochondria in cells [48] (Fig. 8.2).

2.1.2 Environmental Factors Regulate Mitochondrial Autophagy

The occurrence of sporadic PD is also related to adverse environmental factors that include MPTP, 6-hydroxydopamine, pesticides (e.g., paraquat and rotenone), heavy metal ions (including manganese and iron), and milk consumption, and pathological conditions such as chronic anemia and traumatic brain injury [50]. A healthy lifestyle, including regular and appropriate exercise, drinking tea, and taking sirolimus and minocycline, is conducive to suppressing the pathological progression of PD. After MPTP passes through the blood–brain barrier, it is converted into the acute metabolite MPP^+ by astrocytes in the brain tissue, under the action of monoamine oxidase B, and selectively accumulates in dopaminergic cells and inhibits the mitochondrial respiratory chain complex I, thus hindering normal electron transfer. The disruption of membrane potential increases the production and accumulation of reactive oxygen species. Subsequently, the altered activity of mitochondrial complexes II–IV can also lead to an increase in the production of reactive oxygen species and reduced production of ATP. Because oxidative stress can trigger mitochondrial autophagy [51], neurotoxins such as MPTP may cause mitochondrial autophagy through oxidative stress. In addition, rotenone and 6-hydroxydopamine can cause alternative splicing of α-synuclein mRNA resulting in the deletion of exon 5, thus

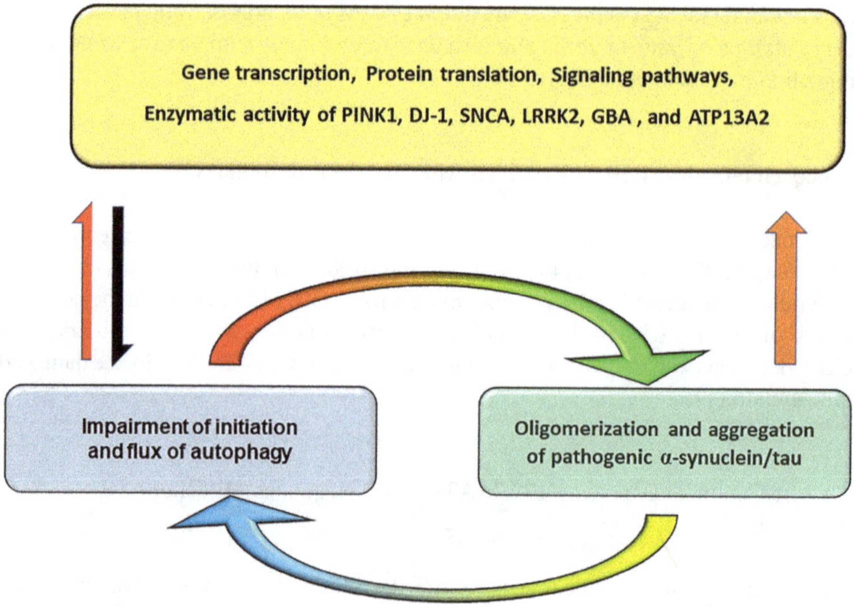

Fig. 8.2 Multidirectional loop of PD-related genes, autophagy function, and PD pathology. (Modified from Hou [49])

leading to the shortening of α-synuclein, changing its localization, and inhibiting proteasome function, thereby indirectly affecting mitochondrial autophagy. At the same time, related studies have reported that 6-hydroxydopamine can cause mitochondrial autophagy, and the overexpression of PINK1 can reverse this effect, suggesting that 6-hydroxydopamine may regulate mitochondrial autophagy by interfering with the expression of PINK1 protein [49].

In vitro experiments have also concluded that some pesticides and herbicides, including those with chlorides such as hexachlorocyclohexane, dieldrin, and rotenone, can increase the occurrence of α-synuclein fibrosis. Insecticides have toxic effects on mitochondria, change the metabolic activity of mitochondria, change the function of the proteasome, or directly interact with α-synuclein to affect the occurrence of intracellular fibrosis [52]. Organochlorine pesticides accumulate in the central nervous system. This leads to the overexpression of α-synuclein aggregates that impair the function of the proteasome, thereby indirectly regulating mitochondrial autophagy through the α-synuclein pathway. In addition to the neurotoxins mentioned above, there is also progressive accumulation of iron in the microglia of brain neurons, which increases with age. Iron can also enhance the expression of α-synuclein, induce the formation of fibrosis, and promote the formation of Lewy bodies [53]. High levels of intracellular iron can also cause oxidative stress, thereby changing the function of mitochondrial autophagy, while the expression of Parkin protein can reverse this effect [54].

In mammals, sirolimus can reduce the release and localization of cytochrome C in mitochondria, and the occurrence of apoptosis, by inactivating its target mTOR, thereby enhancing autophagy and degrading aggregated proteins. Sirolimus is used to treat nervous system diseases and can perform a neuroprotective role. Sirolimus can also induce mitophagy in a dependent manner through other autophagy-related proteins, including Atg11, Atg20, and Atg24 [55]. As a kinase inhibitor, minocycline inhibits nitric oxide synthase involved in cell apoptosis, and blocks the activation of 6-hydroxydopamine and microglia in MPTP-induced Parkinson's animal model, thereby indirectly affecting mitochondrial autophagy and achieving a neuroprotective effect [56].

2.2 *The Relationship Between ALP and PD*

Both UPS and ALP are involved in the degradation pathway of α-synuclein, which plays a very important role in the degeneration of neurons. ALP is a lysosomal-dependent degradation pathway widely observed in eukaryotic cells, and it is closely correlated with the pathogenesis of PD [27]. Based on substrates from different signal pathways entering lysosomes, ALP can be divided into three types: macromolecular spontaneous phagocytosis (macroautophagy), which is typically the default classification for autophagy; molecular chaperone-mediated spontaneous phagocytosis (chaperone-mediated autophagy, CMA); and microautophagy. Autophagy can be induced after short-term nutrient deficiency, and CMA is activated after prolonged nutrient deficiency, while microautophagy is not activated by nutrient deficiency. Compared with UPS, autophagy is likely the major degradation mechanism for some aggregated proteins together with damaged organelles, and it is the only mechanism for recycling damaged organelles such as mitochondria. Some large membrane proteins and protein complexes (including oligomers and aggregates) cannot pass through the pores of the narrow proteasome channel, but can only be degraded through the autophagy pathway. Autophagy is a multistep reaction process, including the formation of autophagosomes with double-layered membrane structures. Autophagolysosomes are then generated after autophagosomes fuse with lysosomes, followed by degradation of the contents in autophagolysosomes by hydrolytic enzymes. Many components involved in macromolecular autophagy have been confirmed experimentally, including the members in the Atg family, and these play various roles in the formation and development of autophagosomes.

Multiple studies have confirmed that the important role of ALP in PD is the degradation of α-synuclein through macromolecular autophagy and CMA [57]. Since α-synuclein is the major component of Lewy bodies, both UPS and macromolecular autophagy are involved. During the removal of α-synuclein, CMA is also necessary for the degradation of wild-type α-synuclein. Moreover, the inhibition of CMA leads to the formation of high molecular weight and insoluble α-synuclein [58], indicating that healthy neurons rely on CMA to clear α-synuclein. Therefore, α-synuclein has an important rate-limiting effect on the oligomerization reaction. Even more important are the A53T and A30P mutations of α-synuclein, which can cause familial PD, and inhibit the CMA pathway by producing a high affinity for Lamp2A [59]. However, the mutant α-synuclein is still partially trapped in the lysosome and is degraded by macromolecular autophagy instead of CMA. Macromolecular autophagy is a compensatory mechanism accompanied by α-synuclein mutations for the inhibition of CMA. The physiological significance of this response is not fully understood. However, this phenomenon suggests that A53T and A30P mutations of α-synuclein may induce α-synuclein aggregation by inhibiting the CMA pathway, thus leading to decreased clearance of α-synuclein in cells. On the other hand, the missense mutation 193M in UCH-L1 has been previously observed in PD familial patients and is also involved in the regulation of the CMA pathway [49]. The mutated UCH-L1 can produce an unusually high affinity for substances that play key roles in the CMA pathway, including Lamp2A and HSC70. More importantly, mutant UCH-L1 increases the levels of α-synuclein [49], suggesting that the mutation of UCH-L1 may also participate in the pathogenesis of PD by regulating the level of α-synuclein. These findings may describe an important mechanism in the occurrence of PD involving inhibition of CMA-mediated degradation of α-synuclein.

2.3 The Role of microRNAs in PD

microRNAs (miRNAs) are small, noncoding, single-stranded RNAs with 21–23 nucleotides that can regulate target gene expression in a sequence-specific manner. miRNA is produced by a single-stranded RNA precursor (pre-miRNA) with a hairpin loop structure and a length of 70–80 nucleotides, after being processed by Dicer enzyme. The combination of mature miRNA and RNA-induced gene silencing complex (RISC) can form an asymmetric RISC complex.

Previous studies have shown that miRNAs are closely related to PD. For example, the expression levels of miR-7, miR-153, miR-34b, and miR-34c in PD are all significantly decreased, and directly regulate α-synuclein mutations [60]. According to a recent report, miR-137, miR-124, and miR-184 are widely expressed in the central nervous system and are essential for neuronal regulation [61]. In addition, miR-137 has a significant regulatory role in Parkin-induced mitochondrial autophagy, which has been confirmed by miR-137 mimics and miR-137 inhibitor transfection in HeLa cells [62]. Similarly, quantitative real-time PCR used to evaluate the expression of miRNAs in the blood samples of PD patients and healthy controls has revealed that miR-137 can regulate hypoxia-induced mitochondrial autophagy by inhibiting the expression of mitophagy receptors such as FUNDC1 and NIX. Moreover, the expression of miR-137 in the blood of PD patients is significantly higher than that of the control group, suggesting that miR-137 is one of the inducements of PD. The inhibitory function of miRNA activity on pathogenic LRRK2 is directly antagonized by the neuronal cell fate determinant TRIM32, indicating that inhibiting miRNA activity may be an important approach to the prevention and treatment of PD [63].

Some scholars have also found that miR-124 is highly expressed in the nervous system, with the expression level higher than other tissues by 100 times. It expression decreases in neurological diseases such as encephalomyelitis and glioma diseases. At the same time, it has a significant neuroprotective effect on ischemic brain injury and stroke. Reduced expression of miR-124 in dopaminergic neurons in the midbrain tissue and substantia nigra of MPTP-treated mice with PD has been reported [64]. Similarly, reduced expression of miR-124 in SH-SY5Y cells is also reported, suggesting that miR-124 may be a biomarker for early diagnosis of PD. In contrast, after exogenous administration of miR-124 agonist through intracerebroventricular catheterization, the loss of dopaminergic neurons in the substantia nigra of MPTP-treated mice with PD is obviously suppressed, with increased dopamine levels in the midbrain tissue, suggesting that miR-124 has a neuroprotective effect on PD.

In another previous study, serum samples have been collected from 109 PD patients and 40 age- and gender-matched healthy volunteers (control group) to evaluate accurate screening analysis of PD [63]. Sample analyses include the extraction and reverse transcribing of RNA from exosome-like microcapsules in serum, the analysis of serum miRNA by quantitative reverse transcription polymerase chain reaction (qRT-PCR), and the establishment of receiver operating characteristic (ROC) curve of miRNA. Furthermore, downregulation of miR-19b and upregulation of miR-195 and miR-24 in PD patients are verified. Compared with the control group, the area under curve (AUC) values of miR-19b, miR-24, and miR-195 are 0.753, 0.908, and 0.697, respectively, for PD patients. Therefore, expression levels of miR-19b, miR-24, and miR-195, in serum, may be beneficial to the diagnosis of PD.

The expression of miRNAs in the cingulate gyrus of patients and controls from the perspective of epigenetics has also been investigated. The 744 well-characterized miRNAs from patients and controls are determined through the cyclotron curve established by TaqMan array miRNA chip. SYBR Green qRT-PCR is used to verify significantly dysregulated miRNAs. In total, 43 miRNAs are identified as being upregulated in the cingulate gyrus of patients during the first screening. Among these, 13 miRNAs are predicted to regulate at least 6 genes that mutate in a single-gene form of PD (DJ-1, PARK2, PINK1, LRRK2, SNCA, and HTRA2). It is also determined that five of these 13 miRNAs (miR-144, miR-199b, miR-221, miR-488, and miR-544) are upregulated based on the evaluation by SYBR Green qRT-PCR and are also predicted to regulate SNCA, PARK2, and LRRK2, or some combination thereof. The expression levels of SNCA, PARK2, and LRRK2 are noticeably reduced in PD patients. Five of the other tested potential target genes are found to be downregulated, specifically miR-221, miR-144, miR-488, tRNA formyltransferase, and Xin actin-binding repeat-containing protein 2 (XIRP2). The five identified miRNA-mediated genes responsible for normal cell function may play critical roles in the pathogenesis of PD by modifying the expression of SNCA, PARK2, and LRRK2. In addition, a PD model is constructed using human SK-N-SH neuroblastoma cells, following treatment with MPP^+ [65], and then, the expression of miR-181a in a PD cell model is analyzed. A gradual decrease in the expression of miR-181a is observed with increasing concentration of MPP^+. However, after cells are transfected with a miR-181a mimic and inhibitor, protein expression is evaluated for LC3-II, LC3-I, and Beclin1, associated with autophagy, and proteins p-p38, p38, p-JNK, and JNK in p38 mitogen-activated protein kinase (MAPK)/c-Jun N-terminal kinase (JNK) signal transduction, associated with apoptosis. Compared with the MPP^+-induced PD cell model, which includes silencing of miR-181a, the overexpression of miR-181a is found to significantly reduce the LC3-II/LC3-I ratio and Beclin1 expression, and promote the expression of p-p38 and p-JNK, thereby stimulating cell apoptosis. These results indicate that miR-181a can regulate apoptosis and autophagy in PD by inhibiting the p38 MAPK/JNK signal pathway. This suggests that miRNAs may be sensitive biomarkers for the early diagnosis of PD and that regulating the expression of miRNAs could be a therapeutic approach to delaying the progression of PD.

3 The Relationship Between Exercise and PD

3.1 Regulation of Exercise on Autophagy

At present, many studies have reported on the relationship between exercise and autophagy-related gene expression. As early as 1984, research reported increased autophagy activity in 5-month-old male mice following treadmill running at various intensities [66]. The presence of mitochondria at different degradation stages in autophagy vesicles is also observed, suggesting that damaged organelles such as mitochondria can be degraded through autophagy, which can provide both energy and synthetic substrates for the regeneration of skeletal muscle fibers during the recovery period after exercise. Similarly, increased expression of autophagy-related proteins has been reported in toe muscle of C57BL/6 mice after 28 days of voluntary wheel running, as evidenced by increased expression levels for Beclin1 (33%), Bnip3 (41.5%), LC3-I (43%), and LC3-II (210%), as well as an increase in the LC3-II/LC3-I ratio [66]. It is therefore believed that autophagy plays an important role in the maintenance of skeletal muscle contraction and function. Similarly, both caloric restriction and lifelong voluntary running exercise, or a combination of both, can upregulate the expression levels of Atg7, Atg9, LC3-I, LC3-II, and Lamp2 proteins, in the toe muscle of older rats [67]. Based on the increase in the LC3-II/LC3-I ratio, caloric restriction combined with lifelong voluntary exercise produces a more obvious improvement, indicating that lifelong voluntary exercise may improve autophagy activity of the digit muscle in older rats, thereby inhibiting cell damage and cell death caused by skeletal muscle aging. Another study has also confirmed that the oxidative stress-induced damage of skeletal muscle is negatively correlated with autophagy activity, and caloric restriction combined with exercise intervention could effectively reduce oxidative stress-induced damage of the cells through autophagy signal pathways, reduce total peroxide content, increase cross-sectional area of toe muscle fibers, and inhibit cell senescence [68]. In addition, the LC3-II/LC3-I ratio of mice with myocardial infarction also exhibits a 2.1-fold increase with treadmill running exercise when compared with the mice without exercise intervention [69]. It is believed that exercise training can increase the degradation rate of damaged proteins by enhancing autophagy activity, promoting the production of new nutrients, reducing myocardial cell damage, inhibiting excessive apoptosis caused by myocardial infarction, and improving cardiac function. Conversely, overtraining can increase the expression of skeletal muscle atrophy-related genes, including Atrogin-1 and MuRF1, and autophagy-related genes, such as Atg7, Beclin1, LC3, and FoxO3, as specific genes for the autophagy of skeletal muscle cells. Therefore, overtraining also can lead to excessive induction of autophagy for degrading proteins and organelles, including mitochondria and the endoplasmic reticulum, as well as other major components in skeletal muscle, thereby causing the atrophy of skeletal muscle, and reducing exercise performance and cellular immunity of skeletal muscle as well as its corresponding functions. From these cited studies, it can be summarized that exercise training at the appropriate intensity can degrade intracellular metabolic waste and damaged organelles by moderately increasing the level of autophagy, thereby maintaining the normal homeostasis of cells. In contrast, overtraining can cause excessive activation of autophagy, and degrade too many proteins and organelles, which in severe cases may lead to excessive apoptosis, thereby resulting in exercise fatigue and abnormal cell damage.

3.2 *The Regulatory Role of Exercise-Mediated Autophagy in PD*

Autophagy is involved in the degradation of α-synuclein, and rapamycin as an autophagy inducer can increase the removal of α-synuclein through the autophagy signal pathway, while autophagy suppression can decrease the removal of α-synuclein and increase its accumulation, which indicates that the dysfunctional degradation of α-synuclein is an important cause of PD [70]. In the PD mouse model, 18 months of treadmill exercise are observed to improve the function of dopamine neurons, and alleviate coordination and motor disorders of PD mice [71, 72]. Similarly, the 8-week exercise in MPTP-induced PD mice reportedly can rescue motor dysfunction, decrease the expression of α-synuclein, and promote the expression of tyrosine hydroxylase and dopamine transporter, as well as plasma dopamine levels [73]. Moreover, regular aerobic exercise is also found to upregulate the expression of Beclin1, LC3-I, and LC3-II; improve the rod-rotating behavior of PD rats; increase the expression of tyrosine hydroxylase; and reduce the accumulation of α-synuclein [73]. Therefore, exercise appears to provide neuroprotective effects on PD animal models. Inducing autophagy using autophagy enhancers can be beneficial to alleviating the corresponding symptoms of PD patients and PD animal models [74].

Furthermore, exercise training at the appropriate intensity can play a central role in the growth and metabolism of tissues by moderately increasing autophagy levels, and can play a constructive role by inhibiting the occurrence and development of autophagy-related diseases, including PD. Overtraining can lead to exercise fatigue and tissue damage because the activation of autophagy exceeds a certain threshold, and excessive apoptosis could be induced in this severe case. Therefore, regulating the functional status of autophagy through exercise with proper intensity may be a determinant for slowing down the initiation and progression of PD.

References

1. Dommershuijsen LJ, Heshmatollah A, Darweesh SKL et al (2020) Life expectancy of parkinsonism patients in the general population. Parkinsonism Relat Disord 77:94–99
2. Marino BLB, de Souza LR, Sousa KPA et al (2020) Parkinson's disease: a review from pathophysiology to treatment. Mini Rev Med Chem 20:754–767
3. Bature F, Guinn BA, Pang D et al (2017) Signs and symptoms preceding the diagnosis of Alzheimer's disease: a systematic scoping review of literature from 1937 to 2016. BMJ Open 7: e015746
4. Krylatov AV, Maslov LN, Voronkov NS et al (2018) Reactive oxygen species as intracellular signaling molecules in the cardiovascular system. Curr Cardiol Rev 14:290–300
5. Bjørklund G, Aaseth J, Dadar M et al (2019) Molecular targets in Alzheimer's disease. Mol Neurobiol 56:7032–7044
6. Mattson MP, Arumugam TV (2018) Hallmarks of brain aging: adaptive and pathological modification by metabolic states. Cell Metab 27:1176–1199
7. Postuma RB, Berg D, Stern M et al (2015) MDS clinical diagnostic criteria for Parkinson's disease. Mov Disord 30:1591–1601
8. Eratne D, Loi SM, Farrand S et al (2018) Alzheimer's disease: clinical update on epidemiology, pathophysiology and diagnosis. Australas Psychiatry 26:347–357
9. Tysnes OB, Storstein A (2017) Epidemiology of Parkinson's disease. J Neural Transm (Vienna) 124:901–905
10. Alves G, Forsaa EB, Pedersen KF et al (2008) Epidemiology of Parkinson's disease. J Neurol 255(Suppl 5):18–32
11. de Lau LM, Breteler MM (2006) Epidemiology of Parkinson's disease. Lancet Neurol 5:525–535
12. Picillo M, Nicoletti A, Fetoni V et al (2017) The relevance of gender in Parkinson's disease: a review. J Neurol 264:1583–1607
13. Smeyne RJ, Breckenridge CB, Beck M et al (2016) Assessment of the effects of MPTP and Paraquat on dopaminergic neurons and microglia in the substantia nigra pars compacta of C57BL/6 mice. PLoS One 11:e0164094
14. Wakabayashi K, Tanji K, Odagiri S et al (2013) The Lewy body in Parkinson's disease and related neurodegenerative disorders. Mol Neurobiol 47:495–508
15. Polymeropoulos MH, Lavedan C, Leroy E et al (1997) Mutation in the alpha-synuclein gene identified in families with Parkinson's disease. Science 276:2045–2047
16. Bendor JT, Logan TP, Edwards RH (2013) The function of α-synuclein. Neuron 79:1044–1066
17. Kawamoto Y, Akiguchi I, Nakamura S et al (2002) Accumulation of 14-3-3 proteins in glial cytoplasmic inclusions in multiple system atrophy. Ann Neurol 52:722–731
18. Niemann N, Jankovic J (2019) Juvenile parkinsonism: differential diagnosis, genetics, and treatment. Parkinsonism Relat Disord 67:74–89

19. Zheng Q, Huang T, Zhang L et al (2016) Dysregulation of ubiquitin-proteasome system in neurodegenerative diseases. Front Aging Neurosci 8:303
20. Bonifati V (2005) Genetics of Parkinson's disease. Minerva Medica 96:175–186
21. Qi Z, Miller GW, Voit EO (2014) Rotenone and paraquat perturb dopamine metabolism: a computational analysis of pesticide toxicity. Toxicology 315:92–101
22. Folke J, Rydbirk R, Løkkegaard A et al (2019) Distinct autoimmune anti-α-synuclein antibody patterns in multiple system atrophy and Parkinson's disease. Front Immunol 10:2253
23. Koutzoumis DN, Vergara M, Pino J et al (2020) Alterations of the gut microbiota with antibiotics protects dopamine neuron loss and improve motor deficits in a pharmacological rodent model of Parkinson's disease. Exp Neurol 325:113159
24. Oikawa S, Kobayashi H, Kitamura Y et al (2014) Proteomic analysis of carbonylated proteins in the monkey substantia nigra after ischemia-reperfusion. Free Radic Res 48:694–705
25. Tiwari HS, Misra UK, Kalita J et al (2016) Oxidative stress and glutamate excitotoxicity contribute to apoptosis in cerebral venous sinus thrombosis. Neurochem Int 100:91–96
26. Dupouy J, Ory-Magne F, Brefel-Courbon C (2017) Other care in Parkinson's disease: psychological, rehabilitation, therapeutic education and new technologies. Presse Med 46:225–232
27. Lu J, Wu M, Yue Z (2020) Autophagy and Parkinson's disease. Adv Exp Med Biol 1207:21–51
28. Bai Y, Wang Y, Yang Y (2018) Hepatic encephalopathy changes mitochondrial dynamics and autophagy in the substantia nigra. Metab Brain Dis 33:1669–1678
29. Marongiu R, Spencer B, Crews L et al (2009) Mutant Pink1 induces mitochondrial dysfunction in a neuronal cell model of Parkinson's disease by disturbing calcium flux. J Neurochem 108:1561–1574
30. Lin J, Prahlad J, Wilson MA (2012) Conservation of oxidative protein stabilization in an insect homologue of parkinsonism-associated protein DJ-1. Biochemistry 51:3799–3807
31. Cookson MR (2012) Parkinsonism due to mutations in PINK1, parkin, and DJ-1 and oxidative stress and mitochondrial pathways. Cold Spring Harb Perspect Med 2:a009415
32. Zeng J, Zhao H, Chen B (2019) DJ-1/PARK7 inhibits high glucose-induced oxidative stress to prevent retinal pericyte apoptosis via the PI3K/AKT/mTOR signaling pathway. Exp Eye Res 189:107830
33. Pozo Devoto VM, Falzone TL (2017) Mitochondrial dynamics in Parkinson's disease: a role for α-synuclein? Dis Model Mech 10:1075–1087
34. Xie W, Chung KK (2012) Alpha-synuclein impairs normal dynamics of mitochondria in cell and animal models of Parkinson's disease. J Neurochem 122:404–414
35. Binukumar BK, Bal A, Kandimalla RJ et al (2010) Nigrostriatal neuronal death following chronic dichlorvos exposure: crosstalk between mitochondrial impairments, α synuclein aggregation, oxidative damage and behavioral changes. Mol Brain 3:35
36. Guerreiro PS, Huang Y, Gysbers A et al (2013) LRRK2 interactions with α-synuclein in Parkinson's disease brains and in cell models. J Mol Med (Berl) 91:513–522
37. Hyun CH, Yoon CY, Lee HJ et al (2013) LRRK2 as a potential genetic modifier of Synucleinopathies: interlacing the two major genetic factors of Parkinson's disease. Exp Neurobiol 22:249–257
38. Luo GR, Le WD (2010) Collective roles of molecular chaperones in protein degradation pathways associated with neurodegenerative diseases. Curr Pharm Biotechnol 11:180–187
39. Sarkar S, Olsen AL, Sygnecka K et al (2021) α-Synuclein impairs autophagosome maturation through abnormal actin stabilization. PLoS Genet 17:e1009359
40. Chinta SJ, Mallajosyula JK, Rane A et al (2010) Mitochondrial α-synuclein accumulation impairs complex I function in dopaminergic neurons and results in increased mitophagy in vivo. Neurosci Lett 486:235–239
41. McKinnon C, Tabrizi SJ (2014) The ubiquitin-proteasome system in neurodegeneration. Antioxid Redox Signal 21:2302–2321
42. Zhu Y, Wang C, Yu M et al (2013) ULK1 and JNK are involved in mitophagy incurred by LRRK2 G2019S expression. Protein Cell 4:711–721

43. Ikuno M, Yamakado H, Akiyama H et al (2019) GBA haploinsufficiency accelerates alpha-synuclein pathology with altered lipid metabolism in a prodromal model of Parkinson's disease. Hum Mol Genet 28:1894–1904
44. Yun SP, Kim D, Kim S et al (2018) α-Synuclein accumulation and GBA deficiency due to L444P GBA mutation contributes to MPTP-induced parkinsonism. Mol Neurodegener 13:1
45. Li H, Ham A, Ma TC et al (2019) Mitochondrial dysfunction and mitophagy defect triggered by heterozygous GBA mutations. Autophagy 15:113–130
46. Wang R, Tan J, Chen T et al (2019) ATP13A2 facilitates HDAC6 recruitment to lysosome to promote autophagosome-lysosome fusion. J Cell Biol 218:267–284
47. Nyuzuki H, Ito S, Nagasaki K et al (2020) Degeneration of dopaminergic neurons and impaired intracellular trafficking in Atp13a2 deficient zebrafish. IBRO Rep 9:1–8
48. Park JS, Blair NF, Sue CM (2015) The role of ATP13A2 in Parkinson's disease: clinical phenotypes and molecular mechanisms. Mov Disord 30:770–779
49. Hou X, Watzlawik JO, Fiesel FC et al (2020) Autophagy in Parkinson's disease. J Mol Biol 432:2651–2672
50. Osellame LD, Duchen MR (2013) Defective quality control mechanisms and accumulation of damaged mitochondria link Gaucher and Parkinson diseases. Autophagy 9:1633–1635
51. Park JS, Koentjoro B, Veivers D et al (2014) Parkinson's disease-associated human ATP13A2 (PARK9) deficiency causes zinc dyshomeostasis and mitochondrial dysfunction. Hum Mol Genet 23:2802–2815
52. Campdelacreu J (2014) Parkinson disease and Alzheimer disease: environmental risk factors. Neurologia 29:541–549
53. Dusek P, Jankovic J, Le W (2012) Iron dysregulation in movement disorders. Neurobiol Dis 46:1–18
54. Hamaï A, Mehrpour M (2017) [Autophagy and iron homeostasis]. Med Sci (Paris) 33:260–267
55. Pan T, Jankovic J, Le W (2003) Potential therapeutic properties of green tea polyphenols in Parkinson's disease. Drugs Aging 20:711–721
56. Dixit A, Srivastava G, Verma D et al (2013) Minocycline, levodopa and MnTMPyP induced changes in the mitochondrial proteome profile of MPTP and maneb and paraquat mice models of Parkinson's disease. Biochim Biophys Acta 1832:1227–1240
57. Frank M, Duvezin-Caubet S, Koob S et al (2012) Mitophagy is triggered by mild oxidative stress in a mitochondrial fission dependent manner. Biochim Biophys Acta 1823:2297–2310
58. Wu JZ, Ardah M, Haikal C et al (2019) Dihydromyricetin and Salvianolic acid B inhibit alpha-synuclein aggregation and enhance chaperone-mediated autophagy. Transl Neurodegener 8:18
59. Galindo MF, Solesio ME, Atienzar-Aroca S et al (2012) Mitochondrial dynamics and mitophagy in the 6-hydroxydopamine preclinical model of Parkinson's disease. Parkinsons Dis 2012:131058
60. Leggio L, Vivarelli S, L'Episcopo F et al (2017) microRNAs in Parkinson's disease: from pathogenesis to novel diagnostic and therapeutic approaches. Int J Mol Sci 18:2698
61. Mendl N, Occhipinti A, Müller M et al (2011) Mitophagy in yeast is independent of mitochondrial fission and requires the stress response gene WHI2. J Cell Sci 124:1339–1350
62. Li W, Zhang X, Zhuang H et al (2014) MicroRNA-137 is a novel hypoxia-responsive microRNA that inhibits mitophagy via regulation of two mitophagy receptors FUNDC1 and NIX. J Biol Chem 289:10691–10701
63. Martinez-Vicente M, Cuervo AM (2007) Autophagy and neurodegeneration: when the cleaning crew goes on strike. Lancet Neurol 6:352–361
64. Gong X, Wang H, Ye Y et al (2016) miR-124 regulates cell apoptosis and autophagy in dopaminergic neurons and protects them by regulating AMPK/mTOR pathway in Parkinson's disease. Am J Transl Res 8:2127–2137
65. Levine B, Klionsky DJ (2004) Development by self-digestion: molecular mechanisms and biological functions of autophagy. Dev Cell 6:463–477
66. Salminen A, Vihko V (1984) Autophagic response to strenuous exercise in mouse skeletal muscle fibers. Virchows Arch B Cell Pathol Incl Mol Pathol 45:97–106

67. Kabuta T, Furuta A, Aoki S et al (2008) Aberrant interaction between Parkinson disease-associated mutant UCH-L1 and the lysosomal receptor for chaperone-mediated autophagy. J Biol Chem 283:23731–23738
68. Kim JH, Kwak HB, Leeuwenburgh C et al (2008) Lifelong exercise and mild (8%) caloric restriction attenuate age-induced alterations in plantaris muscle morphology, oxidative stress and IGF-1 in the Fischer-344 rat. Exp Gerontol 43:317–329
69. Li N, Pan X, Zhang J et al (2017) Plasma levels of miR-137 and miR-124 are associated with Parkinson's disease but not with Parkinson's disease with depression. Neurol Sci 38:761–767
70. Sweet ES, Saunier-Rebori B, Yue Z et al (2015) The Parkinson's disease-associated mutation LRRK2-G2019S impairs synaptic plasticity in mouse hippocampus. J Neurosci 35:11190–11195
71. Minakaki G, Canneva F, Chevessier F et al (2019) Treadmill exercise intervention improves gait and postural control in alpha-synuclein mouse models without inducing cerebral autophagy. Behav Brain Res 363:199–215
72. Shin MS, Jeong HY, An DI et al (2016) Treadmill exercise facilitates synaptic plasticity on dopaminergic neurons and fibers in the mouse model with Parkinson's disease. Neurosci Lett 621:28–33
73. Kanagaraj N, Beiping H, Dheen ST et al (2014) Downregulation of miR-124 in MPTP-treated mouse model of Parkinson's disease and MPP iodide-treated MN9D cells modulates the expression of the calpain/cdk5 pathway proteins. Neuroscience 272:167–179
74. Wang H, Ye Y, Zhu Z et al (2016) MiR-124 regulates apoptosis and autophagy process in MPTP model of Parkinson's disease by targeting to Bim. Brain Pathol 26:167–176

Chapter 9
Exercise-Mediated Autophagy in Cardiovascular Diseases

Shaohui Jia, Hu Zhang, Jiling Liang, Yin Zhang, Yanju Guo, and Ning Chen

1 Introduction

Cardiovascular diseases (CVDs) are a group of common disorders associated with heart and its blood vessels, with the characteristics of high incidence or high mortality and poor prognosis, especially for elderly population. CVD is the leading cause of global mortality and the dominant contributor for reduced quality of life among people with chronic diseases all over the world. CVDs are usually composed of hypertension, atherosclerosis, cardiomyopathy, myocardial ischemia–reperfusion injury, and other complications. According to the statistics from the WHO report in 2017, approximately 17.8 million people are died of CVDs each year, accounting for 31% of all deaths worldwide, and CVDs could result in 35.6 million people lived with disability [1]. Obesity, drug abuse, hyperlipidemia, high blood pressure, reduced physical activity, smoking, and alcohol drinking are important factors for inducing CVDs. The current incidence of CVDs shows a rapidly increasing trend, which could result in the significant impact on the quality of life of these patients and the larger burden on social and medical system.

As an important physiological process for the recycling of intracellular substances through the evolutionarily conserved degradation pathway, autophagy can regulate protein degradation, organelle turnover, and recycling of cytoplasmic components to maintain cellular metabolism and homeostasis, as well as cell survival under the conditions with nutritional deficiency or suffering from cellular stress, thereby helping cells to adapt to the demand of nutrients and energy. Autophagy is

S. Jia · H. Zhang · J. Liang · Y. Guo · N. Chen (✉)
Tianjiu Research and Development Center for Exercise Nutrition and Foods, Hubei Key Laboratory of Exercise Training and Monitoring, College of Health Science, Wuhan Sports University, Wuhan, China

Y. Zhang
School of Physical Education, Hanshan Normal University, Chaozhou, China

N. Chen (ed.), *Exercise, Autophagy and Chronic Diseases*,
https://doi.org/10.1007/978-981-16-4525-9_9

also considered to be involved in the pathogenesis of a series of CVDs and corresponding complications, such as hypertension, atherosclerosis, cardiomyopathy, and myocardial ischemia–reperfusion injury, as well as heart failure [2]. Therefore, autophagy may be a potential target for the prevention, treatment, and rehabilitation of cardiovascular diseases.

Increasing studies have demonstrated that regular exercise at appropriate intensity and duration can largely reduce the incidence and mortality of CVDs. According to previous studies, 15 min of exercise training each day can confer the reduction in mortality by 14% and another 4% reduction with the increase for each additional 15 min of daily exercise training time [3]. It has been confirmed that regular long-term exercise training can induce anti-atherosclerotic adaptation of vascular function and structure, reduce the risk of cardiovascular diseases, and promote myocardial regeneration by stimulating circulating angiogenic cells [4]. Several studies have documented that appropriate volume of exercise can inhibit the apoptosis of cardiomyocytes [5], increase ejection fraction, significantly improve the function of myocardium contraction and relaxation, and ameliorate the function of aging heart, as well as reduce the stroke risk [6, 7]. Currently, exercise is an inducer of autophagy, and exercise-mediated autophagy has gained tremendous attention during the prevention and rehabilitation of cardiovascular diseases.

2 Autophagy in Cardiovascular Diseases

Autophagy is believed to be a major recycling way to eliminate harmful cellular metabolic wastes and modulate cellular homeostasis, in the cardiovascular system, so that the optimal functional status of autophagy can play a positive role in preventing and treating cardiovascular diseases, mainly by improving the intracellular environment, regulating the quality of mitochondria, and alleviating oxidative stress-induced cardiovascular damage, and providing cardiovascular energy. On the other hand, the deficient autophagy, excessive autophagy, impaired autophagy flux, or autophagy genetic defects are not conducive to the normal regulation of cardiovascular system and could result in the damage of tissue structure and function of cardiovascular system, thereby greatly triggering the risks and occurrence of cardiovascular diseases [8].

Cardiovascular diseases mainly include hypertension, atherosclerosis, cardiomyopathy, and myocardial ischemia–reperfusion injury as well as other corresponding complications. Similarly, the damage of cardiovascular endothelial cells is also the basis of many cardiovascular diseases. A certain degree of autophagy activation plays a positive role in regulating the functions of endothelial cells, such as angiogenesis, nitric oxide (NO) production, and hemostatic function [9]. As the biomarker of autophagy, LC3-II is significantly reduced in atherosclerotic plaques of humans and mice [10, 11]. Similarly, in atherosclerotic plaques, the autophagy degradation substrate p62 shows a significant accumulation, thereby exhibiting an obvious increase in apoptosis, and further aggravating atherosclerosis [12]. At the late stage of atherosclerosis, the excessive oxidative stress will cause autophagy to be irreparable and induce apoptosis [13]. Under the condition with Atg5 or Atg7 knockout, the formation of atherosclerosis is accelerated [14]. The abnormal endothelial function and atherosclerosis can further induce hypertension. The increased level of autophagy in arteries exhibits the relief of hypertension symptoms [15]. In addition, the heart is an energy-sensitive organ, and the normal or optimal induction of autophagy plays an important regulatory role in myocardial energy supply and function. However, the overexpression of protein Beclin1 associated with autophagy initiation may also aggravate the increased induction of autophagy and may lead to cardiac hypertrophy, and suppressed autophagy to a certain extent may alleviate this phenomenon [16]. At the low level of autophagy, the activation of autophagy to a certain extent shows the protection from myocardial ischemia and heart remodeling after myocardial ischemia [17]. After 21 days of acute myocardial infarction, the anterior wall of the ventricle became thinner, and myocardial cells are degenerated. The severe loss of myocardial cells and the increasing area of myocardial infarction can induce a significantly synchronous increase in LC3-II and p62, suggesting the impaired autophagy flux [18]. After the inhibition of autophagy, the death and loss of cardiomyocytes will further be aggravated, thereby accelerating the occurrence of progressive heart failure. Based on these results, rescuing the abnormal autophagy flux can alleviate the functional impairment caused by myocardial infarction [19]. The relevant studies have also shown that increased exogenous alcohol consumption induces an increase in autophagy levels and appears to promote myocardial hypertrophy in a feedback manner and promotes myocardial defects [20]. After knocking out Atg5, the heart appears hypertrophy, left ventricle enlargement, and cardiovascular functional decline [21]. The feedback myocardial hypertrophy may be the adaptation of deficient autophagy so that heart failure shows an active state of autophagy, which may be specifically correlated with the level of oxidative stress [22]. The pathological myocardial hypertrophy and hypertension induced by angiotensin II also show a certain degree of dysfunctional autophagy [23]. In addition, selective autophagy such as mitochondrial autophagy also has a certain regulatory role in cardiovascular diseases [24]. Therefore, regulating the functional status of autophagy through exercise interventions or exogenous supplementation of natural products may have the positive effect on the prevention and rehabilitation of cardiovascular diseases [25].

Although current studies have confirmed that cardiovascular diseases show the abnormal functional status of autophagy, the specific mechanisms still need to be further explored, which should be focused on the models and stages of cardiovascular diseases (Fig. 9.1).

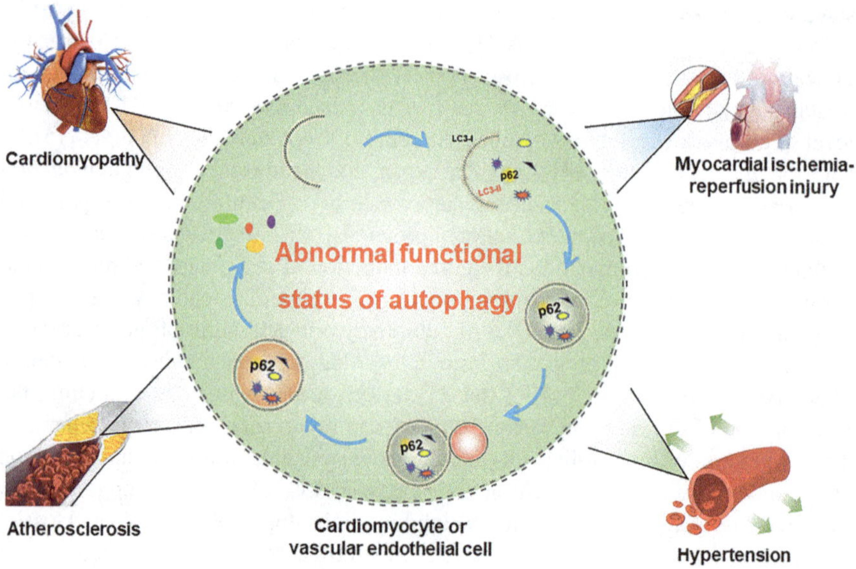

Fig. 9.1 Abnormal functional status of autophagy in cardiomyocytes or vascular endothelial cells can trigger cardiovascular diseases such as hypertension, atherosclerosis, cardiomyopathy, and myocardial ischemia–reperfusion injury

3 Exercise-Mediated Autophagy in CVDs

Exercise, as an important external method for the intervention of cardiovascular diseases, shows great potential value in the prevention and treatment of cardiovascular diseases. The major part of physical benefits of exercise is due to exercise-induced optimal functional status of autophagy. In the exercising state, the changes in energy demand and supply can effectively stimulate the initiation of autophagy so that the material circulation inside the cell occurs to alleviate the oxidative stress in the body. The abnormal functional status of autophagy in cardiovascular diseases may accelerate the progression of the diseases, and it may be a more economical and environment-friendly way to optimize the functional status of autophagy through exercise interventions. Therefore, in this part, the relevant mechanisms of exercise intervention in several common cardiovascular diseases will be discussed from the perspective of autophagy.

3.1 Exercise-Mediated Autophagy in Hypertension

Hypertension is a high blood pressure state of the systemic circulatory system, which causes the body to produce a series of pathological symptoms such as dizziness and headache and a higher risk of cardiovascular or cerebral hemorrhage, as well as the damage of eye, kidney, and other tissues. Correlational analysis shows that approximately one-fourth of the world's population will suffer from hypertension by 2025 [1]. Regular appropriate exercise is an important external intervention method to relieve high blood pressure in the body. Low- and medium-intensity aerobic exercise is a safer exercise intensity to relieve high blood pressure [26]. The underlying mechanism of exercise relieving high blood pressure may be due to regulating the functional status of autophagy.

With the extension of age, the incidence of hypertension also shows a sharply increasing trend. Meanwhile, the functional changes in endothelial cells and vascular smooth muscle cells are also important inducing factors of hypertension [27]. In the rat spontaneous hypertension model, the level of arterial autophagy presents a significant decrease; in contrast, increasing autophagy can improve the function of arterial endothelial cells and vascular smooth muscle cells, thereby alleviating the symptoms of hypertension [15]. In addition, increased expression of mitochondrial autophagy proteins such as PINK1 and Parkin also exhibits an obvious improvement of vascular endothelial cells, thereby revealing the mitigated effect on high blood pressure [28]. The consumption of exogenous *Rhynchophylla* alkaloid can improve the vasodilation function by activating AMPK signaling pathway due to the rescuing of low-level autophagy [29]. Therefore, autophagy may play an important role in relieving hypertension. Activating autophagy through exercise interventions to relieve high blood pressure is a more economical and green strategy. High-intensity interval training for people with hypertension could result in a significant reduction in blood pressure, accompanied by an increase in autophagy biomarkers [30]. In healthy persons, regular grip exercise can effectively increase the level of autophagy and the generation of NO and $O_2^{\cdot -}$ in arterial endothelial cells [31]. In addition, regular appropriate exercise can also relieve high blood pressure by improving cardiac autonomic nerves [32], angiogenesis, and insulin resistance [33]. Although limited studies have explored the regulatory role of exercise-mediated autophagy in hypertension, these studies have not fully clarified the specific role of autophagy in the control of hypertension.

3.2 Exercise-Mediated Autophagy in Atherosclerosis

Atherosclerosis, one of the common cardiovascular diseases, is the major pathological basis of many cardiovascular and cerebrovascular diseases [34], which is characterized by lipid deposition in the intima of large and middle arteries, intima focal fibrosis, and the formation of atherosclerotic plaques, thereby eventually leading to the hardened vascular wall and lumen stenosis, and causing a series of secondary lesions [35]. Although the exact etiology and pathogenesis of atherosclerosis are controversial, hyperlipidemia, hypertension, hyperinsulinemia, diabetes, obesity, and smoking have been clarified as the risk factors for atherosclerosis [35]. Meanwhile, multiple pathological factors such as lipid infiltration, endothelial cell injury, monocyte–macrophage accumulation, vascular inflammation, and smooth muscle cell proliferation still cannot fully explain the occurrence and pathogenesis of atherosclerosis.

As a modifiable risk factor, physical inactivity is a contemporary public health problem with the trigger of diabetes, hypertension, and hyperlipidemia, so that physical inactivity plays a harmful role in the pathogenesis of atherosclerosis. Conversely, epidemiological studies have proved the regulatory role of regular exercise interventions in the improvement of cardiovascular health [36]. In addition, previous studies have documented that early physical exercise can significantly reduce the morbidity and mortality of atherosclerotic cardiovascular diseases [37], which may be correlated with the improvement of atherosclerosis-related risk factors such as hyperlipidemia, type 2 diabetes, and hypertension [38]. Similarly, other studies have also found that moderate activation of autophagy can prevent the onset and progression of diabetes and atherosclerosis [39]. Meanwhile, regular appropriate exercise can enhance adaptive autophagy to execute its positive role in the prevention and treatment of atherosclerosis. However, there is still no perfect treatment system for exercise intervention. Therefore, it is of great significance to actively seek the optimal combinatorial exercise intervention for the prevention and treatment of atherosclerosis.

3.2.1 Autophagy and Atherosclerosis

Autophagy is a biological process of phagocytosis, in which denatured proteins and damaged or aging organelles are transported to lysosomes for degradation, to achieve the own metabolic demands and the renewal of some organelles [40]. Apoptosis refers to the autonomous and orderly death of cells controlled by apoptotic genes to maintain the stability of the internal environment. Meanwhile, autophagy is a nonapoptotic cell death manner, without the increase in inflammation of plaques. Furthermore, autophagy can suppress apoptosis by degrading misfolded proteins and impaired or damaged cellular organelles. The protective function of autophagy involves the regulation of cholesterol metabolism [41]. Therefore, autophagy plays a dual role in atherosclerosis, and moderate autophagy can protect against atherosclerosis, while the excessive autophagy can contribute to atherosclerosis as a form of cell death [42].

Previous studies have demonstrated that the dysregulated autophagy results in the decreased free cholesterol efflux and the accumulation of apoptotic foam cells attributable to lipid overload, thus leading to the increased lipid content in plaques, increased necrotic core, and secondary inflammatory response, which consequently promotes the vulnerability of atherosclerotic plaques [43]. In the models of atherosclerotic mice, high-fat diet inhibits Beclin1-mediated protective effects of macrophage autophagy, thus accelerating the progression of atherosclerosis [44]. Similarly, rapamycin can restore impaired autophagy by inhibiting the mammalian target of rapamycin (mTOR), thereby leading to the selective clearance of macrophages, increased cholesterol efflux, reduced apoptotic cells in atherosclerotic plaques, and achieving the stabilization of atherosclerotic mice [45, 46].

3.2.2 Exercise and Atherosclerosis

Regular appropriate exercise has a fundamental role in the pathogenesis and treatment of atherosclerotic cardiovascular disease. It exerts a protective effect against the development of atherosclerosis irrespective of other cardiovascular risk factors. Additionally, exercise-induced vascular hemodynamic change is helpful for elucidating the presence of obscure vascular involvement. Also, exercise is one of the major interventional modalities in peripheral arterial diseases with accumulating evidences from improving exercise capacity and reducing the symptoms of cardiovascular diseases. As influenced by numerous endogenous and exogenous factors with complex mechanisms, until now, it is assumed that genetic predisposition is the deterministic factor in the occurrence of atherosclerotic manifestations [47]. Nevertheless, current evidence supports the theme that genetic risk can be attenuated by adherence to a healthy lifestyle. A study has reported that people with high genetic risks adopted a favorable lifestyle including exercise present a 50% reduction in cardiovascular risks when compared to people without exercise [48]. Likewise, a research report confirmed that lifestyle changes can contribute to the reduction in cardiovascular risks due to the epigenetic modification [49]. However, the specific mechanism still needs to be further elucidated.

3.2.3 Effect of Exercise on Autophagy in Aorta

Autophagy is typically activated under energy stress conditions such as energy restriction, obesity, and increased physical activity to degrade damaged proteins and aged cells, which plays a crucial role in maintaining intracellular homeostasis and affecting cell survival [50]. In recent years, it has been confirmed that regular appropriate exercise can enhance the activity of adaptive autophagy and plays a positive role in the prevention and treatment of atherosclerosis. In aortic plaques of atherosclerotic mice, the significantly decreased autophagosomes and LC3-II/LC3-I ratio are detected, suggesting the suppressed autophagy, which may be related to the large amount of lipid depositions in cells to block the fusion between autophagosomes and lysosomes [51]. In addition, long-term regular exercise can produce certain mechanical stress and energy stress, thus leading to the damage of a variety of proteins and organelles. Autophagy is properly activated to remove the damaged cytoplasmic contents, inhibit oxidative stress, inflammatory response, and other mediators for regulating apoptosis, and improve the resistance of cells in stress environment [52]. A study has demonstrated that intermittent hemodynamic stimulation induced by exercise could enhance the structure and function of blood vessel wall and inhibit the formation of atherosclerotic plaques [53]. Recent studies have shown that regular physical exercise can promote the production of nitric oxide (NO) and the phosphorylation of endothelial NO synthase (eNOS) in arterial endothelial cells, and also reduce oxidative stress injury and inflammatory response of vascular lesions [54]. At the same time, autophagy-associated biomarkers such as

Beclin1, LC3, Atg3, and Lamp2 are increased and the expression level of p62 is decreased [31]. However, some studies have shown that no significant changes in autophagy-related proteins such as Atg5/12 or LC3-II/LC3-I ratio are observed in rodents after 3 months of exercise training [55], suggesting that the effect of exercise on autophagy is variable and may be also affected by various factors such as exercise intensity, health state, nutritional level, and tissue type.

3.2.4 Exercise-Induced Autophagy Inhibits Atherosclerosis by Regulating Inflammatory Response and Lipid Metabolism

Inflammation and abnormal lipid metabolism are important risk factors for the occurrence and development of atherosclerosis [56]. A series of inflammatory factors including TNF-α, IL-6, IL-1, human monocyte chemotactic protein-1 (MCP-1), MMP-9, soluble intercellular adhesion molecule-1 (sICAM-1), and other inflammatory factors are involved in the pathogenesis of atherosclerosis [57]. In addition, normally, there should also have a delicate balance between the flow of cholesterol in and out of cells, and the breakdown of this balance can cause abnormal lipid accumulation in endothelial cells and form lipid bands as the early stages of atherosclerosis. More recent studies have confirmed that autophagy can affect immune responses in the microenvironment by regulating the secretion of inflammatory factors [58]. In general, autophagy and its related proteins often have obviously regulatory effects on inflammatory responses. Under the Atg7 knockout condition, lipopolysaccharide (LPS)-dependent inflammasomes are activated, accompanied by the release of various inflammatory factors such as IL-1β and IL-18 [59]. Furthermore, morin hydrate has been found to inhibit the formation of atherosclerotic plaques in $ApoE^{-/-}$ mice and reduce the levels of inflammatory cytokines such as TNF-α, sICAM-1, and COX-2, and its anti-inflammatory effect may be correlated with the activation of autophagy induced by cAMP/PRKA/AMPK/SIRT1 signaling pathway [60]. Therefore, exercise-induced autophagy can effectively regulate inflammatory responses and reverse the abnormal lipid metabolism in the atherosclerotic models, thereby rescuing the pathological processes and suppressing the development of atherosclerosis.

3.3 *Exercise-Mediated Autophagy Alleviates Myocardial Ischemia–Reperfusion Injury*

After acute myocardial infarction, the early recovery of myocardial reperfusion through thrombolysis or primary percutaneous coronary intervention can reduce the area of myocardial infarction and improve clinical symptoms of myocardial ischemia–reperfusion injury, but the recovery of blood supply to the ischemic area may further aggravate myocardial injury, that is myocardial ischemia–reperfusion injury [61]. Autophagy is another important mechanism involved in myocardial ischemia–reperfusion injury [62–64]. Under physiological conditions, autophagy maintains the structure and function of the heart; however, under stress conditions from ischemia and hypoxia, autophagy is activated to execute the protective function for cardiovascular system by degrading unnecessary or denatured proteins and damaged mitochondria and reducing oxidative stress, thereby suppressing myocardial injury. On the other hand, excessive autophagy can lead to the self-destruction and death of myocardial cells, thereby finally aggravating myocardial injury during reperfusion [65]. These results suggest that autophagy may play different regulatory roles in myocardial cells at different stages of ischemia and reperfusion.

3.3.1 The Regulation of Autophagy During Ischemia

ATP required by cardiomyocytes is produced by oxidative phosphorylation of mitochondria under normal conditions in the heart [63, 64]. During myocardial ischemia, mitochondrial damage causes the cessation of oxidative phosphorylation, thereby resulting in the reduced ATP production, and the "starvation state" caused by insufficient oxygen supply and metabolic raw materials in cardiomyocytes [66]. At the same time, autophagy is activated to maintain cellular homeostasis of cardiomyocytes. Decker and Wildenthal have conducted the experiments to validate the activated autophagy in Langendorff perfused rabbit hearts under ischemia, and the prolonged ischemia seems to impair the autophagosome–lysosome pathway [67]. The animal model of myocardial infarction caused by ligation of the anterior descending branch of the left coronary artery in mice has revealed the significant upregulation of both LC3-II and p62 proteins, thereby suggesting the significantly enhanced autophagy or impaired autophagy flux in the ischemic myocardium [68]. In addition, autophagy is generally believed as an evolutionarily conserved process of restoring energy under the condition of myocardial ischemia and plays a protective role in the myocardium by clearing damaged mitochondria and inhibiting mitochondrial signal pathway-mediated apoptosis. Several studies have explored the possible mechanisms of autophagy activation in ischemia and hypoxia tissues at the early stage of myocardial ischemia as follows: The decrease in ATP level and the increase in AMP/ATP ratio can directly or indirectly activate AMP-activated protein kinase (AMPK), an energy sensor, and then induce autophagy through the AMPK-mTORC1-ULK1 signaling pathway.

3.3.2 The Regulation of Autophagy During Reperfusion

When myocardial reperfusion is restored, autophagy is still activated, but the signal pathway for autophagy induction is different from that of myocardial ischemia. Since the ischemia and hypoxia in cardiomyocytes are removed during reperfusion, the signal pathway of autophagy induced by AMPK during ischemia is inhibited, while Beclin1 plays an important role in the induction of autophagy during reperfusion at this time. One study has demonstrated that Beclin1$^{+/-}$ mice have reduced autophagy during reperfusion when compared with wild-type mice [69]. However, Beclin1 overexpression could increase autophagy activity during ischemia and reperfusion model in vitro [70]. These studies indicate that Beclin1 is a critical autophagy-related protein-regulating autophagosome formation and processing, and the upregulation of Beclin1 is responsible for the initiation of autophagy during reperfusion. The suppression of autophagy by 3-methyladenine (3-MA) [69, 71] and Beclin1 siRNA [16] can interfere with the initiation process of autophagy, thereby deficient autophagy at the reperfusion stage could aggravate myocardial injury. The possible mechanisms may be as follows: During reperfusion process, the reoxygenation through myocardial blood supply can induce the more production of ROS as a stress signal, and can upregulate Beclin1 expression, thus promoting autophagosome formation, inhibiting the fusion between autophagosomes and lysosomes, and then leading to limited autophagosome removal and excessive accumulation, which in turn results in the damage of normal function of proteins and organelles, and finally accelerates the autophagic death of cardiomyocytes. These studies confirm that autophagy plays a dual role in myocardial ischemia and reperfusion injury. However, the specific role and regulatory mechanism of autophagy in cardiac ischemia and reperfusion injury remain to be further studied, in order to lay a theoretical foundation for the development of novel and effective drugs targeting autophagy in the future, and take full advantage of beneficial roles of autophagy in myocardial ischemia–reperfusion injury, and improve the prognosis of these patients.

3.3.3 Exercise Regulates Autophagy-Mediated Myocardial Ischemia–Reperfusion Injury

Exercise training is a kind of mechanical and stress stimulation. Human epidemiological studies have shown that regular exercise can reduce the risk of death during myocardial ischemia–reperfusion injury [72]. Meanwhile, the studies in animal models have also shown that regular endurance exercise (running or swimming training) may mimic the favorable cardioprotective effect of ischemia preconditioning and attenuate cardiomyocyte death under the circumstance of myocardial ischemia–reperfusion injury [73–76]. Appropriate exercise training could maintain the homeostasis of intracellular environment through moderate activation of autophagy. During cardiac adaptive exercise, autophagy plays an important role in exercise-induced cardiac protection as a stress response and is a necessary process of exercise preconditioning.

A growing body of evidence suggests that appropriate exercise training can improve the level of myocardial autophagy, promote the proliferation of cardiomyocytes and the generation of new blood vessels in myocardial tissue, increase myocardial blood reperfusion, reduce the level of local tissue inflammation, help to remove damaged organelles and longevity proteins, and finally achieve the effect of improving cardiac function. Therefore, exercise training with appropriate intensity can moderately activate autophagy, reduce the release of apoptotic factors such as cytochrome C in myocardial mitochondria, and enhance antioxidant activity and anti-apoptotic capacity in myocardium of rats [77]. Similarly, He et al. have also found that appropriate exercise can promote the enhancement of autophagy in cardiomyocytes for the degradation and elimination of misfolded proteins and damaged organelles, thereby providing certain energy for the synthesis of substrates for the regeneration of muscle fibers, so that the self-stable state of cardiomyocytes can be maintained [78]. The adaptive changes in myocardial mitochondrial autophagy are also correlated with the improvement of myocardial fibrous membrane function and/or mitochondrial ATP-sensitive potassium ion channels as the underlying mechanism of myocardial protection during exercise intervention [79]. Exercise, as a stimulus signal, could simulate myocardial ischemia to induce autophagy through AMPK-mTORC signal pathway, thus achieving the protective effect on heart during ischemia [80].

Therefore, regulating autophagy can reduce the degree of myocardial injury through exercise interventions, but the specific mechanism of whether exercise interventions can protect basic cardiac functions by enhancing or inhibiting autophagy is still controversial. However, many questions related to myocardial protection and potential mechanism of exercise-mediated autophagy against myocardial ischemia–reperfusion injury need to be further addressed.

3.4 Exercise-Induced Autophagy in Cardioprotection

Autophagy is a biological process in eukaryotic cells for degrading aging or damaged organelles and denatured proteins using lysosomes to maintain normal heart function [81]. Dysfunction of autophagy can lead to the accumulation of misfolded proteins due to the weak clearance capability, thus resulting in a large amount of metabolic waste accumulation in cells, and impacting normal metabolism of the cells, and triggering a series of chronic diseases. In recent years, increasing studies have indicated that autophagy not only plays a critical role in metabolic stress state and the choices of cell survival and death, but also plays a key role in prevention and rehabilitation of cardiovascular diseases. As an endogenous cell protection mechanism, a moderate level of autophagy may play a protective role in cardiomyocytes, but insufficient autophagy or excessive autophagy may lead to cell death of cardiomyocytes.

Appropriate intensity of exercise training can enhance the level of autophagy, accelerate the degradation of metabolic waste in cells, and is essential for maintenance of cellular homeostasis and functions [78]. Autophagy can stimulate the production of new amino acids for generation and formation of new cellular contents or organelles under hypoxia condition after degrading damaged organelles and abnormal proteins, thereby maintaining myocardial energy metabolism. Under the condition of exercise and hypoxia, AMPK, an energy sensor, can induce autophagy to enhance aerobic respiration efficiency of cardiomyocytes [82], thereby promoting cell survival and reducing apoptosis of myocardium [83]. Activated AMPK can enhance glucose uptake through the translocation of the glucose transporter 4 (GLUT4) on cardiac cell membranes [84], and fatty acid oxidation [85], which increases glucose transport and ATP production to regulate cardiac metabolism [86].

Mitochondrial damage, oxidative stress, and autophagy are involved in myocardial ischemia and injury. Under normal circumstances, autophagy is maintained at a low level in cardiomyocytes. However, 8-week exercise training can re-establish the status of autophagic flux and mitochondrial quality control to enhance cardiac function during the process of heart failure [87]. On the other hand, the mice with autophagy-related gene defects such as *Atg7* knockout do not exhibit the improvement of cardiac functions upon exercise interventions, which further confirms the crucial role of the functional status of autophagy induced by exercise in cardioprotection [88].

Autophagy may suppress myocardial apoptosis by reducing the release of pro-apoptotic substances such as apoptosis-inducing factor and cytochrome C (CytC). Autophagosomes encapsulate damaged mitochondria, prevent the release of CytC into the cytoplasm, and inhibit the formation of apoptotic bodies [89]. As a golden biomarker of autophagy, LC3-II or LC3-II/LC3-I ratio combined with p62 level is usually considered to be a specific indicator for the functional status of autophagy including autophagy activation or improved autophagic flux. The significant increase in LC3-II level in cardiomyocytes around the ischemic area also indicates the activation of autophagy in this region of cardiomyocytes [90]. A previous study has demonstrated that the number of autophagosomes exhibits a rapid and significant increase in lesion areas in ischemic preconditioning heart, and an effectively reduced myocardial infarction area. However, Tat-ATG5 (K130R), as an inhibitor of autophagy, can reduce the protective effect of ischemic preconditioning on cardiomyocytes, thus confirming that autophagy induced by ischemic preconditioning is essential for cardioprotection [91]. Further studies have confirmed that the expression level of LC3-II is significantly increased and p62 is significantly decreased in the infarction–peripheral area after the treatment with rapamycin when compared with the acute myocardial infarction group. At the same time, the size of left ventricular infarction area is significantly decreased in the rapamycin group, but that is significantly increased in the acute myocardial infarction group [18].

Mitophagy also plays an important role in cardiac health. Parkin and BNIP3L are key proteins to regulate mitophagy. It has been found that the number of damaged mitochondria in the infarct area is increased and the level of myocardial apoptosis is significantly increased in the myocardial infarction model of mice with Parkin knockout [92]. Similarly, the significantly pathological cardiac hypertrophy of the mice with *BNIP3L* knockout at the age of 12 months is also observed [93]. A number of studies demonstrate that both acute exercise and long-term exercise may activate myocardial mitophagy, thus exerting a cardioprotective function. Acute exercise for 45 min could significantly upregulate the BNIP3L expression and activate mitophagy in cardiomyocytes in young mice [94]. The sustained short-duration exercise (15-min swimming training) can contribute to the attenuation of cardiac dysfunction after myocardial infarction in aged mice, rather than long-duration exercise (60-min swimming training). This study has also found that 15-min swimming training improves mitochondrial quality control via regulating mitochondrial fission–fusion and increasing the expression of PINK/Parkin to regulate mitophagy, thereby augmenting left ventricular function, suppressing myocardial fibrosis and apoptosis, and increasing survival rate of cardiomyocytes in aged mice [95]. In contrast, exhaustive exercise could cause serious cardiomyofibril injury and result in the failure to eliminate damaged mitochondria, but late exercise preconditioning can significantly attenuate exhaustive exercise-induced injury via Parkin-mediated mitophagy. However, the suppression of mitophagy could lead to the alleviation of cardioprotection induced by late exercise preconditioning [96].

Aging and ischemic heart are often accompanied by the dysregulation of autophagy and decreased and dysfunctional mitochondrial quality control. Exercise at appropriate intensity can execute the maintenance at the optimal level of autophagy, especially mitophagy, within the normal range, thereby playing a protective role on ischemic heart and delaying the development of ischemic heart disease through enhancing mitochondrial quality control. For example, 4-week swimming (60 min/day, 5 days/week) can significantly enhance mitophagy, promote mitochondrial dynamics and biogenesis in cardiomyocytes, and improve mitochondrial quality, thus reducing cardiac injury caused by exhaustive exercise [97]. In addition, the rats subjected to 8 weeks of high-intensity running exercise before the operation for cardiac ischemia–reperfusion injury reveal the significant upregulation of Mitofusion1/2 (MFN1/2) and the significant reduction in myocardial infarction area, indicating that exercise preconditioning could improve the mitochondrial quality and suppress ischemia–reperfusion myocardial injury by modulating mitochondrial fusion and fission, and mitophagy is an underlying mechanism involved in aerobic exercise training-induced cardioprotection against ischemia–reperfusion injury [98]. Moreover, exercise-induced cardioprotection against ischemia–reperfusion injury is highly correlated with exercise-induced adaptive protection of cardiomyocytes, as shown in delaying left ventricular remodeling, promoting cardiac vasodilation and microvascular regeneration, enhancing myocardial contractility, and suppressing arrhythmias [99, 100]. Therefore, exercise intervention for triggering the activation of autophagy or the optimal functional status of autophagy and improving mitochondrial quality control and function may be a novel and effective strategy for the prevention, treatment, and rehabilitation of aging-related ischemic heart disease or other cardiovascular diseases [101, 102].

4 Conclusion

Cardiovascular diseases, as the common and frequent occurrence of chronic diseases, have high incidence or high mortality and poor prognosis, especially for elderly population. Physical inactivity and unhealthy diets are the major leading causes for this kind of chronic diseases. However, once cardiovascular diseases

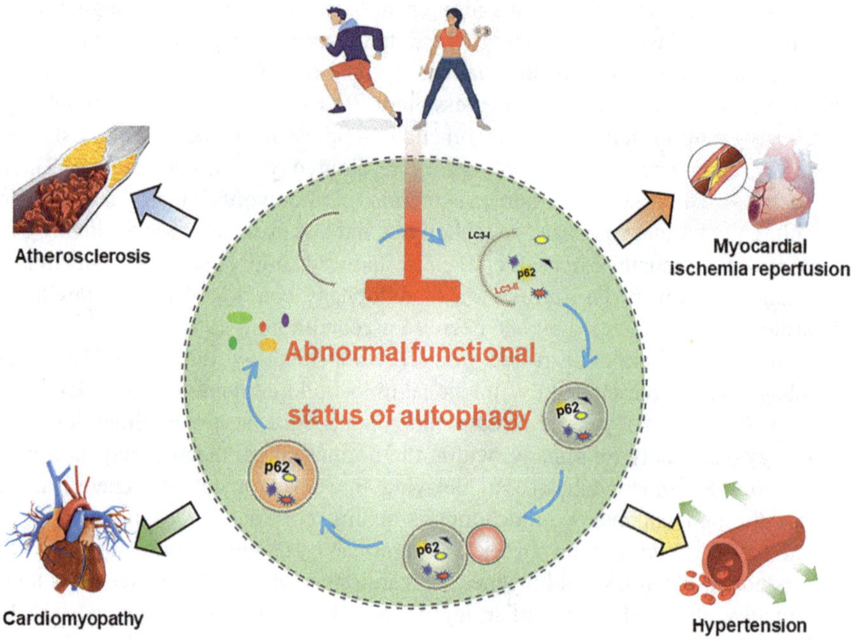

Fig. 9.2 Exercise interventions rescue abnormal functional status of autophagy in cardiomyocytes or vascular endothelial cells to attenuate cardiovascular diseases such as hypertension, atherosclerosis, cardiomyopathy, and myocardial ischemia–reperfusion injury

result in the serious damage for the tissues and organs in cardiovascular system, it is usually irreversible, even lifelong disability or death, especially elderly population. In order to cope with these issues, exercise is a novel and effective strategy for accomplishing cardioprotection, and executing the prevention, treatment, and rehabilitation of cardiovascular diseases including hypertension, atherosclerosis, cardiomyopathy, and myocardial ischemia–reperfusion injury, due to its obvious function of triggering the activation of autophagy or the optimal functional status of autophagy and improving mitochondrial quality control and function, which is also the problem-resolving strategy from the original sources of chronic diseases through exercise interventions (Fig. 9.2).

References

1. Kearney PM, Whelton M, Reynolds K et al (2005) Global burden of hypertension: analysis of worldwide data. Lancet 365:217–223
2. Ren J, Zhang Y (2018) Targeting autophagy in aging and aging-related cardiovascular diseases. Trends Pharmacol Sci 39:1064–1076
3. Wen CP, Wai JP, Tsai MK et al (2011) Minimum amount of physical activity for reduced mortality and extended life expectancy: a prospective cohort study. Lancet 378:1244–1253

4. Fiuza-Luces C, Santos-Lozano A, Joyner M et al (2018) Exercise benefits in cardiovascular disease: beyond attenuation of traditional risk factors. Nat Rev Cardiol 15:731–743
5. Kavazis AN, Mcclung JM, Hood DA et al (2008) Exercise induces a cardiac mitochondrial phenotype that resists apoptotic stimuli. Am J Physiol Heart Circ Physiol 294:H928–H935
6. Choi SY, Chang HJ, Choi SI et al (2009) Long-term exercise training attenuates age-related diastolic dysfunction: association of myocardial collagen cross-linking. J Korean Med Sci 24:32–39
7. Gielen S, Laughlin MH, O'Conner C et al (2015) Exercise training in patients with heart disease: review of beneficial effects and clinical recommendations. Prog Cardiovasc Dis 57:347–355
8. Le W (2020) Autophagy: biology and diseases: clinical science. Springer, Singapore
9. Jiang F (2016) Autophagy in vascular endothelial cells. Clin Exp Pharmacol Physiol 43:1021–1028
10. Swaminathan B, Goikuria H, Vega R et al (2014) Autophagic marker MAP1LC3B expression levels are associated with carotid atherosclerosis symptomatology. PLoS One 9:e115176
11. Razani B, Feng C, Coleman T et al (2012) Autophagy links inflammasomes to atherosclerotic progression. Cell Metab 15:534–544
12. Sergin I, Bhattacharya S, Emanuel R et al (2016) Inclusion bodies enriched for p62 and polyubiquitinated proteins in macrophages protect against atherosclerosis. Sci Signal 9:ra2
13. Grootaert MOJ, Roth L, Schrijvers DM et al (2018) Defective autophagy in atherosclerosis: to die or to senesce? Oxidative Med Cell Longev 2018:7687083
14. Grootaert MO, Da Costa Martins PA, Bitsch N et al (2015) Defective autophagy in vascular smooth muscle cells accelerates senescence and promotes neointima formation and atherogenesis. Autophagy 11:2014–2032
15. Mccarthy CG, Wenceslau CF, Calmasini FB et al (2019) Reconstitution of autophagy ameliorates vascular function and arterial stiffening in spontaneously hypertensive rats. Am J Physiol Heart Circ Physiol 317:H1013–H1027
16. Zhu H, Tannous P, Johnstone JL et al (2007) Cardiac autophagy is a maladaptive response to hemodynamic stress. J Clin Invest 117:1782–1793
17. Dong Y, Chen H, Gao J et al (2019) Molecular machinery and interplay of apoptosis and autophagy in coronary heart disease. J Mol Cell Cardiol 136:27–41
18. Aisa Z, Liao GC, Shen XL et al (2017) Effect of autophagy on myocardial infarction and its mechanism. Eur Rev Med Pharmacol Sci 21:3705–3713
19. Zhang H, Yin Y, Liu Y et al (2020) Necroptosis mediated by impaired autophagy flux contributes to adverse ventricular remodeling after myocardial infarction. Biochem Pharmacol 175:113915
20. Guo R, Hu N, Kandadi MR et al (2012) Facilitated ethanol metabolism promotes cardiomyocyte contractile dysfunction through autophagy in murine hearts. Autophagy 8:593–608
21. Nakai A, Yamaguchi O, Takeda T et al (2007) The role of autophagy in cardiomyocytes in the basal state and in response to hemodynamic stress. Nat Med 13:619–624
22. Li B, Chi RF, Qin FZ et al (2016) Distinct changes of myocyte autophagy during myocardial hypertrophy and heart failure: association with oxidative stress. Exp Physiol 101:1050–1063
23. Zhou L, Ma B, Han X (2016) The role of autophagy in angiotensin II-induced pathological cardiac hypertrophy. J Mol Endocrinol 57:R143–R152
24. Morciano G, Patergnani S, Bonora M et al (2020) Mitophagy in cardiovascular diseases. J Clin Med 9:892
25. Hashemzaei M, Entezari Heravi R, Rezaee R et al (2017) Regulation of autophagy by some natural products as a potential therapeutic strategy for cardiovascular disorders. Eur J Pharmacol 802:44–51
26. Moraes-Silva IC, Mostarda CT, Silva-Filho AC et al (2017) Hypertension and exercise training: evidence from clinical studies. Adv Exp Med Biol 1000:65–84

27. Cao F, Ma S (2018) Autophagy and hypertension. In: Autophagy and cardiometabolic diseases. Elsevier, London, pp 91–99
28. Chen Y, Li S, Guo Y et al (2020) Astaxanthin attenuates hypertensive vascular remodeling by protecting vascular smooth muscle cells from oxidative stress-induced mitochondrial dysfunction. Oxidative Med Cell Longev 2020:4629189
29. Li C, Jiang F, Li YL et al (2018) Rhynchophylla total alkaloid rescues autophagy, decreases oxidative stress and improves endothelial vasodilation in spontaneous hypertensive rats. Acta Pharmacol Sin 39:345–356
30. Fiorenza M, Gunnarsson TP, Ehlers TS et al (2019) High-intensity exercise training ameliorates aberrant expression of markers of mitochondrial turnover but not oxidative damage in skeletal muscle of men with essential hypertension. Acta Physiol (Oxf) 225:e13208
31. Park SK, La Salle DT, Cerbie J et al (2019) Elevated arterial shear rate increases indexes of endothelial cell autophagy and nitric oxide synthase activation in humans. Am J Physiol Heart Circ Physiol 316:H106–H112
32. Herrera NA, Jesus I, Shinohara AL et al (2016) Exercise training attenuates dexamethasone-induced hypertension by improving autonomic balance to the heart, sympathetic vascular modulation and skeletal muscle microcirculation. J Hypertens 34:1967–1976
33. Gambardella J, Morelli MB, Wang XJ et al (2020) Pathophysiological mechanisms underlying the beneficial effects of physical activity in hypertension. J Clin Hypertens (Greenwich) 22:291–295
34. Wu MY, Li CJ, Hou MF et al (2017) New insights into the role of inflammation in the pathogenesis of atherosclerosis. Int J Mol Sci 18:2034
35. Ross R (1993) Atherosclerosis: current understanding of mechanisms and future strategies in therapy. Transplant Proc 25:2041–2043
36. Paffenbarger RS Jr, Hyde RT, Wing AL et al (1986) Physical activity, all-cause mortality, and longevity of college alumni. N Engl J Med 314:605–613
37. Young DR, Hivert MF, Alhassan S et al (2016) Sedentary behavior and cardiovascular morbidity and mortality: a science advisory from the American Heart Association. Circulation 134:e262–e279
38. Pedersen BK, Saltin B (2015) Exercise as medicine—evidence for prescribing exercise as therapy in 26 different chronic diseases. Scand J Med Sci Sports 25(Suppl 3):1–72
39. Luo Z, Xu W, Ma S et al (2017) Moderate autophagy inhibits vascular smooth muscle cell senescence to stabilize progressed atherosclerotic plaque via the mTORC1/ULK1/ATG13 signal pathway. Oxidative Med Cell Longev 2017:3018190
40. Liang J, Zeng Z, Zhang Y et al (2020) Regulatory role of exercise-induced autophagy for sarcopenia. Exp Gerontol 130:110789
41. Liao X, Sluimer JC, Wang Y et al (2012) Macrophage autophagy plays a protective role in advanced atherosclerosis. Cell Metab 15:545–553
42. Luo Y, Lu S, Zhou P et al (2016) Autophagy: an exposing therapeutic target in atherosclerosis. J Cardiovasc Pharmacol 67:266–274
43. Sun W, Lin Y, Chen L et al (2018) Legumain suppresses OxLDL-induced macrophage apoptosis through enhancement of the autophagy pathway. Gene 652:16–24
44. Geng Z, Xu F, Zhang Y (2016) MiR-129-5p-mediated Beclin-1 suppression inhibits endothelial cell autophagy in atherosclerosis. Am J Transl Res 8:1886–1894
45. Pakala R, Stabile E, Jang GJ et al (2005) Rapamycin attenuates atherosclerotic plaque progression in apolipoprotein E knockout mice: inhibitory effect on monocyte chemotaxis. J Cardiovasc Pharmacol 46:481–486
46. Gadioli AL, Nogueira BV, Arruda RM et al (2009) Oral rapamycin attenuates atherosclerosis without affecting the arterial responsiveness of resistance vessels in apolipoprotein E-deficient mice. Braz J Med Biol Res 42:1191–1195
47. White PD (1957) Genes, the heart and destiny. N Engl J Med 256:965–969
48. Khera AV, Emdin CA, Drake I et al (2016) Genetic risk, adherence to a healthy lifestyle, and coronary disease. N Engl J Med 375:2349–2358

49. Whayne TF Jr, Saha SP (2019) Genetic risk, adherence to a healthy lifestyle, and ischemic heart disease. Curr Cardiol Rep 21:1
50. Sanchez AM, Candau R, Bernardi H (2019) Recent data on cellular component turnover: focus on adaptations to physical exercise. Cells 8:542
51. Cao H, Jia Q, Shen D et al (2019) Quercetin has a protective effect on atherosclerosis via enhancement of autophagy in ApoE(−/−) mice. Exp Ther Med 18:2451–2458
52. Mooren FC, Krüger K (2015) Exercise, autophagy, and apoptosis. Prog Mol Biol Transl Sci 135:407–422
53. Green DJ, Hopman MT, Padilla J et al (2017) Vascular adaptation to exercise in humans: role of hemodynamic stimuli. Physiol Rev 97:495–528
54. Hambrecht R, Adams V, Erbs S et al (2003) Regular physical activity improves endothelial function in patients with coronary artery disease by increasing phosphorylation of endothelial nitric oxide synthase. Circulation 107:3152–3158
55. Kim J, Kim YC, Fang C et al (2013) Differential regulation of distinct Vps34 complexes by AMPK in nutrient stress and autophagy. Cell 152:290–303
56. Babar G, Clements M, Dai H et al (2019) Assessment of biomarkers of inflammation and premature atherosclerosis in adolescents with type-1 diabetes mellitus. J Pediatr Endocrinol Metab 32:109–113
57. Ezhov M, Safarova M, Afanasieva O et al (2019) Matrix metalloproteinase 9 as a predictor of coronary atherosclerotic plaque instability in stable coronary heart disease patients with elevated lipoprotein(a) levels. Biomolecules 9:129
58. Grootaert MOJ, Moulis M, Roth L et al (2018) Vascular smooth muscle cell death, autophagy and senescence in atherosclerosis. Cardiovasc Res 114:622–634
59. Nakahira K, Haspel JA, Rathinam VA et al (2011) Autophagy proteins regulate innate immune responses by inhibiting the release of mitochondrial DNA mediated by the NALP3 inflammasome. Nat Immunol 12:222–230
60. Zhou Y, Cao ZQ, Wang HY et al (2017) The anti-inflammatory effects of Morin hydrate in atherosclerosis is associated with autophagy induction through cAMP signaling. Mol Nutr Food Res 61. https://doi.org/10.1002/mnfr.201600966
61. Chen-Scarabelli C, Agrawal PR, Saravolatz L et al (2014) The role and modulation of autophagy in experimental models of myocardial ischemia-reperfusion injury. J Geriatr Cardiol 11:338–348
62. Choi AM, Ryter SW, Levine B (2013) Autophagy in human health and disease. N Engl J Med 368:651–662
63. Kroemer G (2015) Autophagy: a druggable process that is deregulated in aging and human disease. J Clin Invest 125:1–4
64. Mizushima N, Komatsu M (2011) Autophagy: renovation of cells and tissues. Cell 147:728–741
65. Gustafsson AB, Gottlieb RA (2008) Eat your heart out: role of autophagy in myocardial ischemia/reperfusion. Autophagy 4:416–421
66. Santulli G (2018) Cardioprotective effects of autophagy: eat your heart out, heart failure! Sci Transl Med 10:eaau0462
67. Decker RS, Wildenthal K (1980) Lysosomal alterations in hypoxic and reoxygenated hearts. I. Ultrastructural and cytochemical changes. Am J Pathol 98:425–444
68. Fidziańska A, Bilińska ZT, Walczak E et al (2010) Autophagy in transition from hypertrophic cardiomyopathy to heart failure. J Electron Microsc (Tokyo) 59:181–183
69. Matsui Y, Takagi H, Qu X et al (2007) Distinct roles of autophagy in the heart during ischemia and reperfusion: roles of AMP-activated protein kinase and Beclin 1 in mediating autophagy. Circ Res 100:914–922
70. Efeyan A, Zoncu R, Chang S et al (2013) Regulation of mTORC1 by the Rag GTPases is necessary for neonatal autophagy and survival. Nature 493:679–683
71. Liu H, Lei H, Shi Y et al (2017) Autophagy inhibitor 3-methyladenine alleviates overload-exercise-induced cardiac injury in rats. Acta Pharmacol Sin 38:990–997

72. Powers SK, Quindry JC, Kavazis AN (2008) Exercise-induced cardioprotection against myocardial ischemia-reperfusion injury. Free Radic Biol Med 44:193–201
73. Bowles DK, Starnes JW (1994) Exercise training improves metabolic response after ischemia in isolated working rat heart. J Appl Physiol (1985) 76:1608–1614
74. Demirel HA, Powers SK, Zergeroglu MA et al (2001) Short-term exercise improves myocardial tolerance to in vivo ischemia-reperfusion in the rat. J Appl Physiol (1985) 91:2205–2212
75. Lennon SL, Quindry JC, French JP et al (2004) Exercise and myocardial tolerance to ischaemia-reperfusion. Acta Physiol Scand 182:161–169
76. Huang Y, Liu HT, Yuan Y et al (2021) Exercise preconditioning increases Beclin1 and induces autophagy to promote early myocardial protection via intermittent myocardial ischemia-hypoxia. Int Heart J 62:407–415
77. Kavazis AN, Smuder AJ, Min K et al (2010) Short-term exercise training protects against doxorubicin-induced cardiac mitochondrial damage independent of HSP72. Am J Physiol Heart Circ Physiol 299:H1515–H1524
78. He C, Sumpter R Jr, Levine B (2012) Exercise induces autophagy in peripheral tissues and in the brain. Autophagy 8:1548–1551
79. Golbidi S, Laher I (2011) Molecular mechanisms in exercise-induced cardioprotection. Cardiol Res Pract 2011:972807
80. Klionsky DJ, Saltiel AR (2012) Autophagy works out. Cell Metab 15:273–274
81. Lee Y, Kwon I, Jang Y et al (2017) Potential signaling pathways of acute endurance exercise-induced cardiac autophagy and mitophagy and its possible role in cardioprotection. J Physiol Sci 67:639–654
82. Wang C, Wang Y, Mcnutt MA et al (2011) Autophagy process is associated with anti-neoplastic function. Acta Biochim Biophys Sin (Shanghai) 43:425–432
83. Zhao M, Sun L, Yu XJ et al (2013) Acetylcholine mediates AMPK-dependent autophagic cytoprotection in H9c2 cells during hypoxia/reoxygenation injury. Cell Physiol Biochem 32:601–613
84. Yang J, Holman GD (2005) Insulin and contraction stimulate exocytosis, but increased AMP-activated protein kinase activity resulting from oxidative metabolism stress slows endocytosis of GLUT4 in cardiomyocytes. J Biol Chem 280:4070–4078
85. Dyck JR, Lopaschuk GD (2006) AMPK alterations in cardiac physiology and pathology: enemy or ally? J Physiol 574:95–112
86. Young LH, Li J, Baron SJ et al (2005) AMP-activated protein kinase: a key stress signaling pathway in the heart. Trends Cardiovasc Med 15:110–118
87. Campos JC, Queliconi BB, Bozi LHM et al (2017) Exercise reestablishes autophagic flux and mitochondrial quality control in heart failure. Autophagy 13:1304–1317
88. Yan Z, Kronemberger A, Blomme J et al (2017) Exercise leads to unfavourable cardiac remodelling and enhanced metabolic homeostasis in obese mice with cardiac and skeletal muscle autophagy deficiency. Sci Rep 7:7894
89. Melendez A, Talloczy Z, Seaman M et al (2003) Autophagy genes are essential for dauer development and life-span extension in C. elegans. Science 301:1387–1391
90. Zhao R, Xie E, Yang X et al (2019) Alliin alleviates myocardial ischemia-reperfusion injury by promoting autophagy. Biochem Biophys Res Commun 512:236–243
91. Huang C, Yitzhaki S, Perry CN et al (2010) Autophagy induced by ischemic preconditioning is essential for cardioprotection. J Cardiovasc Transl Res 3:365–373
92. Kubli DA, Zhang X, Lee Y et al (2013) Parkin protein deficiency exacerbates cardiac injury and reduces survival following myocardial infarction. J Biol Chem 288:915–926
93. Dorn GW 2nd (2010) Mitochondrial pruning by Nix and BNip3: an essential function for cardiac-expressed death factors. J Cardiovasc Transl Res 3:374–383
94. Li H, Miao W, Ma J et al (2016) Acute exercise-induced mitochondrial stress triggers an inflammatory response in the myocardium via NLRP3 inflammasome activation with mitophagy. Oxidative Med Cell Longev 2016:1987149

95. Zhao D, Sun Y, Tan Y et al (2018) Short-duration swimming exercise after myocardial infarction attenuates cardiac dysfunction and regulates mitochondrial quality control in aged mice. Oxidative Med Cell Longev 2018:4079041
96. Yuan Y, Pan SS (2018) Parkin mediates mitophagy to participate in cardioprotection induced by late exercise preconditioning but Bnip3 does not. J Cardiovasc Pharmacol 71:303–316
97. Dun Y, Liu S, Zhang W et al (2017) Exercise combined with Rhodiola sacra supplementation improves exercise capacity and ameliorates exhaustive exercise-induced muscle damage through enhancement of mitochondrial quality control. Oxidative Med Cell Longev 2017:8024857
98. Ghahremani R, Damirchi A, Salehi I et al (2018) Mitochondrial dynamics as an underlying mechanism involved in aerobic exercise training-induced cardioprotection against ischemia-reperfusion injury. Life Sci 213:102–108
99. Barboza CA, Souza GI, Oliveira JC et al (2016) Cardioprotective properties of aerobic and resistance training against myocardial infarction. Int J Sports Med 37:421–430
100. Powers SK, Smuder AJ, Kavazis AN et al (2014) Mechanisms of exercise-induced cardioprotection. Physiology (Bethesda) 29:27–38
101. Wu NN, Zhang Y, Ren J (2019) Mitophagy, mitochondrial dynamics, and homeostasis in cardiovascular aging. Oxidative Med Cell Longev 2019:9825061
102. Roca-Agujetas V, De Dios C, Leston L et al (2019) Recent insights into the mitochondrial role in autophagy and its regulation by oxidative stress. Oxidative Med Cell Longev 2019:3809308

Chapter 10
Exercise-Induced Autophagy in the Prevention and Treatment of Sarcopenia

Jingjing Fan, Xia Mo, Kai Zou, and Ning Chen

1 The Pathogenesis of Sarcopenia

Sarcopenia is one of the aging-associated diseases with the gradual development over decades, and its development process consists of diverse changes including physiological loss of skeletal muscle mass and strength, decreased repairing function after injury, and increased risk of falls and fractures, as well as the affected prognosis of diseases in the elderly. According to the statistics [1], the global number of people with the age of 60 years or above has increased substantially in recent years and is projected to accelerate in the coming decades, thereby doubling the number by 2050 to an astonishing 2.1 billion, which predicts that the number of sarcopenia will keep an increasing trend. Beyond the progressive atrophy of skeletal muscle, sarcopenia is also closely correlated with several aging-associated diseases such as cardiovascular diseases, cancer, diabetes, and obesity, which will in turn aggravate the onset of sarcopenia and lead to a significant impairment on the quality of life of the elderly in a vicious cycle [2–4]. According to a meta-analysis study of sarcopenia [5], it reveals that the incidence of sarcopenia in patients (65.3 $\pm$ 1.6 years old) who have cardiovascular diseases (CVD), Alzheimer's disease (AD), diabetes, or respiratory diseases is much higher than the control individuals (54.6 $\pm$ 16.2 years old) having similar diseases, reaching 31.4%, 26.4%, 31.1%, and 26.8%, respectively.

Skeletal muscle is an extremely heterogeneous tissue, composed of a large variety of fiber types that are classified as type I and type II fibers. During natural aging

J. Fan · X. Mo · N. Chen (✉)
Tianjiu Research and Development Center for Exercise Nutrition and Foods, Hubei Key Laboratory of Exercise Training and Monitoring, College of Health Science, Wuhan Sports University, Wuhan, China

K. Zou
Department of Exercise and Health Sciences, University of Massachusetts Boston, Boston, MA, USA

N. Chen (ed.), *Exercise, Autophagy and Chronic Diseases*,
https://doi.org/10.1007/978-981-16-4525-9_10

process, it has been shown an decline in the total number of skeletal muscle fibers with the predominant atrophy of type II fibers [6]. Indeed, the atrophy of skeletal muscle fibers caused by a group of complex factors can be described as sarcopenia. These factors include decreased activity of satellite cells, increased systemic inflammatory factors, reduced mitochondrial metabolism and antioxidant capacity, and dysfunctional regulation of hormones [7–9], thereby leading to an imbalance between protein synthesis and protein degradation. The mass maintenance of skeletal muscle is a dynamic process involving protein synthesis and protein degradation. Insufficient protein ingestion, which plays an important role in skeletal muscle mass, cannot satisfy daily requirements and can lead to negative protein balance [10]. In elderly, accelerated protein degradation, coupled with reduced protein intake, results in skeletal muscle atrophy [11, 12]. Recently, increasing evidence supports that the deficient or dysfunctional autophagy can also accelerate the progression of skeletal muscle atrophy [13]. Autophagy is a process of engulfing cytoplasmic proteins or organelles into closed double-membrane vesicles called autophagosomes, eventually fusing with lysosomes to form autophagolysosomes and degrade the contents, thereby fulfilling the metabolic needs of the cells and the renewal of certain organelles [14]. It is noteworthy that many studies have documented the important role of autophagy in the maintenance of skeletal muscle mass and strength [15, 16]. Besides the intervention of internal factors, sarcopenia is also positively correlated with unhealthy lifestyles, such as sedentary lifestyle. A large number of publications have reported that a sedentary lifestyle exacerbated the progression of sarcopenia [17]. In contrast, regular exercise enhances autophagy flux, improves mitochondrial quality, and re-activates quiescent satellite cells, thereby delaying the atrophy of skeletal muscle or sarcopenia.

Currently, the accurate molecular mechanism by which exercise-induced autophagy regulates skeletal muscle atrophy or sarcopenia is still not fully elucidated. Herein, we reviewed the roles and underlying mechanisms of exercise-induced autophagy in the prevention and rehabilitation of sarcopenia from the aspects of autophagy–lysosomal system (ALS) and ubiquitin–proteasome system (UPS), satellite cell activation, mitochondrial quality control, and systemic inflammation (Fig. 10.1). Our review will provide a rational explanation for the prevention and mitigation of sarcopenia upon exercise intervention, and offer a guidance of scientific exercise intervention of sarcopenia based on regulating the functional status of autophagy.

2 The Regulation of Autophagy in Sarcopenia

2.1 ALS and UPS

Aging-related skeletal muscle atrophy is mainly caused by the imbalance between protein synthesis and protein degradation. There are two main systems for the degradation of skeletal muscle during the aging process: ALS and UPS. ALS, also

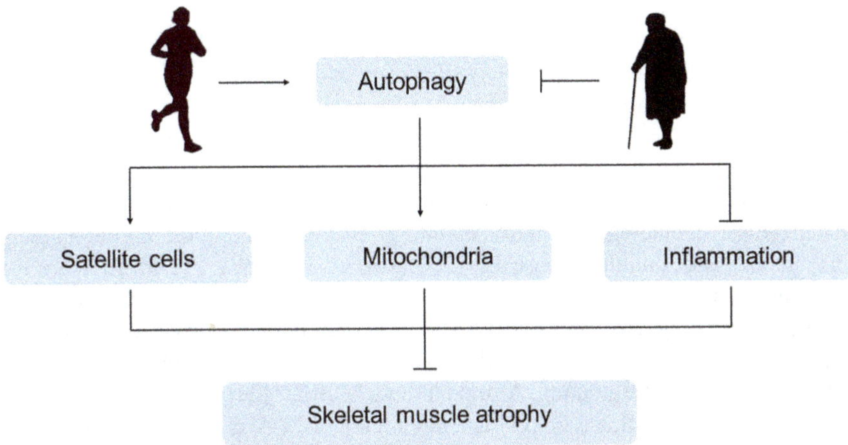

Fig. 10.1 In elderly, the mass and strength loss of skeletal muscle are correlated with autophagy, which can be due to decreased satellite cell activity, reduced mitochondrial function, and increased inflammation. Exercise appears to mediate a significant suppression on the mass and strength loss of skeletal muscle through regulating the functional status of autophagy

known as the autophagolysosomal system, is a highly conserved cell degradation process to execute a significant modulation on maintaining cellular homeostasis of skeletal muscle [18]. It is suggested that autophagy-related proteins such as ATG5 (autophagy-related gene 5), ATG7, and LC3 (microtubule-associated protein light chain 3) are downregulated in aging mice, and as the extension of age, the decline in autophagy level results in a negative impact on skeletal muscle function [17]. Moreover, autophagy defect reduces skeletal muscle function by exacerbating age-related deterioration of neuromuscular synaptic structure and function, and the rescuing of autophagy in aging mice can improve the function of neuromuscular synapticity [19]. These findings suggest that autophagy may play a critical role in the regulation of skeletal muscle atrophy with aging. Phosphoinositide 3-kinase (PI3K)/protein kinase B (AKT)/mammalian target of rapamycin (mTOR) signaling pathway is an important pathway in the regulation of autophagy in skeletal muscle. However, the transduction of PI3K/AKT signaling pathway in skeletal muscle is compromised during aging process, subsequently disrupting the cellular homeostasis in skeletal muscle. The PI3K/AKT signaling pathway also regulates protein synthesis in skeletal muscle mainly through the activation of mTORC1 and the downregulation of FoxO [20]. mTOR is one of the major regulators of autophagy. Rapamycin, an inhibitor of mTOR, can induce autophagy in mammalian cells and yeast [21]. Interestingly, increased activity of mTORC1 is observed during aging process [22], and the inhibitors of mTORC1 can restore autophagy to reverse the regulators of sarcopenia. In addition, FoxO3, a member of the FoxO family and another crucial regulator of autophagy, is necessary for the activation of the lysosome-dependent protein degradation system [23]. In fact, FoxO3 induces autophagy under starvation to maintain the cellular homeostasis of skeletal muscle; however, the overexpression of FoxO3 disrupts the homeostatic state [24], suggesting that overexpression of FoxO3 may induce excessive activation of autophagy and accelerate protein degradation, thereby leading to the atrophy of skeletal muscle.

UPS is the primary proteolytic system for regulating protein degradation and maintaining cellular homeostasis. UPS binds to the target protein covalently, and then, the ubiquitinated protein is tagged and translocated to the proteasome to be degraded by 26S proteasome [25]. The components of UPS mainly involve 3 types of enzymes, including ubiquitin-activating enzyme E1, ubiquitin-conjugating enzyme E2, and ubiquitin ligase E3 [26]. Atrogin-1/MAFbx and MuRF1 are the first identified muscle-specific E3 ligases [27]. FoxO3, as a transcription factor that simultaneously regulates ALS and UPS, plays an important role in the protein degradation of skeletal muscle [28]. FoxO3 could accelerate protein degradation through UPS by upregulating Atrogin-1 and MuRF1 [29], and induce ALS in skeletal muscle [30]. Since FoxO3 is a key regulator in ALS or UPS, more studies are necessary to be conducted to explore its underlying mechanism, which will be a potential therapeutic target of sarcopenia.

Interestingly, the knockout of MuRF1 in mice spares skeletal muscle mass with dexamethasone-induced skeletal muscle atrophy, suggesting its impact on protein synthesis in addition to protein degradation [31]. Similar results have also been reported in fasting animal models [32]. The knockout of Atrogin-1 using shRNA upregulates MyoD, and suppresses the growth differentiation factor (GDF)-8- (myostatin), thereby enhancing muscle cell differentiation and delaying the atrophy of skeletal muscle both in vivo and in vitro.

ALS and UPS are the major signal pathways for protein degradation in skeletal muscle, and the overactivation of both signal pathways will aggravate the atrophy of skeletal muscle. The previous report has demonstrated that autophagy induction can stimulate skeletal muscle atrophy in transgenic mice with the knockout of Nrf2 [33] or specific expression SOD1 ($SOD1^{G93A}$) [34]. In fact, autophagy can contribute to the delaying of skeletal muscle atrophy, yet the excessive activation of autophagy can aggravate the loss of skeletal muscle as an opposite result. This relationship is complicated by the fact that dysfunctional status of autophagy may be deleterious with divergent aging-related diseases. Consistent with above reports, the functional status or level of autophagy is the determinant for the health of skeletal muscle (Fig. 10.2).

2.2 *Mitochondrial Quality Control*

As the indispensable organelles that supply energy and produce ATP, mitochondria play an essential role in the normal morphology and function of tissues and organs, especially in skeletal muscle, which is an energy factory for muscle contraction. Skeletal muscle accounts for about 40% of the body weight and is the most energy-consuming organ. Therefore, the abnormal function of mitochondria is involved in the occurrence of many diseases, such as sarcopenia. Functional decrease in skeletal

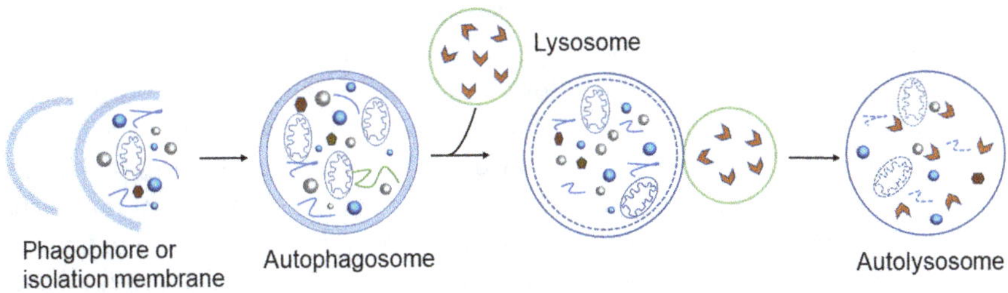

Fig. 10.2 When the cell receives autophagy-induced signals, it first forms a phagophore similar to a double-layer liposome in the cytoplasm, and then, the phagophore continues to extend and wrap damaged organelles or proteins to form an autophagosome. The autophagosome is transported to the lysosome in the cytoplasm and finally fuses with the lysosome. In the acidic environment of the lysosome, the substances in the autophagosome are degraded by various enzymes

muscle can be due to a reduced number of mitochondria and a subsequent reduction in mitochondrial enzymes. During aging process, the abnormal morphology, enlargement, or fragmentation of mitochondria is observed in skeletal muscle [35], and the skeletal muscles of aged rodents and humans have a higher overall rate of fission versus fusion as compared with the skeletal muscle of younger individuals [36]. Based on the free radical theory of mitochondrial aging, reactive oxygen species (ROS) is increased in mitochondrial senescence during aging process, which is associated with impaired mitochondrial function and mitochondrial DNA (mtDNA), thus resulting in the reduction in redox capacity. ROS causes irreversible impairment of skeletal muscle, mainly because ROS changes the balance of the mitochondrial respiratory chain, thereby leading to a weakness in the function of the antioxidant defense system [37, 38]. In fact, the atrophy of skeletal muscle during aging process is not only due to the accumulation of ROS in mitochondria, but also due to the downregulation of antioxidant factors. Superoxide dismutase (SOD) is proposed as an important scavenger of active oxygen free radicals by catalyzing the conversion of superoxide radicals to oxygen and hydrogen peroxide. The functional relationship between SOD and ROS suggests that SOD can resist and block the cell damage caused by ROS, and repair damaged cells timely. The SOD family mainly includes SOD1 (CuZnSOD, mainly in the cytoplasm), SOD2 (MnSOD, mainly in the mitochondria of eukaryotic cells), SOD3 (FeSOD, mainly in prokaryotic cells), and they are characterized by converting superoxide anion radical O_2^- to H_2O_2 [39]. Early studies have identified that 35% of SOD is located in the mitochondria of skeletal muscle. The normal activity and function of SOD are responsible for the function of skeletal muscle. In mice lacking SOD1 (*Sod1*$^{-/-}$), the occurrence of overall accelerated aging phenotype in skeletal muscle can stimulate the decline in skeletal muscle mass, the loss of skeletal muscle fibers, the reduced number of motor units, the loss of motor function, partial denervation, and increased oxidative stress [40]. Actually, an integrated "2-hit" mechanism has proposed that the atrophy of skeletal muscle in global SOD1 knockout mice involves motor neurons and neuromuscular junction (NMJ) dysfunction and increased mitochondrial ROS [41]. In contrast, a mouse model with skeletal muscle-specific knockout of SOD2 exhibits an increase in muscle mass associated with elevated mitochondrial peroxide generation and oxidative stress, but no obvious aging phenotype [42].

Another factor associated with the loss of function in skeletal muscle during aging process is mitochondrial dynamics, including continuous alternation of mitochondrial morphology and distribution controlled by fusion and fission events [43]. Mitochondrial quality control includes mitochondrial biogenesis, mitochondrial fusion and fission, and the clearance of dysfunctional mitochondria via mitophagy [44, 45]. The process of mitochondrial fusion and fission that continuously reshapes the mitochondrial network is defined as mitochondrial dynamics, which plays a critical role in maintaining mitochondrial quality and function. The imbalance of mitochondrial fusion and fission, coupled with the aggregation of damaged mtDNA and abnormal mitochondria, aggravates muscle degradation in aging [46, 47]. Mitophagy plays an indispensable role in the expansion of the mitochondrial reticulum and the degradation of damaged organelles in a quality control process that maintains the function of mitochondria in a balanced state, which is critical for function and homeostasis of skeletal muscle. As a cellular housekeeping mechanism, mitophagy aims to recycle damaged mitochondria to provide future protein synthesis for cell survival. However, the expression of autophagy markers such as LC3, Bnip3, Beclin1, p62, Atg7, Parkin, and LAMP2 exhibits a declining trend as the extension of age in both animals and humans, thus leading to the accumulation of damaged mitochondria [19]. Moreover, the mitochondrial renewal disorder can further stimulate the production of ROS and nondegradable lipofuscin, which is associated with the decreased ATP production and cell metabolism, eventually leading to the irreversible skeletal muscle atrophy [48]. Since the molecular mechanism of mitophagy involved in the regulation of mitochondrial function in sarcopenia is still unclear, many studies have focused on this field. High-temperature demand protein A2 (HtrA2/Omi) is a mitochondrial protease involved in a variety of cellular physiological processes, including autophagy, inflammation, immune activity, and mitochondrial dynamics and biological formation, which contributes to the protein homeostasis of mitochondria [49]. At the same time, the phenotypes associated with sarcopenia can be induced in HtrA2/Omi knockout mice, as shown in decreased skeletal muscle strength, and obvious collagen deposition, irregular arrangement, significantly decreased number of type I and type II fibers in skeletal muscle, as well as the appearance of dysfunctional mitochondria [50]. HtrA2/Omi is suggested to participate in the pathogenesis of sarcopenia based on the regulation of mitophagy. The nucleus adjusted the adaptability of mitochondria in a signal anterograde manner, and PGC-1α may be highly associated with the regulation of this process. As is a key regulator of nuclear-encoded mitochondrial genes, PGC-1α can regulate the biological formation of mitochondria mainly through two major transcription factors NRF1 and NRF2 [51]. This study has demonstrated the decreased expression of NRF1 and PGC-1α and its downstream targets such as GPX1, SOD1/2, catalase, and UCP2 in the HtrA2mnd2(−/−) mouse model [50]. A consequence of increased PGC-1α and AMPK can increase the oxidative capacity and inhibit the degradation of mitochondria, indicating the importance of

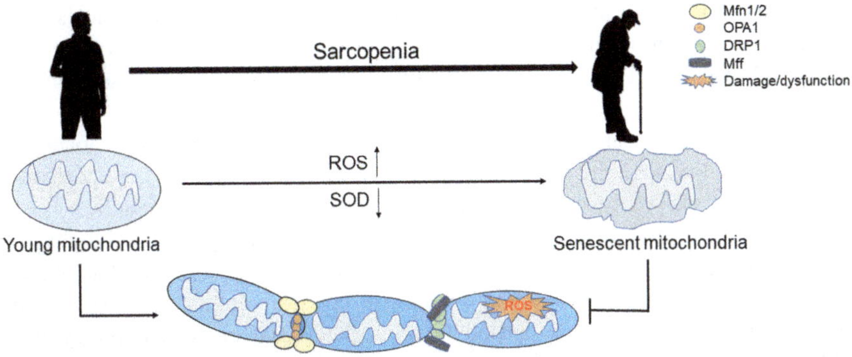

Fig. 10.3 During aging process, the mitochondrial oxidation and antioxidation capacity is unbalanced, thereby resulting in the reduced fission and fusion capacity of mitochondria, and mitochondrial dysfunction. Mfn1/2 (mitofusin1/2) and OPA1 (optic atrophy protein 1) are involved in mitochondrial outer and inner membrane fusion, respectively, to form a functional mitochondrial network. DRP1 (dynamin-related protein 1) and Mff (mitochondrial fission factor) mediate mitochondrial fission. During aging process, the reduced expression of DRP1 and Mff could lead to imbalanced mitochondrial fusion and fission, thereby contributing to skeletal muscle atrophy

mitochondrial dynamics in maintaining mitochondrial biological formation and mitophagy during aging process (Fig. 10.3).

2.3 *Satellite Cells*

In addition to abnormal protein metabolism in skeletal muscle, the regeneration of skeletal muscle is also an essential factor in aging-related skeletal muscle atrophy when subjected to physiological and pathological stimuli. The failure of skeletal muscle regeneration is a major cause of physical incapacity in the elderly. Satellite cells (SCs) are located between the basement membrane and the basal lamina of myofibers, and are necessary for the repairing and regeneration of skeletal muscle in response to stress. SCs are usually divided into two states: the stationary state and the activated state. At the stationary stage, SCs are identified as $Pax7^{+}/Myod^{-}$ cells that normally reside in a quiescent state [52]. When stimulated by physiological (such as muscle contraction during exercise) or pathological (aging, injury, muscle-related diseases) stimuli, the quiescent SCs are capable of further proliferation and differentiation into mature skeletal muscle fibers. Early investigation using eccentric contraction and/or radiation to damage the SCs has shown that the mass of skeletal muscle in Irr (centrifugal + radiation injury)/Inj (eccentric injury) mice exhibits a decrease by 16–21%, which highlights the importance of SCs in the regulation of skeletal muscle mass [53]. Although skeletal muscle maintains a lifelong capacity to regenerate itself, the regenerative response and myogenic potential following injury can be dramatically declined with the reduced quantity and activation of SCs during aging process [54, 55]. Even though SCs develop inherent defects with aging, the host environment may also impact their function. In a study using parabiosis mice, transplanting young tissues to old mice did not recover their function of recipient mice [56], suggesting the aging environment of SCs may compromise their function in repairing capacity.

Although there is abundant evidence indicating that the repairing capacity of skeletal muscle after injury presents a gradual weakness trend in the accelerated aging process, the precise mechanism is less clear. During aging process, SCs lose its regenerative potential, in part due to the deficiency of autophagy [57]. Autophagy is critical for the activation of the static SCs by clearing damaged SCs and acting as a temporary energy source to fuel the initiation of cell proliferation, therefore, delays the atrophy of skeletal muscle caused by aging. The deficiency of autophagy in the elderly SCs causes the accumulation of damaged proteins and organelles, and finally leads to the failure of stem cells. Genetically modified mice with green fluorescent protein (GFP)-LC3 are established to evaluate the functional status of autophagy in silent stem cells of skeletal muscle, which has demonstrated the damaged autophagosomes in senescent SCs. Compared with young mice, the activity of autophagy in SCs exhibits a significant reduction in aging mice. Moreover, based on the observation of the basal level of autophagy in neonatal SCs of hybrid mice, both autophagosomes and SCs are reduced in silent ATG7-null SCs. After skeletal muscle injury, the SC activity and proliferation capacity in young $Atg7^{\Delta Pax7}$ mice (specific deletion of Atg7 in $PAX7^{+}$) are downregulated, with lower regenerated fibers. When treated with tamoxifen for 30 days, $Atg7^{\Delta Pax7}$ mice show a decrease in the number of SCs, indicating that basal autophagy is necessary for the quantity and function of adult stem cells. In addition, blocking autophagy in the resting SCs of young mice presented an aging phenotype. After using rapamycin or upregulating the expression of ATG7 to restore autophagy, the proliferation capacity of SCs was rescued, with a rejuvenating effect and suppressed senescence to some extent [58]. Although the molecular mechanism between SC and autophagy during aging process needs more in-depth discussion, the studies have demonstrated that the inhibition of AMPK/P27 kip1 signaling by the regulation of Notch and Wnt in aging SCs is related to the weakness of autophagy and proliferation, as well as the enhancement of apoptosis and aging [57]. More importantly, when the aging SCs are transplanted into the young recipients, the regenerative capacity of SCs can be restored. These changes can also be mediated by the changes in systemic factors, including synthetic hormones, IGF, and testosterone, suggesting that the host environment may also influence SC function [56]. As age increases, the autophagy in SCs of skeletal muscle is unable to meet energy demand, and is more likely to undergo apoptosis. The activated caspases are significantly overexpressed in aging individuals, while the fragmentation rate of DNA reveals the increase by sevenfold to ninefold when compared with younger groups [59]. Consistent with the above results, the apoptotic rate of SCs is increased during the process of aging. Whether those postulated mechanisms hold true remains largely ambiguous, it is worth noting that substantial evidence also demonstrates inconsistent results and considers an unnecessary role of SCs in fully grown adult mice, thus resulting in bare effects to promote sarcopenia, unloading-induced atrophy, or regrowth [60]. A better understanding of skeletal muscle atrophy is necessary to effectively prevent or minimize this phenomenon.

2.4 *Inflammation*

The immune system is the main defense mechanism against invaders to prevent the body from diseases. In the process of aging, a chronic low-grade inflammation appears in the whole body, which accelerates the damage of mitochondria, promotes ROS production, and in the long term can progressively cause skeletal muscle deterioration [61]. The relationship between inflammation and skeletal muscle performance has demonstrated that poor exercise potential is considered to be the driving force in obese and overweight elderly. A study has shown that this low-grade inflammation is mediated by increased inflammatory factors such as H_2O_2, IL-1β, TNF-α, and hsCRP (high-sensitivity C-reactive protein), along with an alteration of the neurohormonal response such as IGF-1 and $PDGF^{BB}$ (platelet-derived growth factor BB), which results in an imbalance between protein synthesis and protein degradation in skeletal muscle [62]. The key cytokines present in senescence processes, especially high plasma concentration of IL-6 and TNF-α, are critical factors in the deterioration of skeletal muscle mass and function [63, 64]. Accumulating evidences over the past decade have demonstrated that inflammation is a common phenomenon in patients with sarcopenia. Compared with the young group and the healthy control group, the expression of IL-1β, TNF-α, IL-6, and IL-15 in the serum of sarcopenia patients increases significantly [65]. Another study is conducted to compare neutrophil–lymphocyte ratio (NLR) between individuals with sarcopenia and healthy elderly, and clarifies that NLR, white blood cells (WBC), erythrocyte sedimentation rate, and inflammatory marker CRP in the serum of individuals with sarcopenia are significantly higher than those in the serum of healthy elderly [66]. This result suggests that the increased NLP may be an independent predictor of sarcopenia. As a crucial role in allowing cells to adapt to a diverse array of environmental stimuli, NF-κB resides in the cytosol of cells in an inactive state. Given that the major inflammatory factors are mainly controlled by the NF-κB signaling pathway, aberrant persistent activation of NF-κB and its DNA-binding activity are detected in the skeletal muscle of the elderly populations and aged mice [67, 68].

As mentioned above, the cells in the skeletal muscle of the individuals with sarcopenia are probable tendency to apoptosis rather than autophagy. Similarly, inflammation is also involved in the regulation of apoptosis during the aging process of skeletal muscle cells. In 24-month-old mice, the level of apoptosis is 50% higher than 6-month-old young mice [69]; thus, aging-related apoptosis accelerates the atrophy of skeletal muscle in aged mice [70]. In addition, age-related systemic low-grade inflammation boosts insulin resistance and ROS production, thereby stimulating cell apoptosis and leading to protein degradation in skeletal muscle and the subsequent atrophy of skeletal muscle [71]. Consistent with this,

LPS-induced inflammation downregulates the expression of miR-532-3p and upregulates the expression of its target gene *BAK1* (pro-apoptotic factor), thereby facilitating apoptosis of skeletal muscle in a manner of sarcopenia [65]. Autophagy also plays a critical role in regulating inflammatory cells such as monocytes and macrophages. Monocytes are derived from hematopoietic stem cells (HSCs) and can differentiate into macrophages when stimulated by injury or infection. HSC activity decreases during aging process. Autophagy generates ATP and reduces the production of ROS to maintain the static state of HSC. In TSC1-deficient HSC mice, the activity of mTOR is activated, which promotes the formation of mitochondria, the increased ROS production, and the activation of HSC, which could be reversed with activated autophagy induced by rapamycin [72]. Monocytes can be induced to differentiate into pro-inflammatory factor M1 or anti-inflammatory M2 macrophages when stimulated by pro-inflammatory factors such as INF-γ and TNF [52]. The blockade of autophagy in macrophage-specific Atg7 knockout mice can give rise to more differentiation into M1 macrophages with the impairment of glucose homeostasis when fed with high-fat diet [73]. When mTOR is activated, macrophage differentiation stimulated by IL-4 is more trend to the pro-inflammatory M1 phenotype [74].

3 Exercise-Induced Autophagy Regulates Sarcopenia

3.1 Exercise and Sarcopenia

Currently, there is no effective specific medicine to prevent and treat sarcopenia. Interventional strategies for sarcopenia are mainly focused on exercise, nutrition, and medication. Increasing studies have revealed that exercise, high-quality protein intake, and calorie restriction have certain effects on delaying sarcopenia [75, 76]. It has been suggested that regular physical exercise, and particularly resistance training, can promote healthy aging and may be the most effective way to inhibit skeletal muscle wasting.

Sarcopenia is negatively correlated with the adaptability of skeletal muscle to exercise. A recent study has revealed that moderate-intensity resistance training (55–60% of 1 RM) is sufficient to counteract the aging-related muscle mass loss [77]. The unfolded protein response (UPR) regulates the protein quality control of skeletal muscle and physiological adaptability to exercise training. However, aging reduces the activation of the UPR response to exercise stimulation and disrupts the effective self-correction system with the accumulation of misfolded proteins. These maladaptive UPR outputs trigger apoptosis, eventually leading to the damage of skeletal muscle [78]. In the quest to understand whether regular exercise has effects on aging-related dysfunction, a previous study applying lifelong exercise training in rats initiated at the age of 8 months has demonstrated that muscle function, mitochondrial function, and endoplasmic reticulum stress are all ameliorated in comparison with sedentary counterparts [79]. Furthermore, lifelong exercise training also plays a positive role in offsetting lipid peroxidation-induced damage by improving the antioxidant defense with increased activities of superoxide dismutase (SOD), glutathione peroxide (GPX), and glutathione reductase (GR), during aging process [80]. Moreover, exercise regulates mitochondrial proteome in skeletal muscle, rescues the loss of the electron transfer chain (ETC) function in the old sedentary individuals, and suppresses protein damage, indicating its protective role in the maintenance of mitochondrial function during the aging process [81]. In terms of the balance between protein synthesis and protein degradation, it is well known that exercise can affect most of catabolic and anabolic pathways for regulating mitochondrial function and skeletal muscle mass, such as the simultaneous activation of AMPK and AKT-mTOR signaling pathway, to induce an insulin-sensitive phenotype and maintain the quality of skeletal muscle. Moreover, exercise is beneficial to improve the degradation efficiency of UPS and ALS simultaneously, and to stimulate mTOR for protein synthesis in protein turnover [82, 83].

3.2 *Exercise and Autophagy*

As an extremely conserved catabolic procedure in the cytoplasm of eukaryotes, the activation of autophagy by exercise appears to exert a significant suppression on the loss of skeletal muscle mass. It is reported that exercise is not able to rescue the Col6a1$^{-/-}$ mice with impaired autophagic flux, dysfunctional mitochondria, damaged sarcoplasmic reticulum, and excessive apoptosis and degradation of skeletal muscle. Importantly, neither acute exercise nor long-term exercise is able to ameliorate skeletal muscle atrophy in this animal model with impaired autophagy flux, thus illustrating that autophagy plays a vital regulatory role in mediating the positive effects of exercise on the maintenance of skeletal muscle mass [84]. Mechanistically, the relationship between exercise and autophagy has been evaluated by our research group using SAMP8 mice (rapid aging mouse model) in our research group. Exercise combined with spermidine (an autophagy inducer) attenuates aging-related skeletal muscle atrophy through induced autophagy and suppresses apoptosis with enhanced mitochondrial function, which has documented the important roles of AMPK-FoxO3a signaling pathway in the prevention and treatment of aging-induced skeletal muscle atrophy [85]. In a knock-in mouse model with the mutation of Bcl-2 phosphorylation sites, the disruption of Bcl-2-Beclin1 complex can interrupt the activation of autophagy, thus resulting in an impaired maximum exercise capacity. Consistently, the reduced maximum running distance has also been found in ATG6$^{-/-}$ and Beclin1$^{+/-}$ mice [86]. In addition, the increased cross-sectional area and strength of skeletal muscle are observed in aged mice following 9 weeks of resistance exercise training, with concomitant increase in the expression of protein markers associated with autophagy (e.g., Beclin1, ATG5, and ATG7). Meanwhile, the declined p62 and LC3-II/LC3-I ratio indicates that resistance exercise training also upregulates the basal autophagic flux [87]. Therefore, the normal functional status of autophagy or the capacity of autophagy induction is highly required for the adaptation of skeletal muscle to exercise; in other words, autophagy may be required for improved exercise capacity and the prevention and treatment of sarcopenia upon exercise intervention. It is worth noting that the role of autophagy in the prevention and rehabilitation of sarcopenia is highly correlated with the type, intensity, and duration of exercise intervention. In fact, excessive activation of autophagy has been shown to contribute to the loss of skeletal muscle in a plethora of conditions, such as cancer cachexia, fasting, and disuse conditions [88]. Therefore, more studies are still needed to explore the optimal exercise program.

3.3 *Molecular Regulators of Exercise-Induced Autophagy*

In aging, the decrease in skeletal muscle mass is caused by multifactors called sarcopenia. Autophagy is known to protect aging skeletal muscle atrophy by renewing dysfunctional organelles and maintaining skeletal muscle homeostasis, which reveals gradual decline with aging. Exercise is considered to be an important intervention strategy to prevent sarcopenia by restoring autophagy. In recent years, the mechanism of exercise-induced autophagy to maintain the metabolism of aging skeletal muscle has attracted the attentions. Several key factors including energy receptor AMPK, mitochondrial bioformation factor PGC-1α, autophagy inhibitor mTOR, FoxO family, and the common regulators of UPS and ALS have been proved to be involved in this process.

3.3.1 AMPK

5′-AMP-activated protein kinase (AMPK), a master energy sensor in skeletal muscle in response to exercise training, plays an important role in the regulation of cellar metabolism and energy homeostasis [89]. AMPK can be activated at the threonine phosphorylation site 172 in response to the increased AMP/ATP ratio in skeletal muscle [90], thereby preventing aging-related diseases, such as sarcopenia [91]. Besides, AMPK can indirectly activate SIRT1 by regulating the level of NAD^+, which in turn phosphorylates and translocates PGC-1α to nucleus, thereby promoting its interaction with nuclear respiratory factor 1 (NRF1), cAMP response element-binding protein (CREB), myocyte enhancer factor 2 (MEF2), and histone deacetylase (HDACs) to improve skeletal muscle function [92]. However, with aging, the expression of p-$AMPK^{Thr172}$/AMPK in skeletal muscle decreases significantly, accompanied by the decline of mitochondrial fusion (Mfn2) and fission (DRP1) proteins. After exercise intervention, the phosphorylation level of AMPK rises, which can restore skeletal muscle function to a certain extent [47], indicating that AMPK plays an indispensable role in maintaining energy homeostasis and mitochondrial quality control of skeletal muscle in elderly. Of note, studies have provided evidence that exercise can mediate the activation of autophagy in skeletal muscle through AMPK and its downstream target ULK1 [93]. AMPK also induces autophagy by inhibiting the dissociation of mTORC1, ULK1, ATG13, and FIP200 complexes in skeletal muscle [94]. In the context of exercise, interestingly, acute exercise did not induce autophagy in AMPKα2$^{-/-}$ mice [95], suggesting exercise-induced autophagy may require the participation of AMPKα2. As previously mentioned, BCL-2AAA mutant mice block the formation of PI3K complexes and interrupt exercise-induced autophagy activity, accompanied by a reduction in AMPK phosphorylation [86]. Endurance exercise-induced skeletal muscle adaptation may be regulated by AMPK-induced autophagy, which depends on the characteristics of the exercise. There are other factors to execute the regulatory roles in autophagy by the complex network of signaling and metabolic pathways upon exercise interventions.

3.3.2 PGC-1α

PGC-1α, a common crucial transcriptional co-activator of mitochondrial generation, regulates a variety of factors by translocating into nucleus to interact with NRF1, NRF2, MEF2, and FoxO during muscle contraction, which exerts its regulatory effects on mitochondrial function [92]. In the aging mice without exercise intervention, the mRNA expression of PGC-1α in skeletal muscle is 60% lower than that in the young mice without exercise group. Interestingly, PGC-1α in the aging exercise group from the same study is upregulated by lifelong exercise [96], and p-AKT/AKT, BDNF, and LC3-II/LC3-I ratio are also observed in mice with the initiation of exercise training at the age of 8 months, not 18 months [79], suggesting PGC-1α expression is decreased with aging and involved in the regulation of beneficial adaptations to exercise training, including the activation of autophagy and the preservation of oxidative capacity during aging process [97]. Importantly, the overexpression of muscle-specific PGC-1α is strongly correlated with the increase in BNIP3 protein expression and basal autophagy flux, indicating that PGC-1α plays an important role in regulating autophagy in resting skeletal muscle [98]. Furthermore, in an exercise study, the expression of LC3-II is upregulated in the quadriceps muscle of mice with overexpressed PGC-1α after 1-h treadmill training session [99]. Consistent with these findings, PGC-1α in skeletal muscle is also involved in exercise training-induced adaptations, including autophagic renewal of damaged and aged cell components [99].

3.3.3 mTOR

mTOR, an essential cell growth regulator, senses various environmental and intracellular processes including cell growth, autophagy, survival, and metabolism, and interacts with factors for regulating both hypertrophy and atrophy of skeletal muscle. The expression of mTORC1 has been shown to be surprisingly hyperactivated in the rat model of sarcopenia. After partial inhibition of mTORC1 by RAD001 (everolimus, mTORC1 inhibitor), the ubiquitin proteins MuRF1 (muscle RING finger 1) and MT1 (metallothionein1), cyclins p21 (Cdkn1a) and p16 (Cdkn2a) involved in skeletal muscle atrophy, and neuromuscular junction-related proteins Chrnα, ChrnΣ, MuSK, Myogenin, and Gadd45a are inhibited and activated to protect from skeletal muscle atrophy, respectively [22]. Moreover, this signaling axis to autophagy has been supported by correlative evidence in a TSC1-deficient (TSC1mKO) mouse model, which reveals that both basal autophagy and starvation-induced autophagy are inhibited by continuous activation of mTORC1 even though FoxO3 is activated. Further studies have shown that in the TSC1mKO mice treated with rapamycin, the expression of p62 is decreased, and LC3-II/LC3-I ratio, p-ULK1, and Beclin1 are all upregulated, regardless of the soleus muscle or tibialis anterior muscle, thereby improving skeletal muscle mass and suggesting that mTORC1 inhibition is effective to restore autophagy to protect skeletal muscle [100]. However, after exercise intervention, especially resistance exercise, the expression of p-mTOR^{ser2448}/mTOR is reversed in skeletal muscle of aged rats, and the expression of autophagy-related proteins (Beclin1, LC3) is increased, thus improving aging-related atrophy of skeletal muscle [47]. In addition, the downregulated mTOR signaling can contribute to exercise-induced autophagy, thus ensuring the normal operation of metabolic pathways. This conclusion is also supported by alleviated atrophy of gastrocnemius muscle in 18- to 20-month-old rats upon exercise intervention [87]. As a double-edged sword in the regulation of skeletal muscle mass, the mTOR may play a dual function on protein synthesis and protein degradation in response to exercise. A previous study has applied a voluntary weightlifting training model to explore exercise-induced skeletal muscle contraction and metabolic adaptability. After 8 weeks of voluntary weight training, the mass and strength of skeletal muscle increase significantly. Simultaneously, the autophagic signal and the protein synthesis signal regulated by mTOR signaling pathway are also activated significantly [101]. The reason for the above results may be that mTOR regulates the metabolic synthesis of skeletal muscle after exercise through the AMPK/ULK1 and PI3K/AKT signaling pathways, respectively [102, 103]. Due to the complexity of the roles of mTOR in modulating the hypertrophy and atrophy of skeletal muscle, the diverse involvements of mTOR in maintaining skeletal muscle mass during exercise intervention need to be further explored.

3.3.4 FoxO

FoxO controls the expression of rate-limiting enzymes in both protein degradation systems including UPS and ALS in skeletal muscle and plays a critical role in regulating skeletal muscle atrophy to avoid excessive protein breakdown during sarcopenia. A large volume of evidence indicates that the capability of FoxO responds to nutritional deficiency and oxidative stress in dependence of the post-translational mechanisms including phosphorylation, acetylation, glycosylation, and ubiquitination. By interacting with histone acetyltransferase p300, FoxO3 is acetylated and transferred to the cytosol, and then degraded by ubiquitination

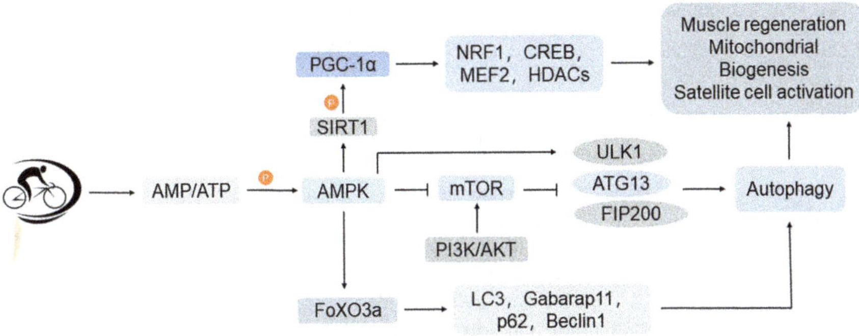

Fig. 10.4 AMPK as the master energy sensor in skeletal muscle response to exercise training can be activated by increased AMP/ATP ratio during exercise. AMPK can also indirectly activate SIRT1 by increased level of NAD^+, and then phosphorylate and transfer PGC-1α to the nucleus, thereby promoting its interaction with NRF1, CREB, MEF2, and HDACs to improve skeletal muscle function. Similarly, AMPK can induce autophagy by inhibiting the dissociation of mTORC, ULK1, ATG13, and FIP200 complexes in skeletal muscle. In addition, AMPK can activate LC3, Gabarap11, p62, and Beclin1 by FoxO3a to induce autophagy, thereby promoting muscle regeneration, mitochondrial biogenesis, and satellite cell activation

[28]. Therefore, AMPK participates in FoxO-mediated autophagy by the deacetylation and phosphorylation of FoxO, accompanied by increased autophagy in the myotubes of mice [104]. Also, autophagy is involved in antioxidant defense and maintaining the steady state of redox reactions. The suppressed activity of antioxidant enzymes and the elevated ROS production are observed in the autophagy-deficient cell model with the loss of FoxO1/3; simultaneously, the accumulated p62 in turn promotes the downregulation of FoxO1/3 expression [105], suggesting the vital role of p62-FoxO1/3 axis in the homeostasis of mitochondria and redox reactions mediated by autophagy. In addition, FoxO family can be inhibited by mTORC2 and activated by SIRT1 to stimulate autophagic process [92]. After acute resistance exercise, the expression of FoxO3 is higher in retired weightlifters than in the untrained control group [106]. During low-intensity endurance exercise under fasting conditions, the insulin-AKT-mTOR signaling pathway and its downstream targets such as FoxO3a and ULK1 play a critical role in the activation of autophagy [107]. These findings suggest the predominant roles of FOXOs as a potential therapeutic target to respond to aging or aging-related diseases upon exercise intervention (Fig. 10.4 and Table 10.1).

4 Exercise Advices

Physical activity is an effective way to delay or prevent skeletal muscle atrophy in older adults. All resistance training, aerobic training, and combinatorial training could improve skeletal muscle quality and physical performance, including improved skeletal muscle mass, body fat mass, appendicular percentage between

Table 10.1 Functional status of autophagy during various exercise intervention modes

Exercise type	Subject	Exercise mode	Levels	Effect
Endurance exercise	Col6a1$^{-/-}$ mice [84]	• Long-term voluntary exercise: running wheels, exercise lasts 3 months. • Acute exercise: treadmill exercise, starting with a speed of 10 cm/s and an increasing interval of 2 cm/s every 2 min until a maximum speed of 40 cm/s, lasting 1 h.	↓ LC3-II/LC3-I ↑ p-AKT	Autophagy ↓ Apoptosis ↑
	Rats [85]	• Swimming: 45 min/day, 5 days in each week within 6-weeks exercise training period.	↑ Beclin-1, LC3-II/LC3-I, AMPK-FoxO3a ↓ Bax, Caspase-3	Autophagy ↑ Apoptosis ↓
	BCL2AAA Mice [86]	• Long-term exercise: exercise treadmill with 10° uphill incline for 50 min/day, 17 m/min, 5 days/week, 8 weeks. • Acute exercise: treadmill exercise starting at the speed of 10 m/min, and 40 min later, increased speed at a rate of 1 m/min for totally 30 min, and then increased at a rate of 1 m/min every 5 min until exhaustion.	↑ p62 ↓ LC3-II/LC3-I, AMPK	Autophagy, Endurance training capacity ↓
	HFD mice AMPKα2$^{-/-}$ [95]	• Chronic exercise: treadmill exercise, 5 days/week, total 6 weeks, intensity of 75% VO_{2max} (12 m/min) for 60 min/day. • Acute exercise: exercise on treadmill for 1 h.	↑ LC3, AMPK, Sestrin2, Sestrin3 ↓ LC3, ULK1	Autophagy ↑ Autophagy ↓
	Mice [96]	• Lifelong voluntary exercise: voluntary wheel running (3–15 months).	↑ PGC-1α, BNIP3, p53, p21, LC3-II ↓ Bax/Bcl-2, Bax, LC3-I	Autophagy ↑ Apoptosis ↓
	Rats [79]	• Exercise initiation at 8 months old; first, 1-min warm-up, running speed of 10 m/min, 35–40% VO_{2max}; 45-min constant running speed at 17 m/min, 75–80% VO_{2max};	↑ p-AKT, p-FoxO1, p-MAPK, SIRT3, PGC-1α, TrkB, BDNF, SOD2, LC3-II/LC3-I ↓ 8-oHdG, 4-HNE	Autophagy ↑ Oxidative stress ↓

(continued)

Table 10.1 (continued)

Exercise type	Subject	Exercise mode	Levels	Effect
		cool-down at speed of 10 m/min, 1 min.		
	Mice [99]	• A single 1-h treadmill running bout at 14 m/min and 10° uphill incline.	↑ PGC-1α, LC3-II/LC3-I, p-AMPK, p-ACC2 ↓ P62	Autophagy ↑
	Mice [107]	• Acute treadmill running in the fasted state: run for 90 min at a speed of 10 m/min, which corresponds to low intensity (55% VO_{2max})	↑ LC3B, BNIP3, Parkin, FoxO3a, ULK1	Autophagy ↑
Resistance exercise	Rats [87]	• Climbing exercise: the initial weight attached is 10% of mouse body weight and gradually increased training period.	↑ Beclin1, Atg5/12, IGF-1, IGF-1R, FoxO3a, AMPK ↓ Cyt-C, Caspase-3, p-AKT, p-mTOR, LC3-II/LC3-I, p62	Autophagy ↑ Apoptosis ↓
	Mice [101]	• Voluntary weightlifting exercise. • Acute exercise: Push against 150% body load. • Long-term exercise: first day 100% with incremental increase in loading by 20% body weight each day until reaching 240%, lasting 8 weeks, 5 sessions/week with 1-day rest after 3 days of training and another.	↑ LC3-II/LC3-I ↓ p62	Autophagy ↑
	Human [106]	• Acute exercise: 3 sets, 70–75% of 1RM until voluntary fatigue.	↑ FoxO3 ↓ Myostatin	Cellular homeostasis ↑
Combination exercise	Rats [47]	• Treadmill exercise: running at a speed of 4.2 m/min with a progressive increase to a final speed of 12 m/min at a speed increment of 1 m/min every 30 s, 60 min/day, 3 days/week, 12 weeks. • Resistance exercise: climbing exercise, beginning with 10% of body weight, progressive loading increased at a rate of	↑ LC3, Beclin1, Bcl-2, PGC-1α, mTOR, AKT, AMPK, FoxO3a ↓ p62, Bax, Atrogin-1, MuRF1	Autophagy ↑ Apoptosis ↓

(continued)

Table 10.1 (continued)

Exercise type	Subject	Exercise mode	Levels	Effect
		10% body weight weekly, until 80% of body weight, 2 sets with 3 repetitions, followed by 1-min rest between each repetition. • Treadmill exercise plus climbing exercise. • Voluntary wheel running exercise.		

skeletal muscle mass and body weight, and visceral fat area (VFA) [108]. It is worth noting that resistance exercise exhibits the most effective performance to delay skeletal muscle atrophy [109, 110], and the greater grip strength and knee extensor strength in the resistance training group are significantly superior to that of the control group even after terminating exercise training for 4 weeks [108]. This may be related to the eccentric contraction-induced skeletal muscle loss during aging process, which is significantly less than concentric and isometric contractions. Therefore, resistance exercise prescription should be the target intervention in an exercise program [111, 112]. Moreover, the elderly is prone to multiple chronic diseases including diabetes, hypertension, obesity, coronary heart disease, and osteoarthritis, which can affect the functions of the entire body during aging process [113, 114]. Therefore, the state of other systems should be considered while offering an exercise prescription for sarcopenia. To sum up, combined exercise is more effective than resistance exercise alone to improve the overall health of the elderly [115]. In addition, nutrient intake is closely related to sarcopenia [75]. In a study of 457 elderly people, 2/3 of the subjects are at the risk of malnutrition, indicating that continuous nutrient supplementation could be helpful to preserve the strength and mass of skeletal muscle, and further to prevent and treat sarcopenia [116]. Since nutrients and dietary patterns have been shown to offer protective effects against aging-related decline in strength and function of skeletal muscle, combinatorial intervention of exercise and nutrients (including whey protein, vitamin D, calcium, minerals, and omega-3 fatty acids) seems to be more effective than exercise alone [75]. Based on above research, this chapter suggests exercise prescriptions combined with nutritional supplements to preserve sarcopenia for the elderly. Generally, due to individual difference, the exercise prescription needs to be consistent with personal physique (Table 10.2).

Table 10.2 Exercise interventions and corresponding precautions for sarcopenia

Exercise mode [115, 117, 118]	*Resistance exercise* (elastic band, loaded training): two or more days of muscle-strengthening activities per week; *Aerobic exercise* (walking, jogging, swimming, cycling, and so on): at least 150-min moderate-intensity aerobic exercise or high-intensity aerobic exercise per week; *Balance training* (Tai Chi, Yoga, stand on one leg, heel walking, and so on): two or more days per week; *Flexibility training* (stretching): 2 or 3 days/week, better effect after aerobic exercise or resistance exercise.
Daily diet [75, 76]	*Protein*: Meat, egg, milk; *Vitamins*: Fruits, vegetables; *Trace elements*: Coarse grains, nut, tea.
Nutritional supplements [110, 119, 120]	Whey protein, leucine supplements, vitamin D, vitamin E, calcium
Precautions for patients with sarcopenia [117, 118]	1. Sarcopenia: (a) Exercise is carried out in a progressive manner, and exercise intensity and frequency are selected according to individual differences; (b) resistance exercise, combined with aerobic exercise, balance, and flexibility training, is used to increase cardiopulmonary function and reduce the risk of falls; (c) less oil, salt and sugar, and drinking more water should be conducted to avoid senile constipation; (d) nutritional supplements should be supplemented according to individual needs under the guidance of doctors. 2. Sarcopenia complicated with hypertension: (a) When symptoms such as chest discomfort or pain, abnormal dyspnea, or dizziness appear, exercise should be immediately terminated; (b) avoiding the increase in blood pressure caused by high-intensity weight training; (c) during exercise, if systolic blood pressure higher than 250 mmHg or diastolic blood pressure higher than 115 mmHg, it should stop training and visit their doctors immediately; (d) quitting smoking and alcohol in diet, and controlling sodium intake [121]. 3. Sarcopenia complicated with diabetes: (a) monitoring blood glucose before, during, and after exercise to avoid abnormal blood glucose; (b) high-intensity exercise must be avoided when blood glucose higher than 16.7 mmol/L; (c) avoiding weight-bearing exercise when combined with active foot lesions; (d) avoiding high-intensity running and jumping exercises when combined with retinopathy; (e) patients who use drugs to control blood glucose should consult doctor to adjust the dosage to avoid hypoglycemia; (f) reducing the intake of carbohydrates (such as rice, noodles, and steamed buns), and eating less or no sugary foods (cakes, biscuits, candies, and so on) [122]. 4. Sarcopenia complicated with obesity: (a) Choosing low-load exercises, such as swimming, cycling, and rowing; (b) controlling calorie intake [123]. 5. Sarcopenia complicated with coronary heart disease: (a) there should be someone to accompany during exercise; (b) heart rate should be monitored throughout the exercise process, and if abnormal heart rate or discomfort occurs,

(continued)

Table 10.2 (continued)

	exercise should stop immediately. 6. Sarcopenia complicated with osteoarthritis: choosing low-loaded exercise to avoid increasing joint damage.

5 Conclusion

With the advent of an aging society, the diagnosis, pathogenesis, prevention, and treatments of sarcopenia have received increasing attention. The molecular mechanisms for the onset of sarcopenia are multifactorial, including dysfunctional status of autophagy, excessive apoptosis, persistent systemic low-grade inflammation, inactivated satellite cells, and reduced mitochondrial quality control. Lifestyle is also an important account for accelerating the loss of skeletal muscle during aging process. Insufficient protein intake, a sedentary lifestyle, and high-glucose and high-fat diet can aggravate the wasting of skeletal muscle. In recent years, the strategies for delaying the onset of sarcopenia and alleviating the symptoms of sarcopenia are focused on exercise, calorie restriction, and medication. It appears that resistance training in combination with amino acid supplements is the best candidate to attenuate aging-related skeletal muscle atrophy [124]. Autophagy plays an important role in delaying aging-related loss of skeletal muscle mass and strength through clearing damaged cells or organelles, maintaining mitochondrial function, and reducing inflammation levels. Recent studies have demonstrated that exercise can induce autophagy to delay or reverse the atrophy of skeletal muscle. However, little is known about the optimal exercise program for activating autophagy to delay the atrophy of skeletal muscle. Therefore, the relationship between autophagy and sarcopenia remains elusive. A better understanding of exercise-induced autophagic signal pathways and optimal exercise intervention is highly desired, which will be beneficial for the prevention and treatment of sarcopenia.

References

1. Beard J, Officer A, Cassels AJTG (2016) The world report on ageing and health, pp S163–S166
2. Leenders M, Verdijk LB, Van Der Hoeven L et al (2013) Patients with type 2 diabetes show a greater decline in muscle mass, muscle strength, and functional capacity with aging. J Am Med Dir Assoc 14:585–592
3. Kim K, Park K, Kim M et al (2014) Type 2 diabetes is associated with low muscle mass in older adults. Geriatr Gerontol Int 14:115–121
4. Gullett N, Rossi P, Kucuk O et al (2009) Cancer-induced cachexia: a guide for the oncologist. J Soc Integr Oncol 7:155–169
5. Pacifico J, Geerlings M, Reijnierse E et al (2020) Prevalence of sarcopenia as a comorbid disease: a systematic review and meta-analysis. Exp Gerontol 131:110801
6. Nilwik R, Snijders T, Leenders M et al (2013) The decline in skeletal muscle mass with aging is mainly attributed to a reduction in type II muscle fiber size. Exp Gerontol 48:492–498

7. Morley J, Anker S, Von Haehling S (2014) Prevalence, incidence, and clinical impact of sarcopenia: facts, numbers, and epidemiology-update 2014. J Cachexia Sarcopenia Muscle 5:253–259
8. Lowe DA, Baltgalvis KA, Greising SM (2010) Mechanisms behind estrogen's beneficial effect on muscle strength in females. Exerc Sport Sci Rev 38:61–67
9. Blau HM, Cosgrove BD, Ho AT (2015) The central role of muscle stem cells in regenerative failure with aging. Nat Med 21:854–862
10. Deer R, Volpi E (2015) Protein intake and muscle function in older adults. Curr Opin Clin Nutr Metab Care 18:248–253
11. Houston D, Nicklas B, Ding J et al (2008) Dietary protein intake is associated with lean mass change in older, community-dwelling adults: the Health, Aging, and Body Composition (Health ABC) Study. Am J Clin Nutr 87:150–155
12. Kang H, Lee K, Kim S et al (2011) Autophagy impairment induces premature senescence in primary human fibroblasts. PLoS One 6:e23367
13. Carter HN, Kim Y, Erlich AT et al (2018) Autophagy and mitophagy flux in young and aged skeletal muscle following chronic contractile activity. J Physiol 596:3567–3584
14. Levine B, Kroemer G (2019) Biological functions of autophagy genes: a disease perspective. Cell 176:11–42
15. Paré MF, Baechler BL, Fajardo VA et al (2017) Effect of acute and chronic autophagy deficiency on skeletal muscle apoptotic signaling, morphology, and function. Biochim Biophys Acta 1864:708–718
16. Marzetti E, Calvani R, Tosato M et al (2017) Physical activity and exercise as countermeasures to physical frailty and sarcopenia. Aging Clin Exp Res 29:35–42
17. Mcmullen CA, Ferry AL, Gamboa JL et al (2009) Age-related changes of cell death pathways in rat extraocular muscle. Exp Gerontol 44:420–425
18. Jiao J, Demontis F (2017) Skeletal muscle autophagy and its role in sarcopenia and organismal aging. Curr Opin Pharmacol 34:1–6
19. Carnio S, Loverso F, Baraibar MA et al (2014) Autophagy impairment in muscle induces neuromuscular junction degeneration and precocious aging. Cell Rep 8:1509–1521
20. Cohen S, Nathan JA, Goldberg AL (2015) Muscle wasting in disease: molecular mechanisms and promising therapies. Nat Rev Drug Discov 14:58–74
21. Lum JJ, Deberardinis RJ, Thompson CB (2005) Autophagy in metazoans: cell survival in the land of plenty. Nat Rev Mol Cell Biol 6:439–448
22. Joseph G, Wang S, Jacobs C et al (2019) Partial inhibition of mTORC1 in aged rats counteracts the decline in muscle mass and reverses molecular signaling associated with sarcopenia. Mol Cell Biol 39:e00141-19
23. Warr M, Binnewies M, Flach J et al (2013) FOXO3A directs a protective autophagy program in haematopoietic stem cells. Nature 494:323–327
24. Mammucari C, Schiaffino S, Sandri M (2008) Downstream of Akt: FoxO3 and mTOR in the regulation of autophagy in skeletal muscle. Autophagy 4:524–526
25. Jana NR (2012) Protein homeostasis and aging: role of ubiquitin protein ligases. Neurochem Int 60:443–447
26. Kwak KS, Zhou X, Solomon V et al (2004) Regulation of protein catabolism by muscle-specific and cytokine-inducible ubiquitin ligase E3alpha-II during cancer cachexia. Cancer Res 64:8193–8198
27. Bdolah Y, Segal A, Tanksale P et al (2007) Atrophy-related ubiquitin ligases atrogin-1 and MuRF-1 are associated with uterine smooth muscle involution in the postpartum period. Am J Physiol Regul Integr Comp Physiol 292:R971–R976
28. Bertaggia E, Coletto L, Sandri M (2012) Posttranslational modifications control FoxO3 activity during denervation. Am J Physiol Cell Physiol 302:C587–C596
29. Sandri M, Sandri C, Gilbert A et al (2004) Foxo transcription factors induce the atrophy-related ubiquitin ligase atrogin-1 and cause skeletal muscle atrophy. Cell 117:399–412

30. Mammucari C, Milan G, Romanello V et al (2007) FoxO3 controls autophagy in skeletal muscle in vivo. Cell Metab 6:458–471
31. Baehr L, Furlow J, Bodine S (2011) Muscle sparing in muscle RING finger 1 null mice: response to synthetic glucocorticoids. J Physiol 589:4759–4776
32. Cong H, Sun L, Liu C et al (2011) Inhibition of atrogin-1/MAFbx expression by adenovirus-delivered small hairpin RNAs attenuates muscle atrophy in fasting mice. Hum Gene Ther 22:313–324
33. Huang D, Yan X, Fan S et al (2020) Nrf2 deficiency promotes the increasing trend of autophagy during aging in skeletal muscle: a potential mechanism for the development of sarcopenia. Aging (Albany, NY) 12:5977–5991
34. Dobrowolny G, Aucello M, Rizzuto E et al (2008) Skeletal muscle is a primary target of SOD1G93A-mediated toxicity. Cell Metab 8:425–436
35. Iqbal S, Hood DA (2015) The role of mitochondrial fusion and fission in skeletal muscle function and dysfunction. Front Biosci (Landmark Ed) 20:157–172
36. Ibebunjo C, Chick J, Kendall T et al (2013) Genomic and proteomic profiling reveals reduced mitochondrial function and disruption of the neuromuscular junction driving rat sarcopenia. Mol Cell Biol 33:194–212
37. Handy DE, Joseph L (2012) Redox regulation of mitochondrial function. Antioxid Redox Signal 16:1323–1367
38. Conley KE, Marcinek DJ, Villarin J (2007) Mitochondrial dysfunction and age. Curr Opin Clin Nutr Metab Care 10:688–692
39. Aruoma OI, Grootveld M, Bahorun T (2006) Free radicals in biology and medicine: from inflammation to biotechnology. Biofactors 27:1–3
40. Deepa S, Van Remmen H, Brooks S et al (2019) Accelerated sarcopenia in Cu/Zn superoxide dismutase knockout mice. Free Radic Biol Med 132:19–23
41. Romanello V, Sandri MJC (2021) The connection between the dynamic remodeling of the mitochondrial network and the regulation of muscle mass. Cell Mol Life Sci 78:1305–1328
42. Ahn B, Ranjit R, Premkumar P et al (2019) Mitochondrial oxidative stress impairs contractile function but paradoxically increases muscle mass via fibre branching. J Cachexia Sarcopenia Muscle 10:411–428
43. Bleck C, Kim Y, Willingham T et al (2018) Subcellular connectomic analyses of energy networks in striated muscle. Nat Commun 9:5111
44. Valero T (2014) Mitochondrial biogenesis: pharmacological approaches. Curr Pharm Des 20:5507–5509
45. Zhang Y, Oliveira A, Hood D (2020) The intersection of exercise and aging on mitochondrial protein quality control. Exp Gerontol 131:110824
46. Romanello V (2020) The interplay between mitochondrial morphology and myomitokines in aging sarcopenia. Int J Mol Sci 22:91
47. Zeng Z, Liang J, Wu L et al (2020) Exercise-induced autophagy suppresses sarcopenia through Akt/mTOR and Akt/FoxO3a signal pathways and AMPK-mediated mitochondrial quality control. Front Physiol 11:583478
48. Marzetti E, Calvani R, Cesari M et al (2013) Mitochondrial dysfunction and sarcopenia of aging: from signaling pathways to clinical trials. Int J Biochem Cell Biol 45:2288–2301
49. Cilenti L, Ambivero C, Ward N et al (2014) Inactivation of Omi/HtrA2 protease leads to the deregulation of mitochondrial Mulan E3 ubiquitin ligase and increased mitophagy. Biochim Biophys Acta 1843:1295–1307
50. Zhou H, Yuan D, Gao W et al (2020) Loss of high-temperature requirement protein A2 protease activity induces mitonuclear imbalance via differential regulation of mitochondrial biogenesis in sarcopenia. IUBMB Life 72:1659–1679
51. Barcena C, Mayoral P, Quiros PM (2018) Mitohormesis, an Antiaging Paradigm. Int Rev Cell Mol Biol 340:35–77
52. Lee D, Bareja A, Bartlett D et al (2019) Autophagy as a therapeutic target to enhance aged muscle regeneration. Cells 8:183

53. Rathbone CR, Wenke JC, Warren GL et al (2003) Importance of satellite cells in the strength recovery after eccentric contraction-induced muscle injury. Am J Physiol Regul Integr Comp Physiol 285:R1490–R1495
54. Relaix F, Zammit PS (2012) Satellite cells are essential for skeletal muscle regeneration: the cell on the edge returns centre stage. Development 139:2845–2856
55. Hikida R (2011) Aging changes in satellite cells and their functions. Curr Aging Sci 4:279–297
56. Brack A, Conboy M, Roy S et al (2007) Increased Wnt signaling during aging alters muscle stem cell fate and increases fibrosis. Science 317:807–810
57. White JP, Billin AN, Campbell ME et al (2018) The AMPK/p27(K)(iP 1) axis regulates autophagy/apoptosis decisions in aged skeletal muscle stem cells. Stem Cell Rep 11:425–439
58. García-Prat L, Martínez-Vicente M, Perdiguero E et al (2016) Autophagy maintains stemness by preventing senescence. Nature 529:37–42
59. Jejurikar S, Henkelman E, Cederna P et al (2006) Aging increases the susceptibility of skeletal muscle derived satellite cells to apoptosis. Exp Gerontol 41:828–836
60. Murach K, Fry C, Kirby T et al (2018) Starring or supporting role? Satellite cells and skeletal muscle fiber size regulation. Physiology (Bethesda) 33:26–38
61. Bharath LP, Agrawal M, Mccambridge G et al (2020) Metformin enhances autophagy and normalizes mitochondrial function to alleviate aging-associated inflammation. Cell Metab 32:44
62. Zembron-Lacny A, Dziubek W, Wolny-Rokicka E et al (2019) The relation of inflammaging with skeletal muscle properties in elderly men. Am J Mens Health 13. https://doi.org/10.1177/1557988319841934
63. Fan J, Kou X, Yang Y et al (2016) MicroRNA-regulated proinflammatory cytokines in sarcopenia. Mediators Inflamm 2016:1438686
64. Schaap L, Pluijm S, Deeg D et al (2009) Higher inflammatory marker levels in older persons: associations with 5-year change in muscle mass and muscle strength. J Gerontol A Biol Sci Med Sci 64:1183–1189
65. Chen FX, Shen Y, Liu Y et al (2020) Inflammation-dependent downregulation of miR-532-3p mediates apoptotic signaling in human sarcopenia through targeting BAK1. Int J Biol Sci 16:1481–1494
66. Öztürk Z, Kul S, Türkbeyler İ et al (2018) Is increased neutrophil lymphocyte ratio remarking the inflammation in sarcopenia? Exp Gerontol 110:223–229
67. Cuthbertson D, Smith K, Babraj J et al (2005) Anabolic signaling deficits underlie amino acid resistance of wasting, aging muscle. FASEB J 19:422–424
68. Vasilaki A, Mcardle F, Iwanejko L et al (2006) Adaptive responses of mouse skeletal muscle to contractile activity: the effect of age. Mech Ageing Dev 127:830–839
69. Dirks A, Leeuwenburgh C (2002) Apoptosis in skeletal muscle with aging. Am J Physiol Regul Integr Comp Physiol 282:R519–R527
70. Marzetti E, Wohlgemuth S, Lees H et al (2008) Age-related activation of mitochondrial caspase-independent apoptotic signaling in rat gastrocnemius muscle. Mech Ageing Dev 129:542–549
71. Dalle S, Rossmeislova L, Koppo K (2017) The role of inflammation in age-related sarcopenia. Front Physiol 8:1045
72. Chen C, Liu Y, Liu R et al (2008) TSC-mTOR maintains quiescence and function of hematopoietic stem cells by repressing mitochondrial biogenesis and reactive oxygen species. J Exp Med 205:2397–2408
73. Kang Y, Cho M, Kim J et al (2016) Impaired macrophage autophagy induces systemic insulin resistance in obesity. Oncotarget 7:35577–35591
74. Hallowell R, Collins S, Craig J et al (2017) mTORC2 signalling regulates M2 macrophage differentiation in response to helminth infection and adaptive thermogenesis. Nat Commun 8:14208
75. Ganapathy A, Nieves J (2020) Nutrition and sarcopenia-what do we know? Nutrients 12:1755

76. Peterson M, Gordon P (2011) Resistance exercise for the aging adult: clinical implications and prescription guidelines. Am J Med 124:194–198
77. Létocart A, Mabesoone F, Charleux F et al (2021) Muscles adaptation to aging and training: architectural changes—a randomised trial. BMC Geriatr 21:48
78. Hart CR, Ryan ZC, Pfaffenbach KT et al (2019) Attenuated activation of the unfolded protein response following exercise in skeletal muscle of older adults. Aging (Albany, NY) 11:7587–7604
79. Gao H, Wu D, Sun L et al (2020) Effects of lifelong exercise on age-related body composition, oxidative stress, inflammatory cytokines, and skeletal muscle proteome in rats. Mech Ageing Dev 189:111262
80. Bouzid M, Filaire E, Matran R et al (2018) Lifelong voluntary exercise modulates age-related changes in oxidative stress. Int J Sports Med 39:21–28
81. Alves RM, Vitorino R, Figueiredo P et al (2010) Lifelong physical activity modulation of the skeletal muscle mitochondrial proteome in mice. J Gerontol A Biol Sci Med Sci 65:832–842
82. Bolotta A, Filardo G, Abruzzo P et al (2020) Skeletal muscle gene expression in long-term endurance and resistance trained elderly. Int J Mol Sci 21:3988
83. Liu G, Sabatini D (2020) mTOR at the nexus of nutrition, growth, ageing and disease. Nat Rev Mol Cell Biol 21:183–203
84. Grumati P, Coletto L, Schiavinato A et al (2011) Physical exercise stimulates autophagy in normal skeletal muscles but is detrimental for collagen VI-deficient muscles. Autophagy 7:1415–1423
85. Fan J, Yang X, Li J et al (2017) Spermidine coupled with exercise rescues skeletal muscle atrophy from D-gal-induced aging rats through enhanced autophagy and reduced apoptosis via AMPK-FOXO3a signal pathway. Oncotarget 8:17475–17490
86. He C, Bassik M, Moresi V et al (2012) Exercise-induced BCL2-regulated autophagy is required for muscle glucose homeostasis. Nature 481:511–515
87. Luo L, Lu A, Wang Y et al (2013) Chronic resistance training activates autophagy and reduces apoptosis of muscle cells by modulating IGF-1 and its receptors, Akt/mTOR and Akt/FOXO3a signaling in aged rats. Exp Gerontol 48:427–436
88. Sandri M (2013) Protein breakdown in muscle wasting: role of autophagy-lysosome and ubiquitin-proteasome. Int J Biochem Cell Biol 45:2121–2129
89. Röckl K, Witczak C, Goodyear L (2008) Signaling mechanisms in skeletal muscle: acute responses and chronic adaptations to exercise. IUBMB Life 60:145–153
90. Morales-Alamo D, Calbet JAL (2016) AMPK signaling in skeletal muscle during exercise: role of reactive oxygen and nitrogen species. Free Radic Biol Med 98:68–77
91. Bujak A, Crane J, Lally J et al (2015) AMPK activation of muscle autophagy prevents fasting-induced hypoglycemia and myopathy during aging. Cell Metab 21:883–890
92. Ferraro E, Giammarioli AM, Chiandotto S et al (2014) Exercise-induced skeletal muscle remodeling and metabolic adaptation: redox signaling and role of autophagy. Antioxid Redox Signal 21:154–176
93. Martin-Rincon M, Morales-Alamo D, Calbet J (2018) Exercise-mediated modulation of autophagy in skeletal muscle. Scand J Med Sci Sports 28:772–781
94. Sanchez A, Csibi A, Raibon A et al (2012) AMPK promotes skeletal muscle autophagy through activation of forkhead FoxO3a and interaction with Ulk1. J Cell Biochem 113:695–710
95. Liu X, Niu Y, Yuan H et al. (2015) AMPK binds to Sestrins and mediates the effect of exercise to increase insulin-sensitivity through autophagy. Metabolism: clinical and experimental 64: 658–665
96. Dethlefsen M, Halling J, Møller H et al (2018) Regulation of apoptosis and autophagy in mouse and human skeletal muscle with aging and lifelong exercise training. Exp Gerontol 111:141–153

97. Halling J, Jessen H, Nøhr-Meldgaard J et al (2019) PGC-1α regulates mitochondrial properties beyond biogenesis with aging and exercise training. Am J Physiol Endocrinol Metab 317: E513–E525
98. Lira V, Okutsu M, Zhang M et al (2013) Autophagy is required for exercise training-induced skeletal muscle adaptation and improvement of physical performance. FASEB J 27:4184–4193
99. Halling J, Ringholm S, Nielsen M et al (2016) PGC-1α promotes exercise-induced autophagy in mouse skeletal muscle. Physiol Rep 4:e12698
100. Castets P, Lin S, Rion N et al (2013) Sustained activation of mTORC1 in skeletal muscle inhibits constitutive and starvation-induced autophagy and causes a severe, late-onset myopathy. Cell Metab 17:731–744
101. Cui D, Drake J, Wilson R et al (2020) A novel voluntary weightlifting model in mice promotes muscle adaptation and insulin sensitivity with simultaneous enhancement of autophagy and mTOR pathway. FASEB J 34:7330–7344
102. Liu H, Pan S (2019) Late exercise preconditioning promotes autophagy against exhaustive exercise-induced myocardial injury through the activation of the AMPK-mTOR-ULK1 pathway. Biomed Res Int 2019:5697380
103. Yin L, Lu L, Lin X et al (2020) Crucial role of androgen receptor in resistance and endurance trainings-induced muscle hypertrophy through IGF-1/IGF-1R- PI3K/Akt- mTOR pathway. Nutr Metab 17:26
104. Kim J, Kundu M, Viollet B et al (2011) AMPK and mTOR regulate autophagy through direct phosphorylation of Ulk1. Nat Cell Biol 13:132–141
105. Zhao L, Li H, Wang Y et al (2019) Autophagy deficiency leads to impaired antioxidant defense via p62-FOXO1/3 axis. Oxidative Med Cell Longev 2019:2526314
106. Wessner B, Ploder M, Tschan H et al (2019) Effects of acute resistance exercise on proteolytic and myogenic markers in skeletal muscles of former weightlifters and age-matched sedentary controls. J Sports Med Phys Fitness 59:1915–1924
107. Jamart C, Naslain D, Gilson H et al (2013) Higher activation of autophagy in skeletal muscle of mice during endurance exercise in the fasted state. Am J Physiol Endocrinol Metab 305: E964–E974
108. Chen H, Chung Y, Chen Y et al (2017) Effects of different types of exercise on body composition, muscle strength, and IGF-1 in the elderly with sarcopenic obesity. J Am Geriatr Soc 65:827–832
109. Snijders T, Nederveen J, Bell K et al (2019) Prolonged exercise training improves the acute type II muscle fibre satellite cell response in healthy older men. J Physiol 597:105–119
110. Yamada M, Kimura Y, Ishiyama D et al (2019) Synergistic effect of bodyweight resistance exercise and protein supplementation on skeletal muscle in sarcopenic or dynapenic older adults. Geriatr Gerontol Int 19:429–437
111. Roig M, Macintyre D, Eng J et al (2010) Preservation of eccentric strength in older adults: evidence, mechanisms and implications for training and rehabilitation. Exp Gerontol 45:400–409
112. Vandervoort A (2009) Potential benefits of warm-up for neuromuscular performance of older athletes. Exerc Sport Sci Rev 37:60–65
113. Fulop T, Witkowski J, Olivieri F et al (2018) The integration of inflammaging in age-related diseases. Semin Immunol 40:17–35
114. Franceschi C, Garagnani P, Parini P et al (2018) Inflammaging: a new immune-metabolic viewpoint for age-related diseases. Nat Rev Endocrinol 14:576–590
115. Piercy K, Troiano R, Ballard R et al (2018) The physical activity guidelines for Americans. JAMA 320:2020–2028
116. Kaiser M, Bauer J, Rämsch C et al (2010) Frequency of malnutrition in older adults: a multinational perspective using the mini nutritional assessment. J Am Geriatr Soc 58:1734–1738

117. Lee P, Jackson E, Richardson C (2017) Exercise prescriptions in older adults. Am Fam Physician 95:425–432
118. Hansen D, Niebauer J, Cornelissen V et al (2018) Exercise prescription in patients with different combinations of cardiovascular disease risk factors: a consensus statement from the EXPERT Working Group. Sports Med 48:1781–1797
119. Robinson S, Granic A, Sayer A (2019) Nutrition and muscle strength, as the key component of sarcopenia: an overview of current evidence. Nutrients 11:2942
120. Komar B, Schwingshackl L, Hoffmann G (2015) Effects of leucine-rich protein supplements on anthropometric parameter and muscle strength in the elderly: a systematic review and meta-analysis. J Nutr Health Aging 19:437–446
121. Juraschek S, Miller E, Weaver C et al (2017) Effects of sodium reduction and the DASH diet in relation to baseline blood pressure. J Am Coll Cardiol 70:2841–2848
122. Wang L, Wang Q, Hong Y et al (2018) The effect of low-carbohydrate diet on glycemic control in patients with type 2 diabetes mellitus. Nutrients 10:661
123. Kemmler W, Von Stengel S, Kohl M et al (2020) Safety of a combined WB-EMS and high-protein diet intervention in sarcopenic obese elderly men. Clin Interv Aging 15:953–967
124. Argilés J, Busquets S, Stemmler B et al (2015) Cachexia and sarcopenia: mechanisms and potential targets for intervention. Curr Opin Pharmacol 22:100–106

Chapter 11
Prospective Advances in Exercise-Induced Autophagy on Health

Jiling Liang, Michael Kirberger, and Ning Chen

1 Introduction

Physical inactivity and a sedentary lifestyle have become a growing public health problem [1]. Approximately 10% of all deaths annually are attributed to physical inactivity. The optimal approach to prevent morbidity and mortality associated with various diseases involves lifestyle modification, including regular physical activity as the primary strategy for people at any age [2]. Regular physical activity at an appropriate intensity is a preventative strategy with few adverse effects and with less cost than medication. Therefore, practical strategies to increase physical activity should be a public health priority, and to address this, the World Health Organization, in concert with national governments and other agencies, has developed evidence-based physical activity guidelines to promote an active lifestyle for people of all ages [3].

Metabolic disorders are frequently reported in conjunction with many chronic diseases. The critical regulators of major energy biomolecules, such as glucose, lipids, and proteins, are also involved in regulating the functional status of autophagy. Common pathophysiological changes in chronic diseases result from the accumulation of harmful substances, such as reactive oxygen species (ROS), damaged organelles, protein aggregates, lipid droplets, and senescent cells. In mammals, the autophagosome membrane is first generated during autophagy initiation, and the autophagosome is sequentially formed after encapsulating damaged materials for degradation, and upon fusing with lysosomes [4]. Autophagy in cells is

J. Liang · N. Chen (✉)
Tianjiu Research and Development Center for Exercise Nutrition and Foods, Hubei Key Laboratory of Exercise Training and Monitoring, College of Health Science, Wuhan Sports University, Wuhan, China

M. Kirberger
School of Science and Technology, Georgia Gwinnett College, Lawrenceville, GA, USA

N. Chen (ed.), *Exercise, Autophagy and Chronic Diseases*,
https://doi.org/10.1007/978-981-16-4525-9_11

a conserved stress response to environmental changes and plays a vital role in cellular metabolism. Autophagy is also an adaptive response to exogenous stimuli, including nutrient deficiency, oxidative stress, and infection. As a consequence of cellular metabolic pressure, autophagy can execute clearance of dysfunctional organelles, and optimal sustained autophagy plays a crucial role in maintaining homeostasis in normal tissues. Alternatively, autophagy can act as a cellular defense mechanism to remove dysfunctional organelles and denatured macromolecules in the cytoplasm for the protection of damaged cells, and can also induce cell death [5].

Furthermore, autophagy is not only critical in cells for the regulation of multiple diseases, but it is also required for exercise-related physiological adaptations (i.e., angiogenesis, mitochondrial biogenesis, insulin sensitivity, and muscle hypertrophy) [6, 7]. Exercise-induced autophagy/mitophagy can also improve autophagic flux, and the improved functional status of autophagy may degrade harmful substances to alleviate chronic diseases such as cardiovascular diseases and metabolic syndromes, sarcopenia, osteoporosis, and neurological diseases. The focus of this chapter will therefore review the numerous beneficial effects of exercise-induced autophagy on health promotion and activity performance, as well as prevention and treatment of chronic diseases.

2 Autophagy Pathways

In macroautophagy (hereafter autophagy), cellular materials are engulfed in the double-membrane structure of the autophagosome, which ultimately fuses with a lysosome. In the autolysosome, autophagic cargo is degraded in an acidic milieu by lysosomal hydrolases. Autophagy serves to recycle cellular components for energy and building blocks to restore nutrient shortage, and maintains cellular homeostasis by degrading damaged or superfluous proteins and organelles. Under basal physiological conditions, autophagy is ubiquitous in eukaryotic cells at a basal level, but can be activated by various stress states including starvation, oxidative stress, growth factor deficiency, endoplasmic reticulum stress, and exercise [8]. To date, more than 30 autophagy-related (Atg) genes whose protein products regulate and execute the critical steps of autophagy have been identified in species ranging from yeast to mammals [9]. A series of complexes composed of the products encoded by Atg genes are involved in coordinating the formation of autophagosomes. The Atg1/Unc-51-like autophagy-activating kinase 1 (ULK1) complex is an essential positive regulatory factor for correct localization of ULK1 to autophagy precursors, and maintaining the stability of ULK1 protein, and this complex is mainly composed of ULK1-Atg13-FIP200 [10]. For mammals receiving adequate nutrition, mechanistic target of rapamycin (mTOR) complex 1 (mTORC1) and the ULK1 complex bind to inhibit autophagy. mTORC1 is an important regulator of cell growth and metabolism, consisting of five subunits, including Raptor (regulatory associated protein of mTOR, interacting with ULKl) and mTOR. mTORC1 inhibits the initiation of autophagy by phosphorylating Atg1/ULK1 and Atg13. Under starvation

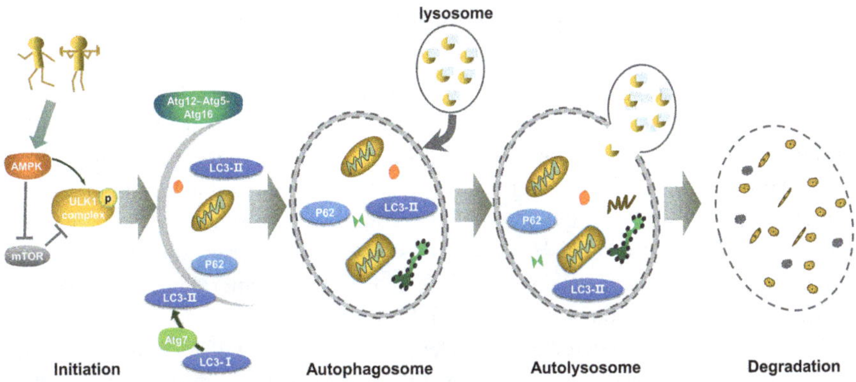

Fig. 11.1 Overview of basic molecular mechanisms involved in the initiation of exercise-induced autophagy

conditions, mTORC1 is separated from the Atg1/ULK1 complex, thereby inducing nucleation and elongation of autophagosomes [11]. The nucleation of autophagy vesicles requires a complex containing Atg6 (mammalian homologous protein Beclin1), which forms a supercompound with the class III phosphoinositide 3-kinase vacuolar protein sorting 34 (Vps34), and activates the production of phosphatidylinositol-3-phosphate (PI3P) [12]. The extension of the autophagosome membrane is primarily controlled by Atg7 and Atg10. As an important biomarker for autophagic vasculogenesis and elongation, Atg7 is responsible for regulating the formation of autophagic bubbles and can further catalyze the conversion of the key protein microtubule-associated protein light chain 3-I (LC3-I) to LC3-II. The expression of LC3-II and the LC3-II/LC3-I ratio are distinguishable by electrophoretic mobility to evaluate the autophagosome content and autophagic flux in cells and tissues [13]. Finally, autophagosomes combine with lysosomes to form autophagosomes. In the autophagosomes, the inner membrane and contents of autophagy are degraded in lysosomes. Lysosomal permease releases degraded products into the cytoplasm for biosynthesis and metabolism involving amino acids, lipids, nucleotides, and carbohydrates [14]. Autophagic adaptor p62/SQSTM1 interacts with both ubiquitin and LC3 to execute specific identification, separation, and transport of p62 or its substrate degradation product, thereby releasing nutrients or ATP for cellular recycling [15] (Fig. 11.1).

Thus, the maturation and degradation of autophagosomes and autolysosomes are a complicated process that is regulated by many autophagy-related proteins. As previously noted, the normally low occurrence of autophagy in eukaryotic cells can become increasingly elevated in response to external stresses, like starvation, in order to maintain cellular homeostasis, or it can be a pathological condition related to diseases. Interestingly, physical exercise has also been identified as a stressor capable of inducing elevated autophagy, which is discussed in the next section.

3 Autophagy Is Activated During Exercise

Exercise-induced autophagy caused by strenuous exercise was first discovered in skeletal muscle of mice by Salminen [16] in 1984, while autophagic degradation of proteins in liver, induced by endurance exercise, was first reported by Dohm [17] in 1987. More recently, in 2011, Grumati et al. [18] have reported that ultra-endurance exercise activates autophagy in skeletal muscle, as confirmed by the increased conversion of LC3-I to LC3-II. In human studies, the level of autophagy and the expression of autophagy-related proteins in the lateral femoral muscle fibers are upregulated by superendurance exercise [19].

Currently, increasing evidence indicates that the recycling of cellular constituents by autophagy is an essential process of the body's adaptive response to exercise [20, 21]. During exercise, the demand for oxygen and glucose in the body increases significantly, and skeletal muscle accounts for the largest proportion of the body. In this low-oxygen and low-sugar environment, autophagy plays a vital role in preventing excessive fatigue and damage to skeletal muscle as it adapts to exercise by improving metabolic levels, maintaining the stability of metabolism, and facilitating the elimination and recycling of damaged or denatured proteins or organelles in skeletal muscle [22]. On the other hand, recent data suggest that the regulation of skeletal muscle metabolism and the sustainment of internal environment stability by exercise-induced autophagy/mitophagy have been reported to improve exercise function [23, 24]. Activation of autophagy in response to exercise has now been identified in multiple metabolism-related organs such as skeletal muscle, heart, liver, pancreas, and adipose tissue [25, 26], as well as in the brain, which may partially explain the neuroprotective effects of exercise [6].

4 Molecular Mechanisms of Exercise-Induced Autophagy

The mass of skeletal muscle accounts for about 40% of adult body weight and is composed of myocytes with contractile function. Skeletal muscle as the major organ involved in exercise response can complete contraction functions under different exercise stresses. During exercise, increases are observed in the levels of calcium, NAD^+/NADH ratio, AMP/ATP ratio, insulin-like growth factor 1 (IGF-1), and ROS for activating downstream effectors that can induce autophagy through different signal pathways such as adenosine monophosphate-activated protein kinase (AMPK)-mTOR-ULK1, protein kinase B (Akt or PKB)-mTOR, Beclin1-B-cell lymphoma 2 (Bcl-2) complex, FoxO family, and other signal pathways (Fig. 11.2).

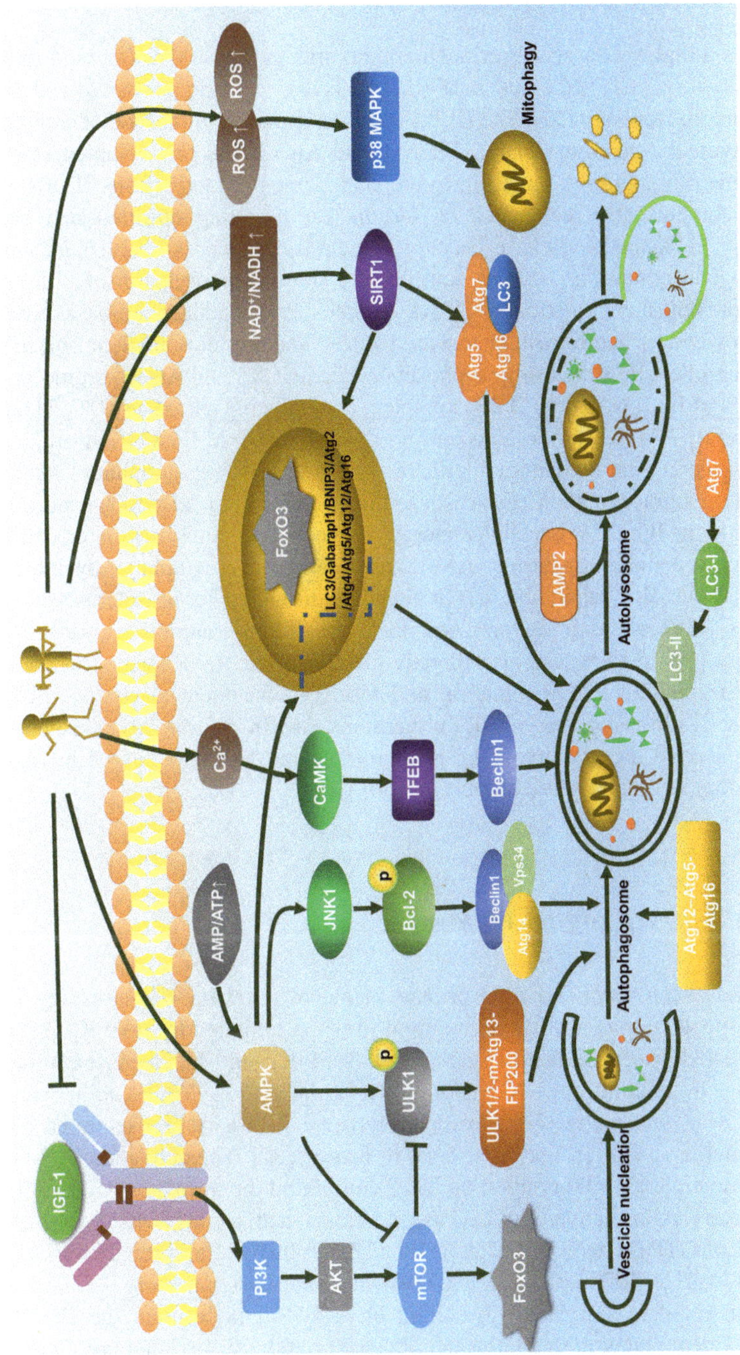

Fig. 11.2 Potential exercise-induced autophagy signaling pathways

4.1 AMPK-mTOR-ULK1 Signal Pathway

mTOR is a highly conserved serine/threonine protein kinase that can bind to other ligands to form mTORC1 and mTORC2. mTORC1 is primarily involved in the inhibition of autophagy [27]. The ULK1 complex is mainly composed of autophagy-related proteins including ULK1, FIP200, and Atg13, and this complex acts as a node converting multiple signals into autophagosome formation [28]. mTORC1 and AMPK are currently recognized as sensors for detecting nutrition and energy efficiency in skeletal muscle, especially in exercise responses [29]. Under normal physiological conditions, mTOR can promote the phosphorylation of ULK1 and Atg13, prevent the formation of ULK1 complex, and maintain the basal environment. However, when growth factors are reduced and nutrients are deficient, mTOR can be inhibited, thus releasing its phosphorylated ULK1 and Atg13, promoting the formation of ULK1-Atg13-FIP200 complex, and inducing autophagy [30, 31]. During exercise, AMPK can be activated by the increased AMP/ATP ratio [32, 33], which is also a vital autophagy regulator, thereby activating autophagy. AMPK is known as an intracellular energy sensor so that exercise-induced autophagy occurs in response to the limited intracellular energy charge. In addition to activating catabolic processes in cellular organisms, AMPK also can induce autophagy by inhibiting mTOR [34]. AMPK can inhibit mTOR signaling through tuberous sclerosis complex 1/2 (TSC1/2)/Ras homolog enriched in brain (Rheb) pathway, thus inducing autophagy [35]. AMPK can also directly phosphorylate the Raptor at Ser722 and Ser792 sites, inhibit mTOR signaling, and deactivate the phosphorylation of ULK1, thereby activating autophagy [29]. Furthermore, AMPK can directly phosphorylate ULK1, enhance the activity of ULK1, promote the formation of ULK1-Atg13-FIP200 complex, and induce autophagy [36].

4.2 Akt-mTOR Signal Pathway

mTOR has been identified as a primary negative regulator of autophagy in an AMPK-opposing manner [31]. In addition to being inhibited by AMPK, mTOR is also regulated by several factors such as IGF-1, PI3K, and Akt. Exercise can induce autophagy by inhibiting the IGF/Akt/mTOR signal pathway. IGF can activate the Akt/mTOR signal pathway via receptor tyrosine kinase (RTK) recruited to the plasma membrane, after binding to growth factors [37]. The activation of Akt can also promote the phosphorylation of TSC2 and inhibit the formation of TSCl/TSC2 heterodimers. As an activator of GTPase, TSC2 can activate mTORC1 by inhibiting the Rheb of GTPase, which phosphorylates ULK1, thereby inhibiting the formation of ULK1-Atg13-FIP200 complex and attenuating autophagy activity [38, 39]. Also, PI3K can regulate autophagy by acting on mTOR and forming the PI3K/Akt/mTOR2 signal pathway to control the phosphorylation of forkhead box O (FoxO) transcription factors [40].

4.3 Beclin1-Bcl-2 Complex

As a homolog of yeast autophagy gene Atg6, Beclin1/(Atg6) can interact with Vps34, Vps15, ultraviolet (UV) radiation resistance-associated gene (UVRAG), and activating molecule in BECN1-regulated autophagy protein 1 (Ambra1), to form a complex that facilitates the production and maturation of autophagosomes, thereby enhancing autophagy activity. This signal pathway is also regulated by the anti-apoptotic Bcl-2 [41]. Under normal physiological conditions, Bcl-2 binds strongly to Beclin1, which inhibits the formation of Beclin1 complex with autophagy-related proteins such as Vps34 and Vps15, and maintains the basal level of autophagy. Studies have reported that exogenous stimuli such as oxidative stress and nutrient deprivation can activate c-Jun N-terminal kinase (JNK) and phosphorylate Bcl-2 in skeletal muscle to stimulate the efficient release of Beclin1 from the Bcl-2-Beclin1 complex, and simultaneously induce the autophagic signal pathway [42, 43]. Moreover, AMPK is also an essential regulator of the Bcl-2-Beclin1 complex, and under conditions of glucose deprivation, AMPK can phosphorylate Vps34 and Beclin1 to promote the formation of the Beclin1-Vps34 complex involved in the activation of autophagy [44].

4.4 FoxO Family

The evolutionarily conserved transcription factors from FoxO family were the first discovered genes associated with the regulation of autophagy [45]. Evidence indicates that FoxO3 mainly reduces the expression of atrophy-associated ubiquitin ligase (Atrogin-1) and muscle-specific ubiquitin ligase muscle RING-finger protein-1 (MuRF1) by activating atrophy-related genes and decomposing proteins in skeletal muscle through the ubiquitin signaling pathway [46]. However, several stress-induced transcription factors have also been identified that may regulate the expression of FoxO family transcription factors involved in autophagy. In particular, FoxO3 reportedly can transcribe a series of autophagy-related genes such as LC3, Gamma-aminobutyric acid receptor-associated protein-like 1 (Gabarapl1), BCL2 interacting protein 3 (BNIP3), ULK2, Atg4, Atg5, Atg12, and Atg16, thereby inducing autophagy [19, 45, 47], and posttranslational modification of FoxO3 also plays a vital role in regulating autophagic flux in skeletal muscle [48]. AMPK also plays a crucial role in the FoxO3-mediated autophagy signaling pathway. In addition, AMPK can promote the phosphorylation of FoxO3 at the site of Ser588/Ser413, which enhances transcriptional activity and promotes autophagy [49]. Furthermore, FoxO1 can specifically bind to Atg7 after acetylation to promote the initiation of autophagy [50]. Collectively, these findings suggest that FoxO3-induced autophagy in skeletal muscle is also an important activation pathway, independent of the mTOR/ULK1 signal pathway.

Exercise, as a factor to stimulate the expression of genes associated with energy metabolism, can increase mitochondrial respiration and redox reaction, promote the oxidation of NADH to NAD^+, and increase the proportion of NAD^+/NADH [51]. It can also activate various downstream signal kinases and stress reactants such as silent information regulator 1 (SIRT1) and FoxO family transcription factors [52]. Alternatively, exercise can stimulate the increase in AMP/ATP ratio, while NADH is oxidized as NAD^+. AMPK can also indirectly activate SIRT1 deacetylase by changing the intracellular NAD^+/NADH ratio during exercise [52, 53]. In the cytosol, SIRT1 can activate autophagy-related proteins such as Atg5, Atg7, Atg12, and LC3, thereby inducing autophagy [54]. Therefore, cellular energy metabolism during exercise can also regulate the autophagy process by changing the NAD^+/NADH ratio.

4.5 Other Signal Pathways

Changes in intracellular and extracellular ions caused by exercise can also affect the occurrence of autophagy through downstream kinases. The significantly increased concentration of Ca^{2+} during exercise can enhance the activity of downstream calmodulin phosphatase and calcium/calmodulin-dependent protein kinase (CaMK) [55]. CaMK can promote the expression of cellular transcription factor EB (TFEB) [56]. TFEB is a critical factor in autophagy and lysosome function through transcriptional regulation of genes encoding autophagy and lysosome proteins [57]. Similarly, TFEB can upregulate Beclin1 expression during autophagosome formation and promote intracellular material recovery by increasing the rate of autophagosome–lysosomal degradation [58]. Moreover, relevant studies have reported that free cytosolic Ca^{2+} can also act as a potent inducer for autophagy. Ca^{2+} agonists inhibit mTOR activity and activate autophagy via CaMKKβ-mediated AMPK activation and promote the formation of autophagosomes in a manner dependent upon Beclin1 or Atg7 [59]. Furthermore, Ca^{2+} also can induce autophagy in an AMPK-independent pathway, while Ca^{2+} agonists can increase the localization of Atg18 to autophagosomes [60].

5 Mitophagy

Mitophagy refers to the depolarization damage to mitochondria during starvation, oxidative stress, and aging, and damage from ROS. This results in encapsulation of the damaged mitochondria into the autophagosome, where it fuses with intracellular lysosomes and then is degraded to maintain homeostasis of the internal environment [61]. Autophagy can be divided into three types: chaperone-mediated autophagy, microautophagy, and macroautophagy. Mitophagy is generally considered a form of macroautophagy that results in selective degradation of dysfunctional or damaged mitochondria. It plays a vital role in the maintenance of cellular function by decreasing oxidative stress and restoring cellular homeostasis [62]. At the same time, the gradual decrease in physical activity, and increase in inflammation and oxidative stress, can cause abnormal accumulation of damaged mitochondria, higher production of ROS, and initiation of mitochondria-mediated apoptosis, ultimately leading to a series of mitochondria-related diseases. In response to these changes, mitophagy can selectively identify damaged and dysfunctional mitochondria to degrade them, thus generating a self-protective coping mechanism against mitochondrial dysfunction [63]. Phosphatase and tensin homologue-induced kinase 1 (PINK1), E3 ubiquitin ligase (Parkin), and BNIP3/Nix are specific proteins believed to be involved in mitophagy [64]. PINK1/Parkin is one of the components in the classic signaling pathway of mitophagy. Under conditions of stress, injury, or aging, mitochondrial kinase PINK1 can aggregate outside the membrane of the dysfunctional mitochondrion and selectively translocate across the barrier by phosphorylating and recruiting Parkin. It then ubiquitinates outer mitochondrial membrane proteins such as mitofusin1/2 (Mfn1/2) and dynamin-related protein 1 (Drp1) to participate in mitophagy [65]. BNIP3/Nix is mainly located in the mitochondrial outer membrane and interacts with autophagic LC3 to digest and degrade the mitochondrial targets into autophagosomes [66]. Research suggests that Parkin can not only prolong life, but also maintain mitochondrial mass and muscle flight function. Additionally, the overexpression of Parkin can downregulate Mfn2, thereby accelerating the degradation of polyubiquitinated proteins and reducing protein toxicity to mitochondria [67].

The molecular details of the PINK1/Parkin signal pathway are still a subject of debate, further complicated by the multiple roles that PINK1 and Parkin seem to play in regulating mitochondrial biogenesis, dynamics, and transport [68]. Furthermore, additional mechanisms have been described to activate mitophagy. For instance, ER-associated E3 ligase Gp78 (glycoprotein 78), like Parkin, ubiquitinates mitofusins and activates mitophagy of depolarized mitochondria [69]. Under hypoxic conditions, the mitochondrial membrane protein FUN14 domain containing 1 (FUNDC1) interacts with LC3 to promote clearance of mitochondria [69]. In neurons, the externalization of cardiolipin and its interaction with LC3 have been confirmed to mediate mitophagic degradation of malfunctioning mitochondria [70]. Finally, Ambra1, a crucial molecule involved in the regulation of autophagy, has been recently shown to promote mitophagy through a novel PINK1/Parkin-independent mechanism [71, 72]. The uncertainty of these results indicates the necessity of further exploration of the cellular mechanism of mitochondrial turnover.

6 Exercise Adaptation for Health Promotion Through Induced Autophagy/Mitophagy

Several animal and human experiments and clinical evidence have documented the beneficial effects of exercise and physical activity on disease prevention or rehabilitation. In addition, exercise training and physical activity are necessary to maintain normal physiological function of musculoskeletal, cardiovascular, nervous,

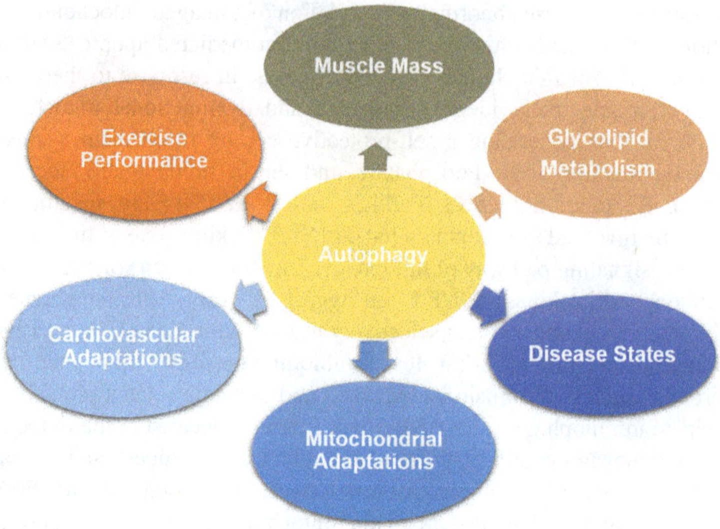

Fig. 11.3 Exercise-induced autophagy/mitophagy for molecular adaptation during the progression of diseases

endocrine, and respiratory systems. Over the last several decades, many studies have described biological adaptation and revealed the modification of intra- and inter-organ communications after physical activity and exercise training [73–75]. Exercise induces several physiological adaptive processes by modulating cellular and molecular regulatory mechanisms. The modification of molecular signaling pathways, including intracellular and extracellular signaling, may be attributed to the alteration of gene/protein expression, thereby leading to cellular/tissue phenotypic changes. In addition, mitophagy is one of the most important factors affecting exercise-induced physiological adaptive processes. Thus, exercise-induced autophagy/mitophagy as an adaptive mechanism could result in supramolecular changes in biological systems, as summarized in Fig. 11.3.

6.1 Exercise Performance

Increasing evidence has demonstrated that autophagy defects may affect athletic performance, mainly due to pathological changes in skeletal muscle. These include the loss of muscle strength, decreased specific tension, increased oxidative stress, pathological changes in mitochondrial structure and function, and reduced capability to cope with muscle damage caused by eccentric exercise and deficits in exercise adaptation [76]. Nonpathological changes can also appear in other tissues lacking autophagy, although skeletal muscle is more sensitive to autophagy defects [77]. Conversely, recent studies have reported that exercise-induced autophagy can regulate skeletal muscle metabolism and maintain cellular homeostasis, thereby improving the function of skeletal muscle [23, 24, 78]. It is also worth noting that maximal running distance is reportedly reduced in the deficient exercise-induced autophagy in BCL2AAA mice. Similar to BCL2AAA mice, Becn1$^{+/-}$ mice also exhibit deficient exercise-induced autophagy in skeletal muscle, normal muscle strength, and decreased maximal treadmill running distance [21].

Previous studies have documented the importance of basal autophagy in skeletal muscle, including recent reports indicating that insufficient increase in basal autophagy in skeletal muscle can impair metabolic adaptation induced by exercise training, such as mitochondrial biogenesis and angiogenesis, without affecting contractile adaptation. Also, the lack of basal autophagy can affect the performance improvements induced by endurance exercise training [7]. Further studies have shown that peroxisome proliferator-activated receptor γ coactivator1-α (PGC-1α) knockout mice exhibit reduced mitochondrial volume and lower mitochondrial turnover due to decreased mitophagy and mitochondrial biogenesis, with an associated reduction in exercise capacity [79]. At the same time, the deletion of the *Atg7* gene in mice has been found to compromise muscle mass, causing profound mitochondrial membrane depolarization in skeletal muscle during downhill running (eccentric muscle contraction), which suggests that autophagy influences exercise performance during more damaging muscle contractions [80]. Moreover, another interesting observation is the lack of improvement in endurance capacity in *Atg6*$^{+/-}$ mice, despite their participation in similar voluntary running activity, when compared with wild-type (WT) littermates. One possible explanation for this is that blunted mitochondrial biogenesis in the skeletal muscle of *Atg6*$^{+/-}$ mice is more functionally limiting for continuous and maximal exercise performance than for the intermittent submaximal exercise, which is a characteristic of the voluntary running model. It is also possible that the reduced Atg6 expression in other tissues, such as heart, brain, and liver, can contribute to the loss of endurance exercise performance in *Atg6*$^{+/-}$ mice. Nevertheless, this is a very important observation because exercise capacity is a major general predictor of mortality from all causes [81]. It also demonstrates that a failure to increase basal autophagy in skeletal muscle, and possibly other tissues, could be associated with individual limitations to enhance endurance capacity in response to exercise training [82].

6.2 Skeletal Muscle Mass

Under catabolic conditions, skeletal muscle protein is mobilized to maintain the energy requirements of multiple organs, but excessive protein degradation induces a decrease in skeletal muscle mass, which causes skeletal muscle atrophy, and rapid deterioration of strength and endurance. Protein degradation in skeletal muscle is mainly controlled by two proteolytic systems: the ubiquitin–proteasome system (UPS) and autophagy–lysosomal system (ALS) [83]. Two E3 ubiquitin ligases, Atrogin-1 and MuRF1 in the UPS, have been identified as the critical factors for regulating protein ubiquitination in a variety of skeletal muscle atrophy models [84]. Recent studies have confirmed that the skeletal muscle of *Atg7*$^{-/-}$ mice displays a significant atrophy, as shown in ultrastructural disorders, including mitochondrial swelling, sarcoplasmic reticulum swelling, sarcomere disorder, protein aggregates in the cytoplasm, increased oxidative stress level, and decreased strength [85]. Similarly, both excessive autophagy and deficient autophagy can induce the atrophy, weakness, or fiber degeneration of skeletal muscle [86]. Thus, the functional status or level of autophagy is the determinant for the health of skeletal muscle.

Skeletal muscle is one of the tissues with highest basal autophagy flux and greater capacity to increase autophagy flux [87]. The maintenance of skeletal muscle health and function requires an adequate level of autophagy and a finely tuned balance between mitophagy and mitochondrial biogenesis [88]. Recent data also suggest that the regulation of skeletal muscle metabolism and the sustainment of internal environment stability can be controlled by exercise-induced autophagy/mitophagy [89]. Similarly, recent reports have verified that autophagy induced by amino acid supplementation promotes a hypertrophic response to resistance exercise and that autophagic signaling regulates resistance exercise-induced protein synthesis in skeletal muscle [78]. Additionally, in comparison with untrained mice, the mice subjected to resistance exercise training exhibit a higher cross-sectional area, stronger skeletal muscle, higher levels of Beclin1, Atg5, and Atg7 proteins, and lower levels of p62 protein, suggesting that resistance exercise can activate basal autophagy [78]. In terms of its mechanism, resistance exercise training can activate autophagy and reduce apoptosis via the IGF-1/Akt/mTOR and IGF-1Akt/FoxO3 signaling pathways, thereby improving the quality and strength of skeletal muscle. Furthermore, several changes such as impairment of autophagic flux in collagen VI null (Col6a1$^{-/-}$) mice after exercise intervention are observed, including impaired autophagy, a slow increase in the LC3-II/LC3-I ratio accompanied by the accumulation of dysfunctional mitochondria, changes in mitochondrial networks, degradation of skeletal muscle fibers, and loss of skeletal muscle strength [18]. These results have further confirmed that exercise-induced autophagy is an important protective process. However, it is also reported that endurance exercise could activate AMPK and phosphorylate ULK1 in skeletal muscle, thereby inhibiting mTOR-mediated phosphorylation of ULK1; enhancing AMPK/FoxO3-mediated mitochondrial E3 ubiquitin ligase 1 (MUL1), MuRF1, and LC3-II expression; and ultimately degrading the protein in skeletal muscle [90], suggesting that endurance exercise could induce autophagy via the AMPK/ULK1 and AMPK/FoxO3 signaling pathways, promote the degradation of dysfunctional proteins in skeletal muscle, and produce energy substrates for skeletal muscle contraction. At the same time, other studies have confirmed that endurance exercise may activate PGC-1α through the phosphorylation of AMPK and p38 MAPK, and the deacetylation of SIRT1, to improve mitochondrial quality control in skeletal muscle [91]. These studies suggest that exercise-induced autophagy is essential for maintaining the mass, metabolism, and function of skeletal muscle.

6.3 Cardiovascular Adaptation

Autophagy is a reparative, life-sustaining process involving the degradation of cytoplasmic constituents enclosed in double-membrane vesicles after fusion with lysosomes. Emerging evidence indicates that basal autophagy is an important process of regulating proper cardiovascular homeostasis and function, as either excessive or insufficient levels of autophagic flux may lead to the pathogenesis of cardiovascular diseases (CVD) [92]. Moreover, environmental stress-related stimuli, like exercise, could potently induce autophagy in the arterial wall and heart, thus conferring cardiovascular protection. As reported previously, acute exercise enhances autophagy activity in skeletal muscle, heart, and other tissues or organs involved in energy homeostasis in normal mice, while transgenic mice deficient in stimulus-induced autophagy exhibit decreased endurance and altered glucose metabolism [21]. Exercise-induced autophagy facilitates the clearance of protein aggregates and alleviates cardiac proteinopathy, as evidenced by enhanced cardioprotective effect of exercise in mice with an overexpression of autophagy-related proteins. Furthermore, exercise training attenuates the impairment of autophagic flux to improve mitochondrial bioenergetics in the failing hearts of mice [93]. Myocardial mitochondrial function has a key role in regulating cellular death and supplying energy. Pathological conditions such as ischemia may induce the fragmentation and fission of cardiomyocyte mitochondria, which results in mitophagy and cellular death. Therefore, mitochondrial fission and fragmentation play adaptive roles in the maintenance of optimal cardiac function during exercise. Beta1-adrenergic receptor-mediated physiological fragmentation of mitochondria during submaximal exercise training may improve mitochondrial function. In this type of fragmentation, Drp1 is also activated; however, membrane potential is maintained, and regulators of mitophagy are downregulated, suggesting that fission is important for maintaining cardiac and mitochondrial bioenergetic homeostasis [94]. One of the common complications of allogeneic hematopoietic stem cell transplantation is postoperative chronic graft-versus-host disease (cGVHD), which leads to morbidity and disability. The beneficial effects of 11 weeks of moderate exercise on cGVHD mice have been reported. In trained animals, an increase is reported for cardiomyocyte biomarkers of autophagy, including Atg12, p-ULK1, p62, and LC3-II [95]. Moreover, myocardial glutathione reductase, catalase, and α-tubulin are higher in the exercise group when compared to the sham group. These results suggest that exercise protects against debilitating cardiac disease [95]. In summary, regular exercise training confers cardiovascular protection and plays an important role in the prevention and treatment of cardiovascular diseases. The effect of exercise on different human physiological systems and even within each system (e.g., cardiovascular system) may vary based on exercise modality, intensity, duration, and repetition [74]. Further elucidation of the mechanisms underlying exercise-afforded cardiovascular benefits holds promise for the discovery of novel therapeutic targets and determining optimal amounts of physical activity to improve outcomes.

6.4 Glycolipid Metabolism Regulation

Autophagy is also an important regulator of lipid metabolism (lipophagy) [96]. In the liver, autophagosomes encapsulate lipid droplets, transport them to lysosomes, and hydrolyze them into free fatty acids (FFAs) and glycerol. The inhibition of autophagy inhibits the β-oxidation of triglycerides (TG), thus causing the accumulation of TG in hepatocytes, and an increase in the number and volume of lipid droplets formed by TG and cholesterol [97]. During starvation, the number of lipid-encapsulated autophagosomes increases. In contrast, a 16-week high-fat diet can significantly increase the number and volume of lipid droplets in hepatocytes, with a significant decrease in the number of lipid-encapsulated autophagosomes [96]. Thus, the gluconeogenesis of fatty acids in the liver is an important source of blood glucose at the late end of long-term endurance exercise and an important guarantee for maintaining the efficiency of ATP supply to skeletal muscle. Another study has reported that mitochondrial dysfunction induced by autophagy deficiency can stimulate fibroblast growth factor 21 (FGF21) expression through upregulating recombinant activating transcription factor 4 (ATF4), thereby promoting the browning of white adipose tissue (WAT) and strengthening protection from obesity and insulin resistance [98].

Skeletal muscle has been established as the major organ targeted for insulin-dependent glucose uptake and utilization, and plays an important role in insulin resistance associated with obesity and type II diabetes. At the same time, autophagy plays a vital role in regulating glucose utilization in skeletal muscle and plasma insulin resistance. Recent animal studies have found that, in a transgenic mouse model with impaired inducible autophagy, exercise exacerbates symptoms such as insulin resistance and glucose metabolism disorders, which may be due to the absorption and utilization of glucose in skeletal muscle, caused by exercise [21]. In contrast, the *Atg7*$^{-/-}$ mice, after exercise, are not affected by plasma lactic acid, glucose, or free fatty acid concentrations [80], indicating that autophagy may be not a necessary condition for exercise-induced regulation of circulating substrates and metabolites. However, under drug-induced inhibition of autophagy, glucose uptake is reduced in C2C12 myotubes stimulated by insulin, thus providing further evidence of the role of autophagy in glucose regulation [57]. Similarly, in BCL2AAA mutant and *Beclin1* knockout mice, exercise-induced autophagy is attenuated, accompanied by reduced translocation of GLUT4 to the sarcolemma, and reduced glucose uptake in skeletal muscle, suggesting that glucose consumption in skeletal muscle is highly correlated with exercise-mediated autophagy. In BCL2AAA mutant mice, a weaker protective effect is observed in exercise training-mediated autophagy against glucose intolerance caused by a high-fat diet [21]. Additionally, endurance exercise can increase the basal level of autophagy by promoting the interaction between AMPK and Sestrin2/3, and reverse the occurrence of low glucose and glucose tolerance caused by a long-term high-fat diet, and the accumulation of lipid droplets [99]. Further studies on exercise-induced autophagy activation will be required to assess the beneficial effects of exercise on glycolipid metabolism.

6.5 Mitochondrial Adaptation

As a double-membrane sealed organelle, mitochondria play an important role in biological oxidation and energy exchange. They can produce ATP through the tricarboxylic acid cycle and oxidative phosphorylation to provide energy and biosynthetic substrates [74]. During exercise, energy requirements increase sharply, and mitochondria can increase the source of available energy. For this reason, it is necessary to improve the biogenesis of mitochondria and maintain its quality control system, stabilize mitochondrial function, and ensure sufficient energy during muscle contraction. Several studies have documented that both acute exercise and chronic exercise training can mediate mitochondrial biogenesis of skeletal muscle [100, 101]. The transcriptional coactivator PGC-1α acts as a mitochondrial biogenesis-regulating gene [102], which is itself regulated by multiple upstream factors such as p38 MAPK, AMPK, and SIRT1 [103, 104]. However, exercise promotes mitochondrial biogenesis and stimulates the generation of excessive ROS. As a double-edged sword, ROS not only serves as the key signal molecules in cell signal transduction, but also regulates cell growth, differentiation, and survival. Excessive ROS will attack DNA, proteins, and lipids; destroy the structure and integrity of mitochondria; cause a peroxidation chain reaction; and induce mitochondrial-mediated apoptosis [91]. In order to maintain the normal state of cellular function, mitochondria need to maintain dynamic changes in the mitochondrial junction structure, controlled by a series of proteins involved in mitochondrial fusion and fission [105]. Mitochondrial dynamics are crucial for cellular homeostasis through regulating the balance between fusion and fission in mitochondria. Mitochondrial fusion is regulated by outer and inner membrane proteins such as Mfn1/2 and optic atrophy 1 (Opa1), which are responsible for regulating the fusion of the outer mitochondrial membrane to maintain mitochondrial function and integrity [106]. Similarly, Drp1 and mitochondrial fission 1 protein (Fis1) are mainly involved in mitochondrial fission to regulate the mitochondrial redistribution process by removing damaged mitochondria [107].

On the other hand, mitochondrial dysfunction, especially related to metabolic activity, has been found to contribute to many disorders, including metabolic diseases, and the aging process [61]. Exercise can remarkably improve the quantity and quality of mitochondria through several adaptive processes, including mitochondrial dynamics and mitophagy [108]. Mitochondrial dynamics not only regulate mitochondrial morphology, but also play a crucial role in controlling mitochondrial function, by responding to apoptosis stimuli and mitochondrial quality control [109]. Recent studies have confirmed that after long-term aerobic exercise training in healthy individuals, the expression of mitochondrial Mfn2 and Drp1 in skeletal muscle cells increases significantly, and the level of mitochondrial oxidative phosphorylation increases, indicating that metabolic changes caused by exercise can regulate the mitochondrial fusion and fission processes [110]. Under normal physiological conditions, the dynamic changes in mitochondria eventually split into two uneven offspring. Mitophagy can selectively reduce the membrane potential and degrade mitochondrial progeny that exceed their repairing capability. Mitochondria responding to depolarization under stress stimuli (such as ROS, nutrient deficiency, and cell aging) are wrapped into autophagosomes and fused with lysosomes to degrade damaged mitochondria, thereby maintaining mitochondrial metabolism [111]. The results in another study have reported that endurance exercise training could significantly upregulate the expression of autophagy-related proteins such as Atg7 and Beclin1; increase the LC3-II/LC3-I ratio; downregulate the expression of p62; and increase the expression of mitophagy-related protein BNIP3 in skeletal muscle. These changes can improve the levels of autophagy and mitophagy, thereby reducing damaged or aged mitochondria, and stabilizing the metabolic adaptability of mitochondria [7]. The results of these studies have demonstrated that mitophagy induced by endurance exercise is an important built-in mechanism that promotes the mitochondrial renewal cycle and functional improvement.

6.6 Disease Status

Increasing evidence has demonstrated that autophagy plays a vital role in the occurrence and development of many diseases such as aging, atherosclerosis, Alzheimer's disease (AD), Parkinson's disease (PD), cardiovascular diseases, and type II diabetes. During the aging process, the number of autophagosomes in tissues decreases with the reduction in physical activity, which promotes the accumulation of senescent organelles and misfolded proteins that cannot be eliminated in a timely and effective manner, thereby reducing cellular defenses and environmental adaptability, and leading to a series of diseases [112]. In contrast, restricted diets or exercise intervention are effective interventional strategies that increase autophagy activity, which may delay aging and prolong lifespan [112].

AD is a common neurological disease in the elderly, and its pathological manifestation includes the aggregation of abnormally folded β-amyloid (Aβ) in cells. As the disease develops, dysfunctional mitophagy is a trigger for apoptosis, where neuronal death is a component of AD pathology [113]. Abnormal autophagy–lysosomal degradation may lead to an increase in toxic Aβ, thus forming extracellular plaques and damaging neurons [114]. PD is a disease characterized by the inclusion of α-synuclein–Lewy bodies in dopaminergic neurons in the substantia nigra compact and striatal dopaminergic neurons [115]. Studies have shown that autophagy is involved in the degradation of α-synuclein, and the autophagy inducer rapamycin can increase the clearance of α-synuclein. The suppression of autophagy can reduce the clearance of α-synuclein and accelerate its accumulation, suggesting that autophagy degradation disorder is an essential cause of PD [115]. Similarly, autophagy is also closely correlated with cardiovascular diseases. When the autophagy function is reduced or impaired, it can cause cardiac insufficiency and even heart failure, thereby leading to a series of cardiovascular diseases [116].

Some recent studies have verified that exercise training can increase the expression of PGC-1α, SIRT1, citrate synthase, and mtDNA, in most brain regions of mice, indicating higher mitochondrial biogenesis in the brain [117]. Furthermore, endurance exercise training, but not voluntary wheel running activity, can improve autophagy, mitophagy, apoptosis signaling, and mitochondrial function in the cortex and cerebellum of the brain tissues of rats. These outcomes are associated with better locomotive capacity and increased willingness to explore new environments [25]. Similarly, autophagy is also necessary for a great glial cell function to modulate inflammatory responses by releasing cytokines, presenting antigens, and producing antibodies, such as anti-Aβ. The studies of AD have consistently reported the accumulation of Aβ plaques followed by a disruption in microglial function. Previous studies have found that mouse microglia with the deficiency of Beclin1, a key autophagy-related protein, are unable to clear Aβ aggregates by phagocytosis [118], indicating that autophagy is necessary for microglia to complete the phagocytosis of Aβ. A more recent study has reported that exercise could induce autophagy in microglia to promote the clearance of Aβ proteins, and suppress the activation of inflammasomes, thereby reducing the release of pro-inflammatory cytokines [119]. Conversely, resistance exercise training has been shown to improve skeletal muscle strength in chloroquine-treated rats, and prevent chloroquine-induced autophagic suppression or autophagic flux impairment, based on the increase in Beclin1 and p62 [120]. This suggests that resistance exercise training can modulate autophagy in atrophying skeletal muscle with potential protective effects on the function of skeletal muscle. In addition to the bona fide autophagy-dependent beneficial effects of exercise, exercise training can adjust the functional status of autophagy during a diseased state, which may alleviate the progression of atherosclerosis, AD, PD, aging, cardiovascular diseases, and other autophagy-related diseases (Fig. 11.3).

7 Conclusion and Future Perspective

Exercise training at an appropriate intensity can play an active role in the growth and metabolism of skeletal muscle, myocardium, brain, and other tissues, by appropriately increasing the level of autophagy or optimizing the functional status of autophagy. It also has a constructive role in inhibiting the occurrence and development of autophagy-related diseases. However, overtraining can cause autophagy activation to exceed the threshold, thus resulting in the fatigue and damage of skeletal muscle, myocardium, brain, and other tissues. In severe cases, it may stimulate excessive autophagy or apoptosis to induce or aggravate various autophagy-related diseases.

Future studies on exercise-mediated health promotion will likely have access to more complex methods to fully understand the beneficial effects of appropriate exercise or regular physical activity from potential aspects such as genes, molecules, cells, and organs. Further studies will be critical to elucidate the diversity and complexity of the cell network involved in the exercise response by analyzing changes in the genome, transcriptome, proteome, and metabolome. Determining the exact mechanisms for optimal and scientific exercise regimens to promote overall health will improve our understanding of the health-promoting factors of exercise, and the pathogenesis of many diseases mediated by physical inactivity, and will provide a new theoretical basis and intervention strategies for the prevention, treatment, and rehabilitation of chronic diseases.

References

1. Blair SN (2009) Physical inactivity: the biggest public health problem of the 21st century. Br J Sports Med 43:1–2
2. Lee IM, Shiroma EJ, Lobelo F et al (2012) Effect of physical inactivity on major non-communicable diseases worldwide: an analysis of burden of disease and life expectancy. Lancet 380:219–229
3. Anonymous (2010) WHO guidelines approved by the guidelines review committee. In: Global recommendations on physical activity for health. World Health Organization, Geneva
4. Yang Z, Klionsky DJ (2009) An overview of the molecular mechanism of autophagy. Curr Top Microbiol Immunol 335:1–32
5. Liu X, Niu Y, Yuan H et al (2015) AMPK binds to Sestrins and mediates the effect of exercise to increase insulin-sensitivity through autophagy. Metabolism 64:658–665
6. He C, Sumpter R Jr, Levine B (2012) Exercise induces autophagy in peripheral tissues and in the brain. Autophagy 8:1548–1551
7. Lira VA, Okutsu M, Zhang M et al (2013) Autophagy is required for exercise training-induced skeletal muscle adaptation and improvement of physical performance. FASEB J 27:4184–4193
8. Carter HN, Kim Y, Erlich AT et al (2018) Autophagy and mitophagy flux in young and aged skeletal muscle following chronic contractile activity. J Physiol 596:3567–3584
9. Ravikumar B, Sarkar S, Davies JE et al (2010) Regulation of mammalian autophagy in physiology and pathophysiology. Physiol Rev 90:1383–1435
10. Ganley IG, Du HL, Wang J et al (2009) ULK1·ATG13·FIP200 complex mediates mTOR signaling and is essential for autophagy. J Biol Chem 284:12297–12305
11. Levine B, Sinha S, Kroemer G (2008) Bcl-2 family members: dual regulators of apoptosis and autophagy. Autophagy 4:600–606
12. Maiuri MC, Zalckvar E, Kimchi A et al (2007) Self-eating and self-killing: crosstalk between autophagy and apoptosis. Nat Rev Mol Cell Biol 8:741–752
13. Wong E, Cuervo AM (2010) Autophagy gone awry in neurodegenerative diseases. Nat Neurosci 13:805–811
14. Morell C, Bort A, Vara-Ciruelos D et al (2016) Up-regulated expression of LAMP2 and autophagy activity during neuroendocrine differentiation of prostate cancer LNCaP cells. PLoS One 11:e0162977
15. Johansen T, Lamark T (2011) Selective autophagy mediated by autophagic adapter proteins. Autophagy 7:279–296
16. Salminen A, Vihko V (1984) Autophagic response to strenuous exercise in mouse skeletal muscle fibers. Virchows Arch B Cell Pathol Incl Mol Pathol 45:97–106
17. Dohm GL, Tapscott EB, Kasperek GJ (1987) Protein degradation during endurance exercise and recovery. Med Sci Sports Exerc 19:S166–S171
18. Grumati P, Coletto L, Schiavinato A et al (2011) Physical exercise stimulates autophagy in normal skeletal muscles but is detrimental for collagen VI-deficient muscles. Autophagy 7:1415–1423
19. Jamart C, Francaux M, Millet GY et al (2012) Modulation of autophagy and ubiquitin-proteasome pathways during ultra-endurance running. J Appl Physiol (1985) 112:1529–1537

20. Ogura Y, Iemitsu M, Naito H et al (2011) Single bout of running exercise changes LC3-II expression in rat cardiac muscle. Biochem Biophys Res Commun 414:756–760
21. He C, Bassik MC, Moresi V et al (2012) Exercise-induced BCL2-regulated autophagy is required for muscle glucose homeostasis. Nature 481:511–515
22. Kim YA, Kim YS, Song W (2012) Autophagic response to a single bout of moderate exercise in murine skeletal muscle. J Physiol Biochem 68:229–235
23. Lenhare L, Crisol BM, Silva VRR et al (2017) Physical exercise increases Sestrin 2 protein levels and induces autophagy in the skeletal muscle of old mice. Exp Gerontol 97:17–21
24. Fan J, Yang X, Li J et al (2017) Spermidine coupled with exercise rescues skeletal muscle atrophy from D-gal-induced aging rats through enhanced autophagy and reduced apoptosis via AMPK-FOXO3a signal pathway. Oncotarget 8:17475–17490
25. Marques-Aleixo I, Santos-Alves E, Balça MM et al (2015) Physical exercise improves brain cortex and cerebellum mitochondrial bioenergetics and alters apoptotic, dynamic and auto (mito)phagy markers. Neuroscience 301:480–495
26. Zhang L, Hu X, Luo J et al (2013) Physical exercise improves functional recovery through mitigation of autophagy, attenuation of apoptosis and enhancement of neurogenesis after MCAO in rats. BMC Neurosci 14:46
27. Wullschleger S, Loewith R, Hall MN (2006) TOR signaling in growth and metabolism. Cell 124:471–484
28. Zachari M, Ganley IG (2017) The mammalian ULK1 complex and autophagy initiation. Essays Biochem 61:585–596
29. Gwinn DM, Shackelford DB, Egan DF et al (2008) AMPK phosphorylation of raptor mediates a metabolic checkpoint. Mol Cell 30:214–226
30. Hosokawa N, Hara T, Kaizuka T et al (2009) Nutrient-dependent mTORC1 association with the ULK1-Atg13-FIP200 complex required for autophagy. Mol Biol Cell 20:1981–1991
31. Kim J, Kundu M, Viollet B et al (2011) AMPK and mTOR regulate autophagy through direct phosphorylation of Ulk1. Nat Cell Biol 13:132–141
32. Toyoda T, Hayashi T, Miyamoto L et al (2004) Possible involvement of the alpha1 isoform of 5'AMP-activated protein kinase in oxidative stress-stimulated glucose transport in skeletal muscle. Am J Physiol Endocrinol Metab 287:E166–E173
33. Hardie DG, Hawley SA, Scott JW (2006) AMP-activated protein kinase—development of the energy sensor concept. J Physiol 574:7–15
34. Liu X, Niu Y, Yuan H et al (2015) AMPK binds to Sestrins and mediates the effect of exercise to increase insulin-sensitivity through autophagy. Metab Clin Exp 64:658–665
35. Inoki K, Li Y, Xu T et al (2003) Rheb GTPase is a direct target of TSC2 GAP activity and regulates mTOR signaling. Genes Dev 17:1829–1834
36. Sanchez AM, Bernardi H, Py G et al (2014) Autophagy is essential to support skeletal muscle plasticity in response to endurance exercise. Am J Physiol Regul Integr Comp Physiol 307: R956–R969
37. Alers S, Loffler AS, Wesselborg S et al (2012) Role of AMPK-mTOR-Ulk1/2 in the regulation of autophagy: cross talk, shortcuts, and feedbacks. Mol Cell Biol 32:2–11
38. Huang J, Manning BD (2009) A complex interplay between Akt, TSC2 and the two mTOR complexes. Biochem Soc Trans 37:217–222
39. Zhang H, Cicchetti G, Onda H et al (2003) Loss of Tsc1/Tsc2 activates mTOR and disrupts PI3K-Akt signaling through downregulation of PDGFR. J Clin Invest 112:1223–1233
40. Ziaaldini MM, Marzetti E, Picca A et al (2017) Biochemical pathways of sarcopenia and their modulation by physical exercise: a narrative review. Front Med (Lausanne) 4:167
41. Russell RC, Tian Y, Yuan H et al (2013) ULK1 induces autophagy by phosphorylating Beclin-1 and activating VPS34 lipid kinase. Nat Cell Biol 15:741–750
42. Wei Y, Pattingre S, Sinha S et al (2008) JNK1-mediated phosphorylation of Bcl-2 regulates starvation-induced autophagy. Mol Cell 30:678–688

43. He C, Zhu H, Li H et al (2013) Dissociation of Bcl-2-Beclin1 complex by activated AMPK enhances cardiac autophagy and protects against cardiomyocyte apoptosis in diabetes. Diabetes 62:1270–1281
44. Mackenzie MG, Hamilton DL, Murray JT et al (2009) mVps34 is activated following high-resistance contractions. J Physiol 587:253–260
45. Mammucari C, Milan G, Romanello V et al (2007) FoxO3 controls autophagy in skeletal muscle in vivo. Cell Metab 6:458–471
46. Neel BA, Lin Y, Pessin JE (2013) Skeletal muscle autophagy: a new metabolic regulator. Trends Endocrinol Metab 24:635–643
47. Milan G, Romanello V, Pescatore F et al (2015) Regulation of autophagy and the ubiquitin-proteasome system by the FoxO transcriptional network during muscle atrophy. Nat Commun 6:6670
48. Tam BT, Siu PM (2014) Autophagic cellular responses to physical exercise in skeletal muscle. Sports Med 44:625–640
49. Sanchez AM, Csibi A, Raibon A et al (2012) AMPK promotes skeletal muscle autophagy through activation of forkhead FoxO3a and interaction with Ulk1. J Cell Biochem 113:695–710
50. Ying Z, Jing Y, Wenjuan L et al (2010) Cytosolic FoxO1 is essential for the induction of autophagy and tumour suppressor activity. Nat Cell Biol 12:665–675
51. Canto C, Gerhart-Hines Z, Feige JN et al (2009) AMPK regulates energy expenditure by modulating NAD+ metabolism and SIRT1 activity. Nature 458:1056–1060
52. Schwer B, Verdin E (2008) Conserved metabolic regulatory functions of sirtuins. Cell Metab 7:104–112
53. Imai S, Armstrong CM, Kaeberlein M et al (2000) Transcriptional silencing and longevity protein Sir2 is an NAD-dependent histone deacetylase. Nature 403:795–800
54. Lee IH, Cao L, Mostoslavsky R et al (2008) A role for the NAD-dependent deacetylase Sirt1 in the regulation of autophagy. Proc Natl Acad Sci U S A 105:3374–3379
55. Chin ER (2005) Role of Ca2+/calmodulin-dependent kinases in skeletal muscle plasticity. J Appl Physiol (1985) 99:414–423
56. Medina DL, Di Paola S, Peluso I et al (2015) Lysosomal calcium signalling regulates autophagy through calcineurin and TFEB. Nat Cell Biol 17:288–299
57. Settembre C, Di Malta C, Polito VA et al (2011) TFEB links autophagy to lysosomal biogenesis. Science 332:1429–1433
58. Vainshtein A, Hood DA (2016) The regulation of autophagy during exercise in skeletal muscle. J Appl Physiol 120(1985):664–673
59. Hoyer-Hansen M, Bastholm L, Szyniarowski P et al (2007) Control of macroautophagy by calcium, calmodulin-dependent kinase kinase-beta, and Bcl-2. Mol Cell 25:193–205
60. Grotemeier A, Alers S, Pfisterer SG et al (2010) AMPK-independent induction of autophagy by cytosolic Ca2+ increase. Cell Signal 22:914–925
61. Coen PM, Musci RV, Hinkley JM et al (2018) Mitochondria as a target for mitigating sarcopenia. Front Physiol 9:1883
62. Scherz-Shouval R, Elazar Z (2011) Regulation of autophagy by ROS: physiology and pathology. Trends Biochem Sci 36:30–38
63. Dombi E, Mortiboys H, Poulton J (2018) Modulating mitophagy in mitochondrial disease. Curr Med Chem 25:5597–5612
64. Bingol B, Sheng M (2016) Mechanisms of mitophagy: PINK1, Parkin, USP30 and beyond. Free Radic Biol Med 100:210–222
65. Chen CCW, Erlich AT, Crilly MJ et al (2018) Parkin is required for exercise-induced mitophagy in muscle: impact of aging. Am J Physiol Endocrinol Metab 315:E404–E415
66. Kluge MA, Fetterman JL, Vita JA (2013) Mitochondria and endothelial function. Circ Res 112:1171–1188

67. Rana A, Rera M, Walker DW (2013) Parkin overexpression during aging reduces proteotoxicity, alters mitochondrial dynamics, and extends lifespan. Proc Natl Acad Sci U S A 110:8638–8643
68. Scarffe LA, Stevens DA, Dawson VL et al (2014) Parkin and PINK1: much more than mitophagy. Trends Neurosci 37:315–324
69. Fu M, St-Pierre P, Shankar J et al (2013) Regulation of mitophagy by the Gp78 E3 ubiquitin ligase. Mol Biol Cell 24:1153–1162
70. Chu CT, Ji J, Dagda RK et al (2013) Cardiolipin externalization to the outer mitochondrial membrane acts as an elimination signal for mitophagy in neuronal cells. Nat Cell Biol 15:1197–1205
71. Strappazzon F, Nazio F, Corrado M et al (2015) AMBRA1 is able to induce mitophagy via LC3 binding, regardless of PARKIN and p62/SQSTM1. Cell Death Differ 22:419–432
72. Van Humbeeck C, Cornelissen T, Hofkens H et al (2011) Parkin interacts with Ambra1 to induce mitophagy. J Neurosci 31:10249–10261
73. Hughes DC, Ellefsen S, Baar K (2018) Adaptations to endurance and strength training. Cold Spring Harb Perspect Med 8:a029769
74. Egan B, Zierath JR (2013) Exercise metabolism and the molecular regulation of skeletal muscle adaptation. Cell Metab 17:162–184
75. Kelly RS, Kelly MP, Kelly P (2020) Metabolomics, physical activity, exercise and health: a review of the current evidence. Biochim Biophys Acta Mol Basis Dis 1866:165936
76. Masiero E, Agatea L, Mammucari C et al (2009) Autophagy is required to maintain muscle mass. Cell Metab 10:507–515
77. Komatsu M, Waguri S, Koike M et al (2007) Homeostatic levels of p62 control cytoplasmic inclusion body formation in autophagy-deficient mice. Cell 131:1149–1163
78. Luo L, Lu AM, Wang Y et al (2013) Chronic resistance training activates autophagy and reduces apoptosis of muscle cells by modulating IGF-1 and its receptors, Akt/mTOR and Akt/FOXO3a signaling in aged rats. Exp Gerontol 48:427–436
79. Vainshtein A, Tryon LD, Pauly M et al (2015) Role of PGC-1α during acute exercise-induced autophagy and mitophagy in skeletal muscle. Am J Physiol Cell Physiol 308:C710–C719
80. Lo Verso F, Carnio S, Vainshtein A et al (2014) Autophagy is not required to sustain exercise and PRKAA1/AMPK activity but is important to prevent mitochondrial damage during physical activity. Autophagy 10:1883–1894
81. Kodama S, Saito K, Tanaka S et al (2009) Cardiorespiratory fitness as a quantitative predictor of all-cause mortality and cardiovascular events in healthy men and women: a meta-analysis. JAMA 301:2024–2035
82. Bouchard C, Rankinen T (2001) Individual differences in response to regular physical activity. Med Sci Sports Exerc 33:S446–S451; discussion S452–S443
83. Sandri M (2013) Protein breakdown in muscle wasting: role of autophagy-lysosome and ubiquitin-proteasome. Int J Biochem Cell Biol 45:2121–2129
84. Rom O, Reznick AZ (2016) The role of E3 ubiquitin-ligases MuRF-1 and MAFbx in loss of skeletal muscle mass. Free Radic Biol Med 98:218–230
85. Masiero E, Sandri M (2010) Autophagy inhibition induces atrophy and myopathy in adult skeletal muscles. Autophagy 6:307–309
86. Sandri M (2010) Autophagy in health and disease. 3. Involvement of autophagy in muscle atrophy. Am J Physiol Cell Physiol 298:C1291–C1297
87. Mizushima N, Yamamoto A, Matsui M et al (2004) In vivo analysis of autophagy in response to nutrient starvation using transgenic mice expressing a fluorescent autophagosome marker. Mol Biol Cell 15:1101–1111
88. Hood DA, Tryon LD, Carter HN et al (2016) Unravelling the mechanisms regulating muscle mitochondrial biogenesis. Biochem J 473:2295–2314
89. Lee DE, Bareja A, Bartlett DB et al (2019) Autophagy as a therapeutic target to enhance aged muscle regeneration. Cells 8:183

90. Pagano AF, Py G, Bernardi H et al (2014) Autophagy and protein turnover signaling in slow-twitch muscle during exercise. Med Sci Sports Exerc 46:1314–1325
91. Ferraro E, Giammarioli AM, Chiandotto S et al (2014) Exercise-induced skeletal muscle remodeling and metabolic adaptation: redox signaling and role of autophagy. Antioxid Redox Signal 21:154–176
92. Bravo-San Pedro JM, Kroemer G, Galluzzi L (2017) Autophagy and mitophagy in cardiovascular disease. Circ Res 120:1812–1824
93. Campos JC, Queliconi BB, Bozi LHM et al (2017) Exercise reestablishes autophagic flux and mitochondrial quality control in heart failure. Autophagy 13:1304–1317
94. Coronado M, Fajardo G, Nguyen K et al (2018) Physiological mitochondrial fragmentation is a normal cardiac adaptation to increased energy demand. Circ Res 122:282–295
95. Fiuza-Luces C, Delmiro A, Soares-Miranda L et al (2014) Exercise training can induce cardiac autophagy at end-stage chronic conditions: insights from a graft-versus-host-disease mouse model. Brain Behav Immun 39:56–60
96. Singh R, Kaushik S, Wang Y et al (2009) Autophagy regulates lipid metabolism. Nature 458:1131–1135
97. Ueno T, Komatsu M (2017) Autophagy in the liver: functions in health and disease. Nat Rev Gastroenterol Hepatol 14:170–184
98. Kim KH, Jeong YT, Oh H et al (2013) Autophagy deficiency leads to protection from obesity and insulin resistance by inducing Fgf21 as a mitokine. Nat Med 19:83–92
99. Little JP, Safdar A, Bishop D et al (2011) An acute bout of high-intensity interval training increases the nuclear abundance of PGC-1α and activates mitochondrial biogenesis in human skeletal muscle. Am J Physiol Regul Integr Comp Physiol 300:R1303–R1310
100. Scalzo RL, Peltonen GL, Binns SE et al (2014) Greater muscle protein synthesis and mitochondrial biogenesis in males compared with females during sprint interval training. FASEB J 28:2705–2714
101. Islam H, Edgett BA, Gurd BJ (2018) Coordination of mitochondrial biogenesis by PGC-1α in human skeletal muscle: a re-evaluation. Metabolism 79:42–51
102. Vega RB, Huss JM, Kelly DP (2000) The coactivator PGC-1 cooperates with peroxisome proliferator-activated receptor alpha in transcriptional control of nuclear genes encoding mitochondrial fatty acid oxidation enzymes. Mol Cell Biol 20:1868–1876
103. Chalkiadaki A, Igarashi M, Nasamu AS et al (2014) Muscle-specific SIRT1 gain-of-function increases slow-twitch fibers and ameliorates pathophysiology in a mouse model of duchenne muscular dystrophy. PLoS Genet 10:e1004490
104. Kang C, Chung E, Diffee G et al (2013) Exercise training attenuates aging-associated mitochondrial dysfunction in rat skeletal muscle: role of PGC-1α. Exp Gerontol 48:1343–1350
105. Tanaka T, Nishimura A, Nishiyama K et al (2020) Mitochondrial dynamics in exercise physiology. Pflugers Arch 472:137–153
106. Bertholet AM, Delerue T, Millet AM et al (2016) Mitochondrial fusion/fission dynamics in neurodegeneration and neuronal plasticity. Neurobiol Dis 90:3–19
107. Losón OC, Song Z, Chen H et al (2013) Fis1, Mff, MiD49, and MiD51 mediate Drp1 recruitment in mitochondrial fission. Mol Biol Cell 24:659–667
108. Hood DA, Memme JM, Oliveira AN et al (2019) Maintenance of skeletal muscle mitochondria in health, exercise, and aging. Annu Rev Physiol 81:19–41
109. Yoo SM, Jung YK (2018) A molecular approach to mitophagy and mitochondrial dynamics. Mol Cells 41:18–26
110. Drummond MJ, Addison O, Brunker L et al (2014) Downregulation of E3 ubiquitin ligases and mitophagy-related genes in skeletal muscle of physically inactive, frail older women: a cross-sectional comparison. J Gerontol A Biol Sci Med Sci 69:1040–1048
111. Pickles S, Vigié P, Youle RJ (2018) Mitophagy and quality control mechanisms in mitochondrial maintenance. Curr Biol 28:R170–R185
112. Wohlgemuth SE, Seo AY, Marzetti E et al (2010) Skeletal muscle autophagy and apoptosis during aging: effects of calorie restriction and life-long exercise. Exp Gerontol 45:138–148

113. Fang EF, Scheibye-Knudsen M, Chua KF et al (2016) Nuclear DNA damage signalling to mitochondria in ageing. Nat Rev Mol Cell Biol 17:308–321
114. Nowicki M, Zabirnyk O, Duerrschmidt N et al (2007) No upregulation of lectin-like oxidized low-density lipoprotein receptor-1 in serum-deprived EA.hy926 endothelial cells under oxLDL exposure, but increase in autophagy. Eur J Cell Biol 86:605–616
115. Rott R, Szargel R, Haskin J et al (2008) Monoubiquitylation of alpha-synuclein by seven in absentia homolog (SIAH) promotes its aggregation in dopaminergic cells. J Biol Chem 283:3316–3328
116. Golbidi S, Laher I (2011) Molecular mechanisms in exercise-induced cardioprotection. Cardiol Res Pract 2011:972807
117. Steiner JL, Murphy EA, Mcclellan JL et al (2011) Exercise training increases mitochondrial biogenesis in the brain. J Appl Physiol (1985) 111:1066–1071
118. Lucin KM, O'Brien CE, Bieri G et al (2013) Microglial beclin 1 regulates retromer trafficking and phagocytosis and is impaired in Alzheimer's disease. Neuron 79:873–886
119. Zhang J, Guo Y, Wang Y et al (2018) Long-term treadmill exercise attenuates Aβ burdens and astrocyte activation in APP/PS1 mouse model of Alzheimer's disease. Neurosci Lett 666:70–77
120. Kwon I, Lee Y, Cosio-Lima LM et al (2015) Effects of long-term resistance exercise training on autophagy in rat skeletal muscle of chloroquine-induced sporadic inclusion body myositis. J Exerc Nutr Biochem 19:225–234

Chapter 12
Exercise Mimetic Pills for Chronic Diseases Based on Autophagy

Jun Lv, Hu Zhang, and Ning Chen

1 Introduction

Noncommunicable chronic diseases such as metabolic diseases including obesity and diabetes, and cardiovascular diseases including ischemic heart disease have high incidence not only in developed countries, but also in developing countries. Lack of physical activity increases a major risk factor for many chronic diseases such as cardiovascular diseases and diabetes, which is close to the effect of smoking. Physical inactivity causes approximately 9% of premature deaths, especially causing some noncommunicable diseases [1]. The effect of physical inactivity on the triggering of chronic diseases has also been confirmed, especially for fatty liver [2], sarcopenia [3], Alzheimer's disease, and diabetes [4], which is mainly due to the deficiency of autophagy from physical inactivity.

Autophagy is an evolutionary process of intracellular turnover in eukaryotes. The damaged proteins or organelles are wrapped in the autophagy vesicles with double-membrane structure and then sent to lysosomes (animals) or vacuoles (yeast and plants) for degradation and recycling. Autophagy is activated when the original cells are in the starvation state. Now, it has been found that exercise [5], high temperature [6], and low oxygen [7] also can activate autophagy in multiple tissues and organs. Among the above methods of autophagy activation, more and more studies have gradually confirmed that exercise is a safe and reliable way to be beneficial for noncommunicable chronic diseases [8].

The most immediate disadvantage of the current rapid development of economic society is the lack of physical activity for more and more people, which can lead to higher occurrence and development of chronic diseases at varying degrees among

J. Lv · H. Zhang · N. Chen (✉)
Tianjiu Research and Development Center for Exercise Nutrition and Foods, Hubei Key Laboratory of Exercise Training and Monitoring, College of Health Science, Wuhan Sports University, Wuhan, China

N. Chen (ed.), *Exercise, Autophagy and Chronic Diseases*,
https://doi.org/10.1007/978-981-16-4525-9_12

sedentary populations. Therefore, more and more biotechnological companies or medical institutions are focusing on the research and development of new compounds that could be consumed to reap the benefits of exercise, enhance exercise capacity, or mimic exercise effects, which will have important implications not only for the majority of sedentary people, but also for the majority of people living with chronic diseases, bedridden persons, people with disability for exercise performance, or people without willing to exercise. Since exercise is well known to have an obvious effect on the activation of autophagy or the optimization of functional status of autophagy, some exercise mimic products and exogenous supplements with regulating functional status of autophagy should be highly desired. It should be interesting if these products or supplements can be used as the exercise mimics to delay the onset or progression of some chronic diseases or rescue these chronic diseases, which will be a promising strategy for health promotion.

Previous pharmaceutical compounds reveal their relatively limited actions for the single organ and have not resolved the combinatorial actions or the communications on multiple organs like exercise intervention [9–11]. However, autophagy, with an obvious effect on improving the intracellular environment, alleviating inflammation, and reducing adverse protein folding, has a positive significance for a variety of chronic diseases and has gradually attracted extensive attention. Therefore, it has a certain practical significance to alleviate chronic diseases and improve health status of patients with chronic diseases worldwide by activating certain stages of autophagy through exercise simulators.

2 Exercise Mimics Modulate Autophagy to Improve Chronic Diseases

Since functional status of autophagy has been correlated with multiple chronic diseases including metabolic diseases such as obesity, diabetes, and nonalcohol fatty liver disease, cardiovascular diseases including ischemia–perfusion injury, atherosclerosis, and hypertension, and aging-related diseases including brain aging, AD, and PD, as well as sarcopenia, which are discussed in detail in above chapters with defined and clear certain mechanisms upon exercise interventions based on regulating signal pathways of autophagy. Therefore, autophagy could be the promising target for the prevention, treatment, and rehabilitation of chronic diseases through exercise interventions. Meanwhile, the upstream or downstream molecules of autophagy could also be the potential exercise mimetics, which are highly desired for the exploration and development.

2.1 Irisin

Irisin is first discovered to be expressed in skeletal muscle during exercise and currently has been confirmed to be a bioactive peptide expressed in many tissues of the body with multiple functions for regulating the health status and modulating the prevention and rehabilitation of a series of chronic diseases upon exercise stimulation. During the exercise intervention, the transcription factor peroxisome proliferator-activated receptor γ (PPARγ) co-activator 1α (PGC-1α) in skeletal muscle is triggered and then upregulates the expression of fibronectin type III domain-containing protein 5 (FNDC5) stimulated by the contraction of skeletal muscle. Once FNDC5 is cleaved, irisin is formed by the N-terminal segment of FNDC5 and then enters the circulation to execute its multiple physiological functions.

For diabetes, aerobic exercise can alleviate insulin resistance and promote glucose transport in a variety of ways through reducing inflammation and oxidative stress due to the generation and secretion of irisin [12]. The studies involved in obese and diabetic patients have documented that low level of irisin in serum is positively associated with insulin resistance and glucose tolerance [13], suggesting the beneficial effects of irisin on glucose homeostasis and insulin sensitivity, so that the low level of irisin has the promising potential to be used as a diagnostic indicator of obesity and prediabetes. There is a significantly positive correlation among irisin, β-cells, and insulin concentration [14], which suggests that irisin-mediated proliferation of β-cells plays an important role in alleviating the development and progression of type 2 diabetes mellitus (T2DM). In addition, high-intensity interval exercise can trigger more production of irisin in the body, so as to relieve the T2DM-related symptoms. In terms of lipid metabolism, both acute aerobic exercise and chronic aerobic exercise can improve glucose and lipid metabolism disorders through phosphatidylinositol 3-kinase (PI3K)/protein kinase B (Akt) signaling pathway [15]. Surprisingly, irisin can also improve T2DM in this manner. In terms of glucose and lipid metabolism, irisin-mediated phosphoenolpyruvate carboxykinase (PEPCK) and glucose-6-phosphatase (G6Pase) could be downregulated through PI3K/Akt/FoxO1 signal pathway [16], and glucose homogenization could be improved by PI3K/Akt/GSK3-mediated activation of glycogen synthase to reduce glucose production. In the liver tissues, exercise promotes fatty acid oxidation, reduces fatty acid synthesis, and suppresses the release of liver damage-related molecules to prevent the damage of mitochondria and hepatocytes [17]. During the studies on liver cirrhosis, FNDC5 inhibits the occurrence and development of nonalcoholic fatty liver disease by reducing hepatocyte adipogenesis and death [18]. All above evidences indicate that irisin has certain benefits for metabolic diseases and can be used as the standard to judge the occurrence and progression of these metabolic diseases. In vitro model studies have also found that insulin resistance is inhibited by increasing glucose uptake through mitogen-activated protein kinase (MAPK)-PGC-1α signal pathway; in contrast, irisin can enhance the functional status of autophagy and mitochondrial quality control and function upon the activation of PGC-1α [19]. During long-term regular exercise training, the expression of PGC-1α protein and autophagy-related proteins in the myocardium of animals subjected to exercise training reveals a significant increase, thus alleviating myocardial damage [20]. Similarly, exercise-induced autophagy plays an important role in alleviating sarcopenia by regulating Akt/mTOR and Akt/FoxO3a signaling pathways and AMPK-mediated mitochondrial quality control [21]. Irisin can also improve the damage of other tissues through autophagic signal pathway, such as myocardial injury [22], hepatocyte injury [23], and skeletal muscle ischemia [24]. Exercise has been confirmed to enhance the dynamic balance between mitochondrial fission and fusion and selective autophagy through PGC-1α/FNDC5/irisin signaling pathway. However, the specific mechanism of irisin on health promotion of myocardium and skeletal muscle needs to be further explored. In neurodegenerative diseases, exercise can induce autophagy through suppressing the mTOR signal pathway to alleviate cognitive dysfunction and neuronal damage [25]. Irisin may improve AD through reducing insulin resistance, rescuing impaired neurogenesis, reducing oxidative stress, and promoting neurotrophic factor balance based on the exercise-induced autophagy [26]. Therefore, some studies have also attempted to promote the prevention and alleviation of chronic disease by enhancing the generation and secretion of irisin, as confirmed by irisin-mediated autophagy in INS-1 cell model upon metformin treatments [27], thereby suppressing apoptosis and promoting cell proliferation in hyperglycemic environment by activating AMPK/SIRT1/PGC-1α signaling pathway. Other efforts have also been conducted to upregulate irisin expression in the body for alleviating diabetes through vitamin D supplementation [28].

It is well known that autophagy is an evolutionarily conserved process for executing cellular degradation and renewal, and also can be activated by skeletal muscle movement. Both autophagy and irisin can be produced under the stimulation of skeletal muscle movement, so the synergistic correlation between irisin and autophagy is becoming the interesting topic for health promotion, and prevention, treatments and rehabilitation of chronic diseases. Recently, exercise-induced irisin as one of the myokines has also be considered as one of the inducers of autophagy and is involved in the regulating the functional status of autophagy in multiple cells or tissues such as myofibers, cardiomyocytes, hepatocytes, pancreatic β-cells, and hippocampal tissue, thereby contributing to the prevention, treatments and rehabilitation of sarcopenia, cardiovascular diseases, metabolic diseases including obesity and diabetes, and brain aging-related diseases including AD and PD upon exercise intervention (Fig. 12.1). Therefore, irisin could be the inducer of autophagy as the therapeutic target of these chronic diseases; on the other hand, it is also a promising exercise mimetic pill for the prevention and rehabilitation for these chronic diseases.

2.2 *Alpha-Lipoic Acid*

Alpha-lipoic acid (ALA) is a coenzyme existing in mitochondria, which plays an important role in mitochondrial dehydrogenase reaction, with the major focus on

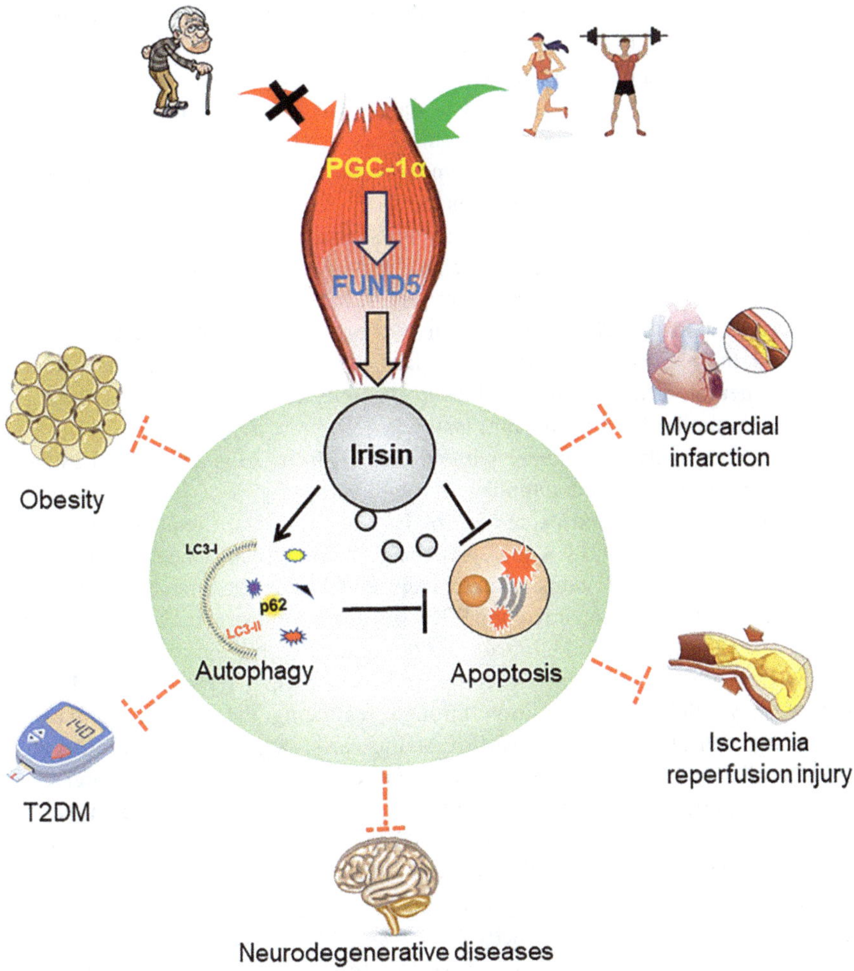

Fig. 12.1 Exercise-induced irisin as an exercise mimetic is beneficial for the suppression or mitigation of chronic diseases such as metabolic diseases, neurodegenerative diseases, and cardiovascular diseases through activating autophagy and inhibiting apoptosis

antioxidant aspects. In addition to acting with glutathione, vitamin C, and vitamin E to protect cell membrane, it can also act as a metal ion chelator to improve oxidative stress [29]. Recently, it has been found to regulate the functional status of autophagy and improve antioxidant capacity during its treatment for chronic diseases, such as diabetes [30], AD [31], and Parkinson's disease (PD) [32].

Oxidative stress occurs mainly because the balance between free radicals and antioxidant systems is broken. The central nervous system is sensitive to oxidative stress [33]. Therefore, ALA has a unique advantage in the treatments of AD and PD. The major causes of the progression of AD are the high phosphorylation of Tau protein, the accumulation of amyloid-beta (Aβ) plaques, and the entanglement of nerve fiber tangles (NFTs). In addition, in the process of oxidative stress, the increase in iron ions will also aggravate the accumulation of free radicals, while binding with Tau protein can further aggravate the formation of NTFs [34]. In a transgenic animal experiment [31], ALA not only increases the activity of antioxidant factors, but also alleviates the neurological dysfunction of experimental animals by inhibiting p38/MAPK signaling pathway and downregulating Tau and p-Tau. Similarly, exogenous supplementation of ALA reveals the apoptotic suppression and the induction of autophagy in a dose-dependent manner. ALA also has the function to optimize the functional status of autophagy in AD and improve mitochondrial quality control through activating PGC-1α. Therefore, ALA has a certain inducing effect on autophagy, but its underlying mechanisms and exogenous supplementation dosages still need to be further explored and clarified. Interestingly, exercise can also improve neural differentiation and neuroimmune activity in patients with AD through the similar MAPK signaling pathway [35].

As for PD, in the cell model with PD, the R-form ALA also could improve mitochondrial function and autophagy through activating PGC-1α signal pathway, as shown in the decreased expression of E3 ubiquitin ligase-related proteins and enhanced expression of LC3 [32]. Similarly, ALA has also been confirmed to optimize the functional status of autophagy and rescue the impaired autophagic flux possibly through rapamycin/mTOR signaling pathway in 6-OHDA-treated SH-SY5Y cells [36]. In addition to neurodegenerative diseases, ALA also has a positive significance for diabetic complications and lipid metabolism. ALA can reduce intracellular fat deposition through regulating the functional status of autophagy and slightly inhibit AMPK at the early stage of lipid differentiation [37]. Similar with the improvement in functional status of autophagy observed in diabetic complications, ALA may also have a protective effect on vascular smooth muscle cells in T2DM by elevating H_2S level and optimizing the functional status of autophagy via the AMPK/mTOR signaling pathway [30].

Exercise has long been thought to alleviate the progression of many chronic diseases, and regular physical activity can significantly prevent the aggravation of AD patients [38]. The mitigation of AD by exercise may involve in the inhibition of mTOR, differentiation of neural factors, activation of AMPK, and regulation of microRNA [25]. Similar to exercise, several studies continue to explore and compare the effects of exercise and ALA on chronic diseases. In streptozotocin (STZ)-induced AD animal experiments, treadmill exercise can alleviate oxidative stress damage and Tau protein accumulation in CA1 region of rats [39]. In another study, in the transgenic mice with AD, ALA and treadmill exercise reach the same effect, but the combinatorial application reveals the better treatment efficacy than the single application [40]. In animal experiments, the treatment of endurance exercise combined with ALA exhibits a unique effect with the increased glucose transport in skeletal muscle through insulin receptor substrate 1 (IRS-1) to alleviate prediabetic insulin resistance [41]. Correspondingly, in human experiments, ALA intervention also can improve insulin sensitivity [42]. Similarly, ALA can increase the expression of PPARβ in C2C12 skeletal muscle cells [43]. These results provide a new possible mechanism for the application of ALA to sarcopenia. Since ALA is one of the endogenous coenzymes in human body to regulate functional status of autophagy in a variety of ways, it eventually executes beneficial functions for chronic diseases. Therefore, ALA may also have the potential as the exercise mimetic for health promotion, and prevention and treatments of chronic diseases based on the similar regulatory role in the functional status of autophagy as exercise (Fig. 12.2).

2.3 *Resveratrol*

Resveratrol (RSV) is a polyphenolic compound found in skins and seeds of grapes so that higher contents of RSV have been determined in some foods such as red wine, berries, and peanuts. The increasing studies on resveratrol are concentrated on delaying aging, improving recognition capacity, reducing blood glucose, mitigating inflammation response, and suppressing cancer initiation and progression.

Neurodegenerative diseases are characterized as the deficiency of autophagy or the impaired autophagic flux [44], and the administration of resveratrol can rescue the abnormal functional status of autophagy, which is beneficial to the improvement of these neurodegenerative diseases such as AD and PD. In earlier studies, resveratrol reveals the alleviation of AD by regulating AMPK-SIRT1-mediated autophagy [45]. In addition, resveratrol can effectively reduce abnormal Aβ deposition by activating AMPK in APP/PS1 mice and maintain Beclin1 and LC3-II levels by activating SIRT1 signaling [46]. Moreover, resveratrol can rescue the impaired autophagy flux in neurons through activating LKB1/AMPK/mTOR/p70S6K signal pathway to mitigate apoptosis in rats with spinal cord injury [47]. Resveratrol also can ameliorate AD by improving mitochondrial function [48] and attenuate Aβ25–35 neurotoxicity in PC12 cells through inducing autophagy partially via the activation of the TyrRS-PARP1-SIRT1 signaling pathway [49]. Therefore, during the process of treating AD and PD based on the functional status of autophagy, resveratrol is mainly dependent on the activation of AMPK signal pathway or the inhibition of mTOR.

Previous studies have also shown that resveratrol can significantly improve exercise endurance and reverse the decline in exercise capacity of the elderly, mainly by improving mitochondrial quality control in skeletal muscle, as exhibited in upregulating mitochondrial transcription factor A, nuclear respiratory factor-1, and PGC-1α. Similarly, the activity of citrate synthase is also significantly enhanced, and the indicators of oxidative stress-induced damage are reduced [50, 51]. In the skeletal muscle wasting model, resveratrol and exercise rescue mitochondrial dysfunction and upregulate anti-apoptotic protein Bcl-2, thereby inhibiting the occurrence of inflammatory factors and apoptosis [52]. Resveratrol can also increase skeletal muscle glycogen reserves and promote muscle hypertrophy to counteract the resistance to exercise capacity [53]. Appropriate intake of resveratrol can delay the fatigue of heart failure and exercise intolerance [54]. In addition, the combinatorial application of exercise and resveratrol may have the synergistic beneficial effect of exercise. In the cardiovascular system, resveratrol also shows the effect on

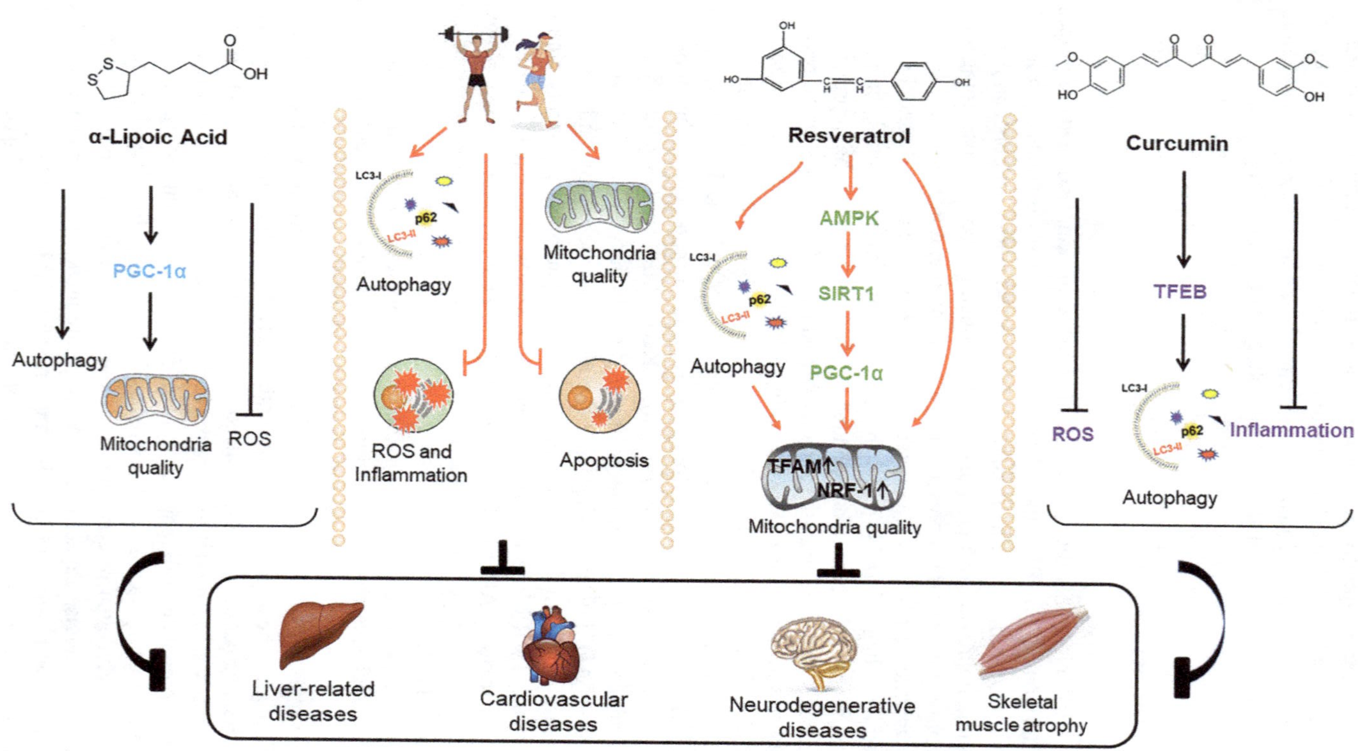

Fig. 12.2 Resveratrol, α-lipoic acid, and curcumin induce autophagy, enhance mitochondrial quality control, and suppress ROS accumulation and inflammation to execute the prevention and treatments of chronic diseases, which exhibits the similar action mechanisms as exercise interventions; therefore, these compounds have the great potential as the promising exercise mimetics for chronic diseases based on regulating functional status of autophagy

promoting autophagy and inhibiting inflammation [55]. Resveratrol and exercise exhibit the upregulated expression of vascular antioxidant enzymes such as endothelial nitric oxide synthase (eNOS) and nonselective phosphodiesterase1–3 (PDE1–3) and cAMP-selective PDE4 in the aorta, and the improved cardiac function and aortic structure in mice [56, 57]. In the nervous system, resveratrol similar to exercise can suppress neuroinflammation, improve neuroplasticity, and upregulate brain-derived neurotrophic factor (BDNF), as well as enhance mitochondrial quality control of neurons to delay brain aging and rescue reduced cognitive capacity [58]. In addition, in models with T2DM nephropathy, exercise is equivalent to inhibiting inflammation [59], reducing lipid peroxidation [60], and increasing mitochondrial biogenesis [61]. At present, the similar molecular mechanisms of exercise simulation and resveratrol consumption have been gradually discovered and validated during improving exercise capacity, promoting exercise adaptation, and accelerating exercise fatigue recovery. Therefore, resveratrol may also have great potential to reveal multiple benefits equivalent to exercise for people with dyskinesias; therefore, it can be also listed in the family of exercise mimics (Fig. 12.2).

2.4 Curcumin

PD is also one kind of aging-related diseases, which is mainly caused by the aggregation of α-synuclein in the substantia nigra and the obvious mitochondrial dysfunction at the early stage. Since exercise-induced autophagy has been shown to delay the development and progression of PD in different ways [62], the screening and identification of novel and effective natural products for activating autophagy are highly desired to slow down the development and progression of PD.

Curcumin, as an acidic polyphenolic compound extracted from the rhizomes of *Zingiberaceae* plants, has been proved to have multiple functions such as antioxidant, anti-apoptotic, and anti-inflammatory functions [63], which also can execute its excellent prevention and treatment effects on chronic diseases including cancers, diabetes, and neurodegenerative diseases such as PD [64]. Increasing volumes of evidences have documented the elucidated role of curcumin in the prevention and treatments of neurodegenerative diseases through regulating the functional status of autophagy. A curcumin analog-based nanoscavenger (NanoCA) can stimulate the nuclear translocation of the transcription factor EB (TFEB), a major autophagy regulator, and trigger autophagic clearance of a-synuclein [65]. Similarly, curcumin also can improve the survival rate of dopamine neurons and upregulate the expression of TFEB and LC3 proteins [66]. Curcumin treatment can reverse the pathogenesis of PD by decreasing the accumulation of A53T α-synuclein and suppressing mTOR/p70S6K signaling [67]. In addition, curcumin derivatives also can induce the activation of TFEB mainly through the inhibition of Akt/mTOR signaling pathway, as validated in animal experiments [68]. Moreover, curcumin has the active role in prolonging lifespan to a certain extent through anti-inflammatory, reducing oxidative stress, and enhancing autophagy [69], with high similarity with exercise training in the body as the promising exercise mimetic.

With the depth research, the relationship between curcumin and exercise has also attracted widespread attention. Intraperitoneal injection of curcumin in rats can promote mitochondrial biogenesis in skeletal muscle [70]. In the chronic obstructive pulmonary disease (COPD) rat model, the administration of curcumin can suppress mitochondrial damage of skeletal muscle by upregulating PGC-1α/SIRT3 signaling pathway [71]. At the same time, the consumption of curcumin shows an increase in skeletal muscle thickness, strength, and mass in sarcopenia mice, thus exhibiting similar treatment efficacy with exercise. Its underlying molecular mechanism is due to the increased and declined expression of the proteins associated with protein synthesis and protein degeneration in skeletal muscle, respectively [72]. Curcumin can reverse hypoxia-induced atrophy by increasing skeletal muscle fibers and inhibiting protein loss of skeletal muscle [73]. In suspension-induced skeletal muscle atrophy, the consumption of curcumin, like exercise intervention, increases the number of activated skeletal muscle satellite cells by activating SIRT1 [74]. In addition, curcumin has the same effect as exercise in its antioxidant effect. Curcumin can improve all oxidative stress biomarkers, such as inhibiting ROS enzyme, and increasing the activity of GSH, catalase, and SOD enzymes [75]. Exercise combined with curcumin also significantly improves cognitive dysfunction caused by diabetes [76] and realizes the improvement in depression, anxiety, metabolic syndromes, diabetes, and other chronic diseases [77]. Furthermore, currently available studies of curcumin are more focused on preventing sports injuries and accelerating sports fatigue recovery [78, 79]. Therefore, curcumin with exercise mimetic benefits shows its particular potential in the prevention, treatments and rehabilitation of a large number of chronic diseases, so that it also should be the promising family member of exercise mimics (Fig. 12.2).

3 Summary and Prospects

As entering the population aging time, more and more elderly people are suffering from chronic diseases with the occupation of a larger number of medical resources, which brings out gradually increasing burdens for their family and society, as well as results in the poor quality of life. Exercise, as a perfect autophagy inducer with the clearance capacity of oxidative stress-induced denatured cellular components and damaged organelles, has the excellent roles in regulating the cellular homeostasis to be beneficial to for the prevention, treatment,s and rehabilitation of a variety of chronic diseases for a large number of people who cannot execute healthy lifestyle including regular physical activity, or the people with disability for incapability to accomplish regular physical activity, or the people without willingness to execute regular physical activity. Therefore, the exercise mimetics could be the gospel for these populations to realize the interventional efficacy similar to exercise.

References

1. Lee IM, Shiroma EJ, Lobelo F et al (2012) Effect of physical inactivity on major non-communicable diseases worldwide: an analysis of burden of disease and life expectancy. Lancet 380:219–229
2. Tang H, Tan X, Zhu L et al (2019) Swimming prevents nonalcoholic fatty liver disease by reducing migration inhibitory factor through Akt suppression and autophagy activation. Am J Transl Res 11:4315–4325
3. Deval C, Calonne J, Coudy-Gandilhon C et al (2020) Mitophagy and mitochondria biogenesis are differentially induced in rat skeletal muscles during immobilization and/or remobilization. Int J Mol Sci 21:3691
4. Gabbouj S, Ryhänen S, Marttinen M et al (2019) Altered insulin signaling in Alzheimer's disease brain—special emphasis on PI3K-Akt pathway. Front Neurosci 13:629
5. He C, Sumpter R Jr, Levine B (2012) Exercise induces autophagy in peripheral tissues and in the brain. Autophagy 8:1548–1551
6. Ganesan S, Pearce SC, Gabler NK et al (2018) Short-term heat stress results in increased apoptotic signaling and autophagy in oxidative skeletal muscle in Sus scrofa. J Therm Biol 72:73–80
7. Monaci S, Aldinucci C, Rossi D et al (2020) Hypoxia shapes autophagy in LPS-activated dendritic cells. Front Immunol 11:573646
8. Peña-Oyarzun D, Bravo-Sagua R, Diaz-Vega A et al (2018) Autophagy and oxidative stress in non-communicable diseases: a matter of the inflammatory state? Free Radic Biol Med 124:61–78
9. Guerrieri D, Moon HY, Van Praag H (2017) Exercise in a pill: the latest on exercise-mimetics. Brain Plast 2:153–169
10. Choi SH, Bylykbashi E, Chatila ZK et al (2018) Combined adult neurogenesis and BDNF mimic exercise effects on cognition in an Alzheimer's mouse model. Science 361:eaan8821
11. Narkar VA, Downes M, Yu RT et al (2008) AMPK and PPARdelta agonists are exercise mimetics. Cell 134:405–415
12. Yaribeygi H, Atkin SL, Simental-Mendía LE et al (2019) Molecular mechanisms by which aerobic exercise induces insulin sensitivity. J Cell Physiol 234:12385–12392
13. Zhang R, Fu T, Zhao X et al (2020) Association of circulating irisin levels with adiposity and glucose metabolic profiles in a middle-aged Chinese population: a cross-sectional study. Diabetes Metab Syndr Obes 13:4105–4112
14. Amri J, Parastesh M, Sadegh M et al (2019) High-intensity interval training improved fasting blood glucose and lipid profiles in type 2 diabetic rats more than endurance training; possible involvement of irisin and betatrophin. Physiol Int 106:213–224
15. Yi XJ, Sun YX, Yao TT et al (2020) Effects of acute and chronic exercise on fat PI3K/AKT/GLUT4 signal pathway in type 2 diabetic rats. Zhongguo Ying Yong Sheng Li Xue Za Zhi 36:12–16
16. Liu TY, Shi CX, Gao R et al (2015) Irisin inhibits hepatic gluconeogenesis and increases glycogen synthesis via the PI3K/Akt pathway in type 2 diabetic mice and hepatocytes. Clin Sci (Lond) 129:839–850
17. Van Der Windt DJ, Sud V, Zhang H et al (2018) The effects of physical exercise on fatty liver disease. Gene Expr 18:89–101
18. Canivet CM, Bonnafous S, Rousseau D et al (2020) Hepatic FNDC5 is a potential local protective factor against non-alcoholic fatty liver. Biochim Biophys Acta Mol Basis Dis 1866:165705
19. Ye X, Shen Y, Ni C et al (2019) Irisin reverses insulin resistance in C2C12 cells via the p38-MAPK-PGC-1α pathway. Peptides 119:170120
20. Tam BT, Pei XM, Yung BY et al (2015) Autophagic adaptations to long-term habitual exercise in cardiac muscle. Int J Sports Med 36:526–534

21. Zeng Z, Liang J, Wu L et al (2020) Exercise-induced autophagy suppresses sarcopenia through Akt/mTOR and Akt/FoxO3a signal pathways and AMPK-mediated mitochondrial quality control. Front Physiol 11:583478
22. Li R, Wang X, Wu S et al (2019) Irisin ameliorates angiotensin II-induced cardiomyocyte apoptosis through autophagy. J Cell Physiol 234:17578–17588
23. Bi J, Yang L, Wang T et al (2020) Irisin improves autophagy of aged hepatocytes via increasing telomerase activity in liver injury. Oxidative Med Cell Longev 2020:6946037
24. He W, Wang P, Chen Q et al (2020) Exercise enhances mitochondrial fission and mitophagy to improve myopathy following critical limb ischemia in elderly mice via the PGC1a/FNDC5/irisin pathway. Skelet Muscle 10:25
25. Kou X, Chen D, Chen N (2019) Physical activity alleviates cognitive dysfunction of Alzheimer's disease through regulating the mTOR signaling pathway. Int J Mol Sci 20:1591
26. Kim OY, Song J (2018) The role of irisin in Alzheimer's disease. J Clin Med 7:407
27. Li Q, Jia S, Xu L et al (2019) Metformin-induced autophagy and irisin improves INS-1 cell function and survival in high-glucose environment via AMPK/SIRT1/PGC-1α signal pathway. Food Sci Nutr 7:1695–1703
28. Safarpour P, Daneshi-Maskooni M, Vafa M et al (2020) Vitamin D supplementation improves SIRT1, Irisin, and glucose indices in overweight or obese type 2 diabetic patients: a double-blind randomized placebo-controlled clinical trial. BMC Fam Pract 21:26
29. Packer L, Witt EH, Tritschler HJ (1995) alpha-Lipoic acid as a biological antioxidant. Free Radic Biol Med 19:227–250
30. Qiu X, Liu K, Xiao L et al (2018) Alpha-lipoic acid regulates the autophagy of vascular smooth muscle cells in diabetes by elevating hydrogen sulfide level. Biochim Biophys Acta Mol basis Dis 1864:3723–3738
31. Zhang YH, Wang DW, Xu SF et al (2018) α-Lipoic acid improves abnormal behavior by mitigation of oxidative stress, inflammation, ferroptosis, and tauopathy in P301S Tau transgenic mice. Redox Biol 14:535–548
32. Zhao H, Zhao X, Liu L et al (2017) Neurochemical effects of the R form of α-lipoic acid and its neuroprotective mechanism in cellular models of Parkinson's disease. Int J Biochem Cell Biol 87:86–94
33. Salim S (2017) Oxidative stress and the central nervous system. J Pharmacol Exp Ther 360:201–205
34. García De Ancos J, Correas I, Avila J (1993) Differences in microtubule binding and self-association abilities of bovine brain tau isoforms. J Biol Chem 268:7976–7982
35. Sun LN, Qi JS, Gao R (2018) Physical exercise reserved amyloid-beta induced brain dysfunctions by regulating hippocampal neurogenesis and inflammatory response via MAPK signaling. Brain Res 1697:1–9
36. Zhou L, Cheng Y (2019) Alpha-lipoic acid alleviated 6-OHDA-induced cell damage by inhibiting AMPK/mTOR mediated autophagy. Neuropharmacology 155:98–103
37. Hahm JR, Noh HS, Ha JH et al (2014) Alpha-lipoic acid attenuates adipocyte differentiation and lipid accumulation in 3T3-L1 cells via AMPK-dependent autophagy. Life Sci 100:125–132
38. Larson EB, Wang L, Bowen JD et al (2006) Exercise is associated with reduced risk for incident dementia among persons 65 years of age and older. Ann Intern Med 144:73–81
39. Lu Y, Dong Y, Tucker D et al (2017) Treadmill exercise exerts neuroprotection and regulates microglial polarization and oxidative stress in a streptozotocin-induced rat model of sporadic Alzheimer's disease. J Alzheimers Dis 56:1469–1484
40. Cho JY, Um HS, Kang EB et al (2010) The combination of exercise training and alpha-lipoic acid treatment has therapeutic effects on the pathogenic phenotypes of Alzheimer's disease in NSE/APPsw-transgenic mice. Int J Mol Med 25:337–346
41. Henriksen EJ (2006) Exercise training and the antioxidant alpha-lipoic acid in the treatment of insulin resistance and type 2 diabetes. Free Radic Biol Med 40:3–12

42. Gosselin LE, Chrapowitzky L, Rideout TC (2019) Metabolic effects of α-lipoic acid supplementation in pre-diabetics: a randomized, placebo-controlled pilot study. Food Funct 10:5732–5738
43. Rousseau AS, Sibille B, Murdaca J et al (2016) α-Lipoic acid up-regulates expression of peroxisome proliferator-activated receptor β in skeletal muscle: involvement of the JNK signaling pathway. FASEB J 30:1287–1299
44. Stacchiotti A, Corsetti G (2020) Natural compounds and autophagy: allies against neurodegeneration. Front Cell Dev Biol 8:555409
45. Kou X, Chen N (2017) Resveratrol as a natural autophagy regulator for prevention and treatment of Alzheimer's disease. Nutrients 9:927
46. Di Meco A, Curtis ME, Lauretti E et al (2020) Autophagy dysfunction in Alzheimer's disease: mechanistic insights and new therapeutic opportunities. Biol Psychiatry 87:797–807
47. Wang P, Jiang L, Zhou N et al (2018) Resveratrol ameliorates autophagic flux to promote functional recovery in rats after spinal cord injury. Oncotarget 9:8427–8440
48. Wang H, Jiang T, Li W et al (2018) Resveratrol attenuates oxidative damage through activating mitophagy in an in vitro model of Alzheimer's disease. Toxicol Lett 282:100–108
49. Deng H, Mi MT (2016) Resveratrol attenuates Aβ25-35 caused neurotoxicity by inducing autophagy through the TyrRS-PARP1-SIRT1 signaling pathway. Neurochem Res 41:2367–2379
50. Muhammad MH, Allam MM (2018) Resveratrol and/or exercise training counteract aging-associated decline of physical endurance in aged mice; targeting mitochondrial biogenesis and function. J Physiol Sci 68:681–688
51. Rodríguez-Bies E, Tung BT, Navas P et al (2016) Resveratrol primes the effects of physical activity in old mice. Br J Nutr 116:979–988
52. Bai CH, Alizargar J, Peng CY et al (2020) Combination of exercise training and resveratrol attenuates obese sarcopenia in skeletal muscle atrophy. Chin J Physiol 63:101–112
53. Kan NW, Lee MC, Tung YT et al (2018) The synergistic effects of resveratrol combined with resistant training on exercise performance and physiological adaption. Nutrients 10:1360
54. Sung MM, Byrne NJ, Robertson IM et al (2017) Resveratrol improves exercise performance and skeletal muscle oxidative capacity in heart failure. Am J Physiol Heart Circ Physiol 312: H842–H853
55. Breuss JM, Atanasov AG, Uhrin P (2019) Resveratrol and its effects on the vascular system. Int J Mol Sci 20:1523
56. Han S, Bal NB, Sadi G et al (2018) The effects of resveratrol and exercise on age and gender-dependent alterations of vascular functions and biomarkers. Exp Gerontol 110:191–201
57. Esfandiarei M, Hoxha B, Talley NA et al (2019) Beneficial effects of resveratrol and exercise training on cardiac and aortic function and structure in the 3xTg mouse model of Alzheimer's disease. Drug Des Devel Ther 13:1197–1211
58. Broderick TL, Rasool S, Li R et al (2020) Neuroprotective effects of chronic resveratrol treatment and exercise training in the 3xTg-AD mouse model of Alzheimer's disease. Int J Mol Sci 21:7337
59. Sun XJ, Feng WL, Hou N et al (2020) Effects of aerobic exercise and resveratrol on the expressions of JAK2 and TGF-β1 in renal tissue of type 2 diabetes rats. Zhongguo Ying Yong Sheng Li Xue Za Zhi 36:202–206
60. Nasiri M, Ahmadizad S, Hedayati M et al (2020) Trans-resveratrol supplement lowers lipid peroxidation responses of exercise in male Wistar rats. Int J Vitam Nutr Res:1–6. https://doi.org/10.1024/0300-9831/a000654
61. Sun J, Zhang C, Kim M et al (2018) Early potential effects of resveratrol supplementation on skeletal muscle adaptation involved in exercise-induced weight loss in obese mice. BMB Rep 51:200–205
62. Hou X, Watzlawik JO, Fiesel FC et al (2020) Autophagy in Parkinson's disease. J Mol Biol 432:2651–2672

63. Aggarwal BB, Harikumar KB (2009) Potential therapeutic effects of curcumin, the anti-inflammatory agent, against neurodegenerative, cardiovascular, pulmonary, metabolic, autoimmune and neoplastic diseases. Int J Biochem Cell Biol 41:40–59
64. Gupta SC, Patchva S, Aggarwal BB (2013) Therapeutic roles of curcumin: lessons learned from clinical trials. AAPS J 15:195–218
65. Liu J, Liu C, Zhang J et al (2020) A self-assembled α-synuclein nanoscavenger for Parkinson's disease. ACS Nano 14:1533–1549
66. Wu Y, Liang S, Xu B et al (2018) Protective effect of curcumin on dopamine neurons in Parkinson's disease and its mechanism. Zhejiang Da Xue Xue Bao Yi Xue Ban 47:480–486
67. Jiang TF, Zhang YJ, Zhou HY et al (2013) Curcumin ameliorates the neurodegenerative pathology in A53T α-synuclein cell model of Parkinson's disease through the downregulation of mTOR/p70S6K signaling and the recovery of macroautophagy. J Neuroimmune Pharmacol 8:356–369
68. Wang Z, Yang C, Liu J et al (2020) A curcumin derivative activates TFEB and protects against parkinsonian neurotoxicity in vitro. Int J Mol Sci 21:1515
69. Bielak-Zmijewska A, Grabowska W, Ciolko A et al (2019) The role of curcumin in the modulation of ageing. Int J Mol Sci 20:1239
70. Ray HRD, Shibaguchi T, Yamada T et al (2021) Curcumin induces mitochondrial biogenesis by increasing cAMP levels via PDE4A inhibition in skeletal muscle. Br J Nutr:1–34. https://doi.org/10.1017/S0007114521000490
71. Zhang M, Tang J, Li Y et al (2017) Curcumin attenuates skeletal muscle mitochondrial impairment in COPD rats: PGC-1α/SIRT3 pathway involved. Chem Biol Interact 277:168–175
72. Lee DY, Chun YS, Kim JK et al (2021) Curcumin attenuates sarcopenia in chronic forced exercise executed aged mice by regulating muscle degradation and protein synthesis with antioxidant and anti-inflammatory effects. J Agric Food Chem 69:6214–6228
73. Chaudhary P, Sharma YK, Sharma S et al (2019) High altitude mediated skeletal muscle atrophy: protective role of curcumin. Biochimie 156:138–147
74. Mañas-García L, Guitart M, Duran X et al (2020) Satellite cells and markers of muscle regeneration during unloading and reloading: effects of treatment with resveratrol and curcumin. Nutrients 12:1870
75. Sahebkar A, Serban MC, Ursoniu S et al (2015) Effect of curcuminoids on oxidative stress: a systematic review and meta-analysis of randomized controlled trials. J Funct Foods 18:898–909
76. Cho JA, Park SH, Cho J et al (2020) Exercise and curcumin in combination improves cognitive function and attenuates ER stress in diabetic rats. Nutrients 12:1309
77. Hewlings SJ, Kalman DS (2017) Curcumin: a review of its effects on human health. Foods 6:92
78. Mallard AR, Briskey D, Richards BA et al (2020) Curcumin improves delayed onset muscle soreness and postexercise lactate accumulation. J Diet Suppl:1–12. https://doi.org/10.1080/19390211.2020.1796885
79. Hu G, Cao H, Zhou HT et al (2018) Regulatory effect of curcumin on renal apoptosis and its mechanism in overtraining rats. Zhongguo Ying Yong Sheng Li Xue Za Zhi 34:513–518

Chapter 13
Exercise-Mediated Functional Status of Autophagy Is Beneficial to Health

Ning Chen

It is well known that a series of chronic diseases can be induced by unhealthy lifestyles, including physical inactivity; therefore, regular and scientific exercise as one part of the active lifestyles can offer multiple benefits to human health or also can be beneficial for the prevention and mitigation of chronic diseases. Physical activity or appropriate regular exercise has a series of positive contributions to improve metabolic homeostasis in various tissues and to reduce the overall risk of chronic diseases. Correspondingly, the gradually increased average human life expectancy over the past decades is highly correlated with the choice of healthy lifestyles and regular and scientific exercise by these people.

Autophagy, an intracellular recycling process for delivering the damaged cellular materials and organelles to lysosomes for degradation in a selective or nonselective manner, is an evolutionarily conserved process for regulating protein degradation, optimizing organelle turnover, and recycling cytoplasmic components to maintain cellular metabolism, homeostasis, and survival under detrimental conditions with nutritional deficiency or cellular stresses including the accumulation of oxidative damage, the loss of proteostasis, genomic instability, and epigenetic alteration, thereby helping cells to adapt to the demand of nutrients and energy, and cellular stresses. Mounting evidence has demonstrated that autophagy can be induced or the functional status of autophagy can be optimized by appropriate exercise interventions in multiple tissues or organs, especially in skeletal muscle and cardiac muscle. As the largest metabolic and endocrine organ, skeletal muscle followed by exercise-mediated cellular homeostasis can trigger the metabolic balance of other organs in the body through eliminating oxidative stress-induced metabolic wastes, aged or damaged organelles, and denatured cellular macromolecules via the selective autophagy, including the catabolism of carbohydrates (glycophagy), iron

N. Chen (✉)
Tianjiu Research and Development Center for Exercise Nutrition and Foods, Hubei Key Laboratory of Exercise Training and Monitoring, College of Health Science, Wuhan Sports University, Wuhan, China

N. Chen (ed.), *Exercise, Autophagy and Chronic Diseases*,
https://doi.org/10.1007/978-981-16-4525-9_13

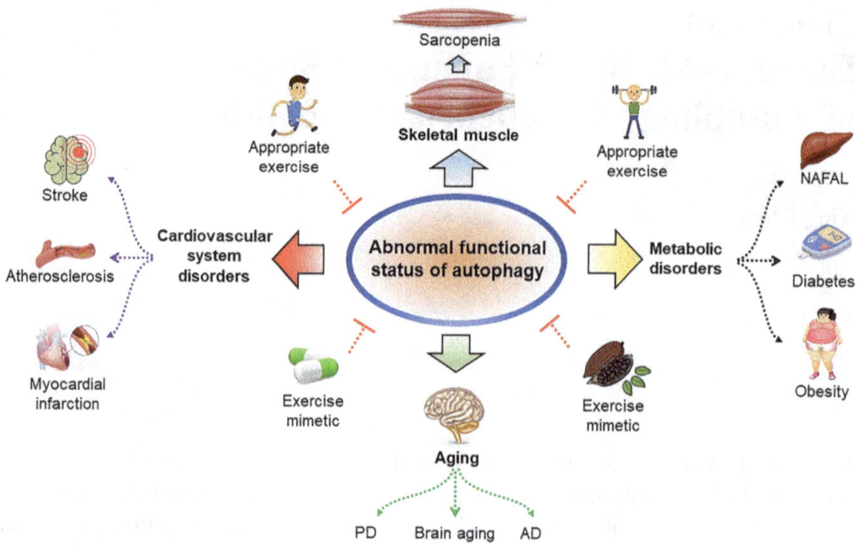

Fig. 13.1 Appropriate exercise or exercise mimetics rescues the abnormal functional status of autophagy for the prevention, treatment and rehabilitation of chronic diseases

(ferritinophagy), and lipids (lipophagy); the clearance of abnormal protein aggregates (aggrephagy); and the elimination of damaged organelles including mitochondria (mitophagy), endoplasmic reticulum (ER-phagy), and ribosomes (ribophagy), thereby rescuing the detrimental cellular environment for accelerating health promotion and the rehabilitation of diseases.

Investigating relationship between chronic diseases and autophagy is beneficial for the development of interventional strategies through optimizing functional status of autophagy. As one of the effective interventional strategies, exercise or exercise mimetic-mediated boosting functional status of autophagy for the recycling of damaged organelles or denatured cellular components may be the optimal choice. Autophagy not only acts as a conserved physiological process involving a defense mechanism for cells in adverse environments, but also is involved in the pathological processes of various diseases. Normal level of autophagy can protect cells from environmental stimuli to maintain the metabolism and cellular homeostasis in organisms; in contrast, the abnormal functional status of autophagy including excessive and deficient autophagy and impaired autophagic flux as well as genetic deficiency of autophagy may be failed to accomplish the above function of maintaining metabolic balance, which may correspondingly trigger the subhealth state or a variety of chronic diseases including metabolic disorders such as obesity, diabetes, and nonalcohol fatty liver disease, cardiovascular system disorders such as cardiomyopathy, cardiac hypertrophy, ischemic heart disease, and heart failure, and aging-related diseases such as brain aging, Alzheimer's disease, and Parkinson's diseases, as well as sarcopenia (Fig. 13.1). Therefore, exercise as the nonpharmaceutical inducer of autophagy, appropriate and scientific exercise

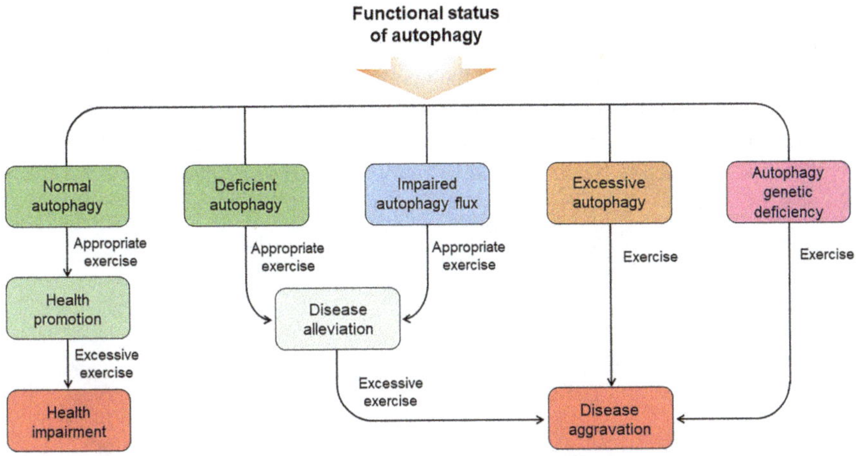

Fig. 13.2 Exercise-mediated functional status of autophagy for the regulation of health and diseases

intervention, or the application of exercise mimetics is critical for rescuing abnormal functional status of autophagy, thereby achieving health promotion, and executing prevention, treatment and rehabilitation of chronic diseases (Fig. 13.2).

Although mounting evidence has been documented for the improvement of health or the mitigation and treatment of chronic diseases by targeting signal pathways of autophagy to optimize functional status of autophagy during the intervention processes by exercise or exercise mimetics, the more complexities including underlying mechanisms, optimal exercise prescriptions, and others are still not uncovered. Increasing concerns for the optimal manipulation of autophagy have been gained extensive attention, which includes the boosted intensity and timing of autophagy for various individuals. In addition, the tissue or organ specificity for the induction of autophagy should also need to be addressed, which could be used for exploring the benefits of autophagy for a single tissue or in a combinatorial style for multiple tissues or organs. Moreover, the major challenges should be the optimal choice for the styles, intensity, duration, and frequency of exercise intervention for the optimization of functional status of autophagy to execute effective health promotion and accomplish the prevention or alleviation of chronic diseases. Furthermore, despite tremendous research progress in exercise-mediated functional status autophagy for chronic diseases, tailoring the interventional steps or processes should also be explored based on the failure in autophagy processes during the onset and progression of chronic diseases, which is also beneficial to develop specific regulators for modulating the functional status of autophagy, such as autophagy-specific drugs or exercise mimetics, for these chronic diseases. It is no doubt that appropriate physical activity or regular and scientific exercise should be the promising nonpharmaceutical interventional strategy with high-efficiency and environment-friendly characteristics for chronic diseases once its underlying mechanisms are further clarified, which will be further beneficial to the development and utilization of novel and effective exercise mimetics for the people that cannot accomplish exercise training or regular physical activity due to disability, or the people without willing to execute exercise training or regular physical activity. Moreover, appropriate physical activity or regular and scientific exercise as the nonpharmaceutical intervention strategy could really realize the active health or healthy aging from passive medication.

CPSIA information can be obtained
at www.ICGtesting.com
Printed in the USA
LVHW081213061121
702606LV00002B/33

9 789811 645242